Transition Metal Carbyne Complexes

NATO ASI Series

Advanced Science Institutes Series

A Series presenting the results of activities sponsored by the NATO Science Committee, which aims at the dissemination of advanced scientific and technological knowledge, with a view to strengthening links between scientific communities.

The Series is published by an international board of publishers in conjunction with the NATO Scientific Affairs Division

A Life Sciences **B Physics**	Plenum Publishing Corporation London and New York
C Mathematical and Physical Sciences **D Behavioural and Social Sciences** **E Applied Sciences**	Kluwer Academic Publishers Dordrecht, Boston and London
F Computer and Systems Sciences **G Ecological Sciences** **H Cell Biology** **I Global Environmental Change**	Springer-Verlag Berlin, Heidelberg, New York, London, Paris and Tokyo

NATO-PCO-DATA BASE

The electronic index to the NATO ASI Series provides full bibliographical references (with keywords and/or abstracts) to more than 30000 contributions from international scientists published in all sections of the NATO ASI Series.
Access to the NATO-PCO-DATA BASE is possible in two ways:

– via online FILE 128 (NATO-PCO-DATA BASE) hosted by ESRIN, Via Galileo Galilei, I-00044 Frascati, Italy.

– via CD-ROM "NATO-PCO-DATA BASE" with user-friendly retrieval software in English, French and German (© WTV GmbH and DATAWARE Technologies Inc. 1989).

The CD-ROM can be ordered through any member of the Board of Publishers or through NATO-PCO, Overijse, Belgium.

Series C: Mathematical and Physical Sciences - Vol. 392

Transition Metal Carbyne Complexes

edited by

F. R. Kreißl

Anorganisch-Chemisches Institut,
Technische Universität München,
Garching, Germany

Springer Science+Business Media, B.V.

Proceedings of the NATO Advanced Research Workshop on
Transition Metal Carbyne Complexes
Wildbad Kreuth, Germany
September 27–October 2, 1992

ISBN 978-94-010-4728-9 ISBN 978-94-011-1666-4 (eBook)
DOI 10.1007/978-94-011-1666-4

Printed on acid-free paper

Table of Contents

Preface

The chemistry of transition metal carbyne complexes has become a highly attractive field during the past twenty years. In recent years its application to aspects of catalysis and metathesis has gained considerable interest from inorganic as well as organic chemists. In addition, organic synthesis by means of metal carbon multiple bond reagents offers the most sophisticated technology currently available. In consideration of these developments some of Professor E. O. Fischer´s former coworkers and colleagues felt obliged to organize this NATO Advanced Research Workshop on Transition Metal Carbyne Complexes in the Bavarian Alps. They have been encouraged by the fact that most of the distinguished scientists in the field of metal-carbon multiple bond chemistry had finally agreed to participate and to present stimulating lectures.

The organizers of the workshop are deeply grateful to the Scientific Affairs Division of the NATO for the generous financial support of the meeting in Wildbad Kreuth and for the preparation of this book. They also feel indebted to acknowledge the generous support from Wacker-Chemie, BASF, Peroxid-Chemie, Hoechst and Bayer. Finally they thank the staff of the Hanns-Seidel-Stiftung in Wildbad Kreuth for providing a pleasant and stimulating atmosphere during the meeting.

Fritz R. Kreißl
Technische Universität München

Foreword

It is not easy these days to put a focus on the rapid development of the chemistry related to metal-carbene and metal-carbyne complexes. This area of scientific research and technology has become so broad in scope and so high in knowledge that it is necessary to apply this focus by means of scientific meetings.

The NATO Series of scientific conferences - another pieceful way of spending public money out of the defense budget - have greatly added to the development of high-technology research areas during the past decades. Organometallic chemistry has developed on this time-scale and has received great benefit from programmes like the NATO Series. Modelled after the success of a previous meeting on "Metal-Carbene Complexes", Professor Kreißl organized the consecutive conference on "Metal-Carbyne Complexes" nearby the cradle of these versatile, successful, chemically as well as industrially useful species: At Technische Universität München, Professor Ernst Otto Fischer was holding the chair of Inorganic Chemistry over a period of twenty years - from 1964 (following Walter Hieber) until 1984.

Ernst Otto Fischer is one of the pioneers of organometallic chemistry. In Europe, he and Franz Hein, Karl Ziegler, Walter Hieber, Geoffrey Wilkinson, and Georg Wittig made the most eminent personal contributions to this highly interdisciplinary field of research. These days, catalysis, organic synthesis and materials science are the major areas of application, with the latter segment being in rapid expansion. The basis of all these applications, however, remains to be *synthesis* and *structure*. Professor Fischer has mastered both areas ever since he entered the exciting field of organometallic chemistry through ferrocene (*structure*) and dibenzene chromium (*synthesis*). In later years he succeeded to synthesize the first organometallic compound exhibiting a metal-to-carbon *triple bond* - predominantly a result of great and long-lasting enthusiasm. Boron trihalides were thought to substitute a halogen atom for the alkoxy group in one of Fischer's metal carbene compounds of general composition $(CO)_5M{=}C(OR')R$ (M = Cr, Mo, W). Quite unexpectedly in those days, the first carbyne complex resulted when my fellow-student Gerhard Kreis, now a chemist at Wacker-Chemie GmbH (Burghausen/Bavaria), performed the very experiments with utmost precision typical of him. The puzzling but correct elemental analysis of his new product, $(CO)_4ClW{\equiv}C(C_6H_5)$, performed by Manfred Barth in our institute is shown in Fig. 1.

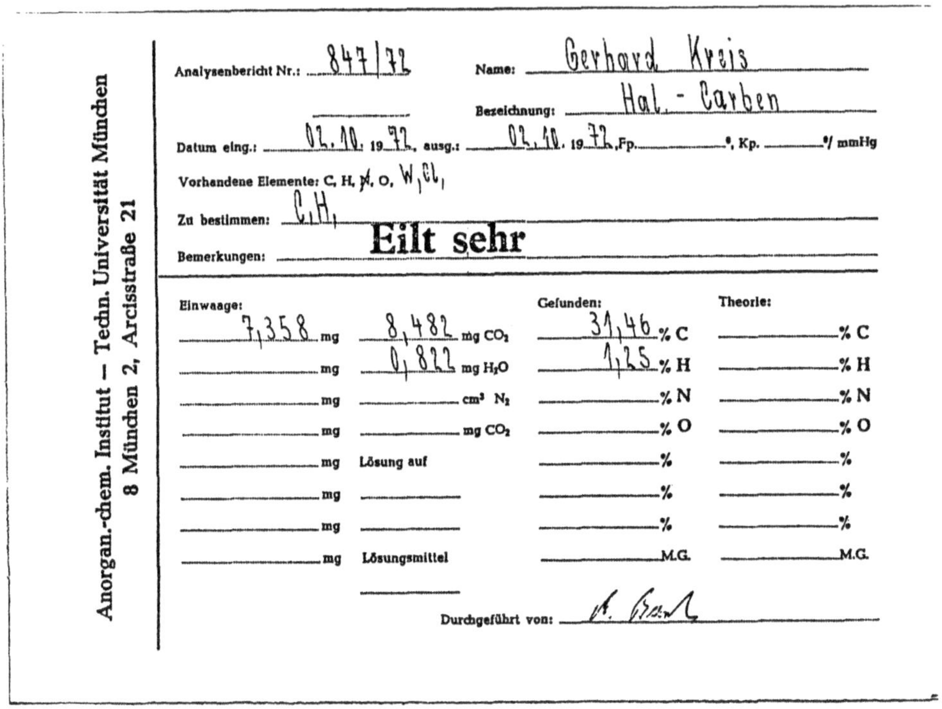

Anorgan.-chem. Institut – Techn. Universität München
8 München 2, Arcisstraße 21

Analysenbericht Nr.: 847/72 Name: Gerhard Kreis

Bezeichnung: Hal. - Carben

Datum eing.: 02.10.1972, ausg.: 02.10.1972, Fp. °, Kp. °/ mmHg

Vorhandene Elemente: C, H, ~~N~~, O, W, Cl

Zu bestimmen: C, H

Bemerkungen: **Eilt sehr**

Einwaage:		Gefunden:	Theorie:
7,358 mg	8,482 mg CO_2	31,46 % C	% C
mg	0,822 mg H_2O	1,25 % H	% H
mg	cm³ N_2	% N	% N
mg	mg CO_2	% O	% O
mg	Lösung auf	%	%
mg		%	%
mg		%	%
mg	Lösungsmittel	M.G.	M.G.

Durchgeführt von: M. Barth

Fig. 1. Analysis certificate of the "expected" chorocarbene tungsten complex "$(CO)_5W{=}C(Cl)C_6H_5$", which compound turned out to be the first "carbyne" complex $(CO)_4ClW{\equiv}C(C_6H_5)$. The analysis was correct, as always for a new compound in E. O.'s laboratory! The Fischer institute has run an excellent microanalysis laboratory, directed by Manfred Barth since the year of 1965. At present, approximately 2000 C,H,N analyses plus numerous oxygen, halogen, sulfur, and metal analyses are performed there every year.

Metal-carbyne complexes have continued to attract broad interest ever since. Organometallic chemistry has gained many new insights as to the bonding properties and reactivity concerning metal-to-carbon bonds in general. According to IUPAC rules, all metal derivatives containing the three-electron carbyne ligands |≡C-R are named "*alkylidyne*" complexes regardless of the nature of both the substituents R and the metal fragment attached to this ligand. In light of the recent development in this area and the understanding of reactivity patterns, a formal discrimination of "Fischer-type" and "Schrock-type" metal-carbyne (alkylidyne) complexes seems no longer justified.

The present book summarizes the lectures held at the symposium that took place in the Hanns-Seidel-Stiftung in Kreuth in Upper Bavaria. The meeting was dedicated to Ernst Otto Fischer. The opening lecture was presented by Dick Schrock who has pioneered the chemical relationship between organometallic compounds, especially alkylidene- and alkylidyne complexes, with the mechanism of the industrially employed process of olefin metathesis. Several other outstanding chemists were present, among them Gordon Stone, Wolfgang Beck, Warren Roper, and Jean-Marie Basset. Most lectures centered around synthesis and reactivity aspects, with highly original contributions also coming from younger researchers who do not have available the machinery of a huge research group.

The excellent personal atmosphere among the conference participants mirrored the high-level quality lectures, intense discussions and excellent organization (which was in the hands of Fritz Kreißl and his crew). One can be sure that this meeting has contributed to a scientific up-date of the present status and the future perspectives of metal-carbon triple bond chemistry.

Wolfgang A. Herrmann
Anorganisch-chemisches Institut
Technische Universität München

LIST OF PARTICIPANTS

J. M.	Basset	Institut de Recherche sur la Catalyse Centre National de la Recherche Scientifique 2, Avenue Albert Einstein 69626 Villeurbanne Cedex F
W.	Beck	Anorganisch-chemisches Institut Universität München Meiserstraße 1 8000 München D
H.	Berke	Anorganisch-chemisches Institut Universität Zürich Winterthurerstr. 190 8057 Zürich CH
R.	Bertani	Centro di Chimica e Technologia dei Composti Metallorganici C. N. R. Via Marzolo 9 35131 Padova I
B.	Bildstein	Institut für Anorganische und Analytische Chemie Universität Innsbruck Innrain 52a 6020 Innsbruck A
S.	Bordoni	Dipartimento di Chimica Fisica ed Inorganica Universita degli Studi di Bologna Viale del Risorgimento 4 40136 Bologna I

L.	Busetto	Dipartimento di Chimica Fisica ed Inorganica Universita degli Studi di Bologna Viale del Risorgimento 4 40136 Bologna I
A.	Choplin	Institut de Recherches sur la Catalyse Centre National de la Recherches Scientifique 2, Avenue Albert Einstein 69626 Villeurbanne Cedex F
V.	Dragutan	Institute of Organic Chemistry Romanian Academy Spl. Independentei 202 B Bucharest R
G.	Erker	Organisch-chemisches Institut Universität Münster Corrensstr. 40 4400 Münster D
A. C.	Filippou	Anorganisch-chemisches Institut Technische Universität München Lichtenbergstr. 4 8046 Garching GR
E. O.	Fischer	Anorganisch-chemisches Institut Technische Universität München Arcisstraße 21 8000 München D
H.	Fischer	Fakultät für Chemie Universität Konstanz Universitätsstraße 10 7750 Konstanz D

G. L.	Geoffroy	Department of Chemistry The Pennsylvania State University 211 Whitemore Laboratory PA 16802 University Park USA
J. W.	Gilje	Department of Chemistry University of Hawaii at Manoa 2545 The Mall Honolulu HI 96822 Honolulu USA
M.	Herberhold	Laboratorium für Anorganische Chemie Universität Bayreuth Postfach 101251 8580 Bayreuth D
W. A.	Herrmann	Anorganisch-chemisches Institut Technische Universität München Lichtenbergstraße 4 8046 Garching D
W.	Hersh	Department of Chemistry and Biochemistry Queens College NY 11367- 0904 Flushing USA
A. F.	Hill	Department of Chemistry Imperial College of Science and Techn. South Kensington SW7 2AY London UK
M.	Hopkins	Department of Chemistry University of Pittsburgh PA 15260 Pittsburgh USA

Y.	Imamoglu	Kimya Fakültesi Inorganik Kimya Bilim D. Hacettepe Universitesi Ankara T
P.	Jaitner	Institut für Anorganische und Analytische Chemie Universität Innsbruck Innrain 52a 6020 Innsbruck A
G.	Kreis	Wacker-Chemie GmbH Postfach 1260 8263 Burghausen D
F. R.	Kreißl	Anorganisch-chemisches Institut Technische Universität München Lichtenbergstraße 4 8046 Garching D
C. G.	Kreiter	Fachbereich Chemie Universität Kaiserslautern Postfach 3049 6750 Kaiserslautern D
C.	Krüger	Röntgenlabor MPI für Kohlenforschung Kaiser-Wilhelm-Platz 1 4330 Mülheim/Ruhr D
E.	Licandro	Dipartimento di Chimica Organica e Industriale Universita degli Studi di Milano Via Venezian 21 20133 Milano I

M.	Lux	Abt. ZKP/NE-M505 BASF AG Postfach 6700 Ludwigshafen D
A.	Mayr	Department of Chemistry Univ.of New York at Stony Brook Stony Brook Stony NY 11794-3400 USA
L.	McElwee-White	Department of Chemistry Stanford University CA 94305 Stanford USA
J. M.	Moretó	Centro de Investigacion Consejo Superior de Investigaciones Cientificas J. Girona 18-26 08034 Barcelona E
D.	Nguyen Quy	Laboratoire de Chimie et Physico-Chimie Moleculaire Ecole Centrale Paris Grande Voie des Vignes 92295 Chatenay-Malabry Cedex F
K.	Öfele	Anorganisch-chemisches Institut Technische Universität München Lichtenbergstraße 4 8046 Garching D
O.	Orama	Department of Inorganic Chemistry University of Helsinki Vuorikatu 20 00100 Helsinki SF

K. A. Ostoja-Starzews
Zentrale Forschung, Wiss. Hauptlaborat.
Bayer AG
Postfach
5090 Leverkusen D

A. Palazzi
Dipartimento di Chimica Fisica ed Inorganica
Universita degli Studi di Bologna
Viale del Risorgimento 4
40136 Bologna I

A. Papagni
Dipartimento Chimica Organica e Industriale
Universita di Milano
Via Golui 19
20131 Milano I

A. Pombeiro
Instituto Superior Tecnico
Centro de Quimica Estrutural
Complexo 1
1096 Lisboa Codex P

H. Raubenheimer
Department of Chemistry and Biochemistry
Rand Afrikaans University
P.O. Box 524
2000 Johannesburg SA

A. Razavi
Fina Research
Zone Industrielle
Seneffe B

S. Ricart
Department de Quimica Organica Biologica
Centro de Investigacion y Desarrollo
c/Jordi Girona 18-26
08034 Barcelona E

W. R.	Roper	Department of Chemistry The University of Auckland Private Bag 92019 Auckland NZ
R.	Schmidt-Radde	Angew. Chem. Int. Ed. Engl. VCH Verlag Pappelallee 3 6940 Weinheim D
R. R.	Schrock	Department of Chemistry Massachusetts Inst.of Technology 77 Massachusetts Avenue MA 02139 Cambridge USA
U.	Schubert	Institut für Anorganische Chemie Universität Würzburg Am Hubland 8700 Würzburg D
F. G. A	Stone	Department of Chemistry Baylor University P.O. Box 97348 TE 76798- 7348 Waco USA
G.	Süss-Fink	Institut de Chimie Universite de Neuchatel Avenue de Bellevaux 51 2000 Neuchatel CH
J. L.	Templeton	Department of Chemistry Univ.of North Carolina Chapel Hill Cb 3290, Venable Hall NC 27599- 3290 Chapel Hill USA

K.	Weiss	Laboratorium für Anorganische Chemie Universität Bayreuth Postfach 101251 8580 Bayreuth D
V.	Zanotti	Dipartimento Fisica ed Inorganica Universita degli Studi di Bologna Viale del Risorgimento 4 40136 Bologna I

OLEFIN METATHESIS BY WELL-CHARACTERIZED RE(VII) ALKYLIDENE/ALKYLIDYNE COMPLEXES

R. R. SCHROCK, R. TOREKI, G. A. VAUGHAN, A. FAROOQ
Department of Chemistry 6-331
Massachusetts Institute of Technology
Cambridge, Massachusetts 02139

ABSTRACT. A convenient one pot synthesis of $Re(NR)_2(py)Cl_3$ from Re_2O_7, trimethyl-chlorosilane, pyridine, and RNH_2 (R = aryl or t-butyl) has been developed. Alkylation of these species with dineopentyl or dineophyl zinc or Grignard reagents affords complexes of the formula $Re(NR)_2(CHR')(CH_2R')$ (R' = CMe_3 or CMe_2Ph). $Re(O)_2(CH$-*t*-$Bu)(CH_2$-*t*-$Bu)$ can be prepared by the acid-catalyzed hydrolysis of $Re(NR)_2(CH$-*t*-$Bu)(CH_2$-*t*-$Bu)$, which upon treatment with HCl in dimethoxyethane produces $[Re(C$-*t*-$Bu)(CH$-*t*-$Bu)Cl_2]_x$. $[Re(C$-*t*-$Bu)(CH$-*t*-$Bu)Cl_2]_x$ is a versatile precursor to a variety of bisalkoxide complexes of the general formula *syn* or *anti*-Re(C-*t*-Bu)(CH-*t*-Bu)$(OR)_2$ (OR = O-*t*-Bu, $OCMe_2(CF_3)$, $OCMe(CF_3)_2$, etc.). Metathetical reactions between Re(C-*t*-Bu)(CH-*t*-Bu)$(OR_{F6})_2$ (OR_{F6} = $OCMe(CF_3)_2$) and 1-decene or methyl-9-decenoate yield the expected new alkylidene complexes, which are unstable in concentrated solution and cannot be isolated. In the presence of THF or dimethoxyethane complexes of the type *syn* or *anti*-Re(C-*t*-Bu)(CHR)$(OR_{F6})_2S_2$ (R = Me, Et, Ph; S = THF or 0.5 DME) could be prepared in high yield from Re(C-*t*-Bu)(CH-*t*-Bu)$(OR_{F6})_2$ and CH_2=CHR. Heteroatom-substituted (O, S, or N) terminal olefins and other electron-rich olefins react more rapidly than ordinary olefins with Re(C-*t*-Bu)(CH-*t*-Bu)$(OR_{F6})_2$ in the presence of THF to yield complexes of the type *syn* or *anti*-Re(C-*t*-Bu)(CHX)$(OR_{F6})_2(THF)_2$ (X = OR, SR, NR_2). 2-Pentene or methyl oleate are metathesized in the presence of Re(C-*t*-Bu)(CH-*t*-Bu)$(OR_{F6})_2$, and intermediate alkylidene complexes can be observed in each case. Ethylene reacts with Re(C-*t*-Bu)(CHX)$(OR)_2$ complexes to give metallacyclobutane complexes first, and then rhenacyclopentene complexes via a "3+2 cycloaddition" across the C≡Re=C unit. X-ray studies of *syn*-$[Re(C$-*t*-$Bu)(CH$-*t*-$Bu)(Naryl)Cl_2]_2$, *syn*-Re(C-*t*-Bu)(CH-*t*-Bu)$(OR_{F6})_2$(THF), *anti*-Re(C-*t*-Bu)(CHferrocenyl)$(OR_{F6})_2$, and *syn*-Re(C-*t*-Bu)(CHOEt)$(OR_{F6})_2(THF)_2$ have been carried out.

Introduction

Approximately ten years ago evidence began to accumulate in favor of the highest possible oxidation state for tungsten metathesis catalysts (d^0 if the alkylidene ligand is viewed as a dianion).[1-4] Since rhenium is one of three metals (molybdenum and tungsten being the other two) that are active for the metathesis of olefins in classical metathesis systems,[5,6] we felt that Re(VII) was the most plausible oxidation state for rhenium metathesis catalysts, and therefore that it should be possible to prepare stable Re(VII) alkylidene complexes. At that time organometallic chemistry of Re(VII) was extremely rare.[7,8] We chose to attempt to synthesize complexes of Re(VII) containing imido

F. R. Kreißl (ed.), Transition Metal Carbyne Complexes, 1–22.

ligands in the belief that imido ligands could support rhenium in its highest oxidation state and might yield more stable catalysts than complexes containing oxo ligands. We soon discovered how to prepare bisimido neopentylidene complexes, and soon thereafter neopentylidene/neopentylidyne complexes.[9] Since then a variety of complexes that contain the neopentylidyne ligand have been synthesized and their chemistry explored. The search for well-characterized soluble rhenium olefin metathesis catalysts ended recently with the discovery that complexes of the type Re(C-*t*-Bu)(CH-*t*-Bu)$(OR_{F6})_2$ (OR_{F6} = $OCMe(CF_3)_2$)[10] will metathesize olefins. In this article I will review some of the developments that led to this discovery, present some recent results concerning metathesis of ordinary and cyclic olefins, and contrast the behavior of rhenium catalysts with related molybdenum and tungsten imido alkylidene catalysts.

Results

SYNTHESIS OF ALKYLIDENE COMPLEXES

Since most tantalum and tungsten neopentylidene complexes had been prepared by α hydrogen abstraction in a d^0 dineopentyl complex,[11] we chose the same approach in order to prepare a rhenium(VII) alkylidene complex. We decided to focus on the synthesis of imido neopentyl complexes, since we felt that metal reduction would be much less likely for imido complexes than for oxo complexes. We found that addition of HCl in ether to Re(N-*t*-Bu)$_3$($OSiMe_3$)[12] yielded one equivalent of *tert*-butyl ammonium chloride and bright orange, highly crystalline Re(N-*t*-Bu)$_2$$Cl_3$ in 83% yield.[13] When Re(N-*t*-Bu)$_2$$Cl_3$ was alkylated by neopentyl regents, Re(N-*t*-Bu)$_2$(CH-*t*-Bu)(CH_2-*t*-Bu), a yellow oil, was formed. Unfortunately, Re(N-*t*-Bu)$_2$(CH-*t*-Bu)(CH_2-*t*-Bu) did not react with olefins, even very reactive olefins such as norbornene.

In an attempt to form complexes that contain less than two imido ligands we tried reacting Re(N-*t*-Bu)$_2$(CH-*t*-Bu)(CH_2-*t*-Bu) with proton sources. An important and surprising finding was that 2,4-lutidine hydrochloride reacts with Re(N-*t*-Bu)$_2$(CH-*t*-Bu)(CH_2-*t*-Bu) in dichloromethane to yield [Re(C-*t*-Bu)(CH-*t*-Bu)(*t*-BuNH_2)Cl_2]$_2$ (equation 1). The initial reaction with HCl was proposed to yield Re(N-*t*-Bu)(NH-*t*-Bu)(CH-*t*-Bu)(CH_2-*t*-Bu)Cl (eq 2). One possibility is that an α proton then transfers from the alkylidene ligand to the imido ligand (eq 3). An amido ligand is then protonated and removed from the system as the ammonium salt (eq 4). A final migration of a proton from a neopentyl α carbon atom to an amido nitrogen atom (eq 5) and dimerization (eq 6) completes the synthesis. The imido ligands can be regarded as protecting groups in these reactions; ultimately they are sacrificed in favor of what are apparently more favorable multiple metal-carbon double and triple bonds. This "exchange" of multiple metal-nitrogen bonds for multiple metal-carbon bonds is the unusual feature of this chemistry, one that may be more characteristic of d^0 metals further to the right in the transition metal series where the ionic component of a metal ligand bond is a less significant fraction of the total bond. Early syntheses of imido alkylidene complexes of tungsten and molybdenum relied on a reaction in which a proton migrated from an amido ligand to a neopentylidyne ligand;[11] the opposite is observed here.

$$\mathrm{Re(NR)_2(CHR)(CH_2R)} \xrightarrow[\substack{-\ \mathrm{RNH_3Cl} \\ -\ 3\ \mathrm{lut}}]{+\ 3\ \mathrm{lutHCl}} 0.5\ \mathrm{[Re(CR)(CHR)(RNH_2)Cl_2]_2} \quad (\mathrm{R} = t\text{-Bu}) \qquad (1)$$

$$\mathrm{Re(NR)_2(CHR)(CH_2R)} \xrightarrow{+\mathrm{HCl}} \mathrm{Re(NR)(NHR)(CHR)(CH_2R)Cl} \quad (2)$$

$$\mathrm{Re(NR)(NHR)(CHR)(CH_2R)Cl} \longrightarrow \mathrm{Re(NHR)_2(CR)(CH_2R)Cl} \quad (3)$$

$$\mathrm{Re(NHR)_2(CR)(CH_2R)Cl} \xrightarrow[-\,t\text{-}\mathrm{BuNH_3Cl}]{2\,\mathrm{HCl}} \mathrm{Re(NHR)(CR)(CH_2R)Cl_2} \quad (4)$$

$$\mathrm{Re(NHR)(CR)(CH_2R)Cl_2} \longrightarrow \mathrm{Re(NH_2R)(CR)(CHR)Cl_2} \quad (5)$$

$$2\,\mathrm{Re(NH_2R)(CR)(CHR)Cl_2} \longrightarrow \mathrm{[Re(NH_2R)(CR)(CHR)Cl_2]_2} \quad (6)$$

Once $[Re(C\text{-}t\text{-}Bu)(CH\text{-}t\text{-}Bu)(t\text{-}BuNH_2)Cl_2]_2$ had been prepared it was relatively straightforward to prepare four-coordinate species such as $Re(C\text{-}t\text{-}Bu)(CH\text{-}t\text{-}Bu)(O\text{-}t\text{-}Bu)_2$, $Re(C\text{-}t\text{-}Bu)(CH\text{-}t\text{-}Bu)(OSiMe_3)_2$, or $Re(C\text{-}t\text{-}Bu)(CH\text{-}t\text{-}Bu)(CH_2\text{-}t\text{-}Bu)_2$. We were disappointed to find that these complexes also did not react with internal olefins. Since the synthetic route to these pseudo four-coordinate species was relatively long and tedious, we were not able to pursue the synthesis of variations at that time.

Two events led us to reevaluate the possibility of metathesis by complexes of the type $Re(CR')(CHR')(OR)_2$. First, facile routes to aryl imido complexes of the type $Re(Naryl)_2(py)Cl_3$ (aryl = 2,6-$C_6H_3X_2$, X = Me, *i*-Pr, Cl) in 70-90% overall yield were developed.[14,15] Second, complexes of the type $M(CH\text{-}t\text{-}Bu)(NAr)(OR)_2$ (M = Mo or W; Ar = 2,6-C_6H_3-i-Pr_2) had been found to be very active metathesis catalysts for ordinary olefins when OR is strongly electron-withdrawing (e.g., $OCMe(CF_3)_2$).[16] Therefore we became convinced that $Re(CR')(CHR')(OR)_2$ complexes would be active for the metathesis of olefins if OR were strongly electron-withdrawing, and moreover, that relatively facile routes to such species could be developed using arylimido ligands as protecting groups in reactions analogous to those that had been outlined previously in *t*-butylimido chemistry.

The goal became more feasible after new syntheses of rhenium starting materials were developed that are analogous to recently developed syntheses of tungsten[17] and molybdenum[18,19] imido complexes. The reaction shown in equation 7 is over in two hours at 25 °C and $Re(NAr')_2(py)Cl_3$ (Ar' = 2,6-$C_6H_3Me_2$) can be isolated in 86% yield. $Re(NAr)_2(py)Cl_3$ can be prepared in a similar fashion in high yield. The potential generality and superiority of this approach is illustrated by the high yield rapid synthesis of $Re(N\text{-}t\text{-}Bu)_3(OSiMe_3)$[12] (eq 8). In this case *tert*-butylamine acts as the trap for HCl. If the reaction is filtered after 20 minutes, $Re(N\text{-}t\text{-}Bu)_3(OSiMe_3)$ can be recovered from the filtrate in >90% yield. Treatment of the crude reaction mixture containing $Re(N\text{-}t\text{-}Bu)_3(OSiMe_3)$ and *tert*-$BuNH_3Cl$ with excess HCl at 0 °C precipitates an additional

$$0.5\ Re_2O_7 + \begin{matrix} 2\ Ar'NH_2 \\ 7\ Me_3SiCl \\ \text{excess py} \end{matrix} \xrightarrow[\substack{-\ 3.5\ (SiMe_3)_2O \\ -\ 4\ pyHCl}]{CH_2Cl_2}$$ 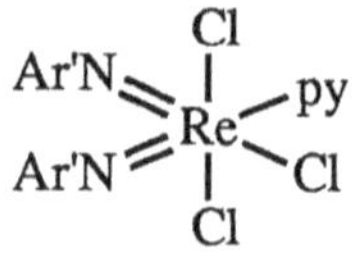 (7)

$$0.5\ Re_2O_7 + 9t\text{-}BuNH_2 + 6\ Me_3SiCl \xrightarrow[\substack{-\ 6\ t\text{-}BuNH_3Cl \\ -\ 2.5\ (SiMe_3)_2O}]{CH_2Cl_2} Re(N\text{-}t\text{-}Bu)_3(OSiMe_3) \quad (8)$$

equivalent of t-$BuNH_3Cl$, and $Re(N\text{-}t\text{-}Bu)_2Cl_3$ then can be isolated from the filtrate as large orange cubes in 85% overall yield (from Re_2O_7). This synthesis makes $[Re(C\text{-}t\text{-}Bu)(CH\text{-}t\text{-}Bu)(t\text{-}BuNH_2)Cl_2]_2$ readily accessible and now the preferred route to reported molecules such as $Re(C\text{-}t\text{-}Bu)(CH\text{-}t\text{-}Bu)(O\text{-}t\text{-}Bu)_2$ or $Re(C\text{-}t\text{-}Bu)(CH\text{-}t\text{-}Bu)(CH_2\text{-}t\text{-}Bu)_2$, but $Re(C\text{-}t\text{-}Bu)(CH\text{-}t\text{-}Bu)(OR)_2$ complexes in which OR is a poor nucleophile (e.g., OR = $OCMe(CF_3)_2$) have not yet been prepared from $[Re(C\text{-}t\text{-}Bu)(CH\text{-}t\text{-}Bu)(t\text{-}BuNH_2)Cl_2]_2$.

$Re(NAr')_2(py)Cl_3$ can be alkylated in high yield with 1.5 equiv of dineopentyl zinc in dichloromethane to give a neopentylidene complex (equation 9). An analogous alkylation

$$Re(NAr')(py)Cl_3 \xrightarrow[-\ 1.5\ ZnCl_2\ -\ py\ -\ CMe_4]{1.5\ Zn(CH_2\text{-}t\text{-}Bu)_2} Re(NAr')_2(CH\text{-}t\text{-}Bu)(CH_2\text{-}t\text{-}Bu) \quad (9)$$

employing $Zn(CH_2CMe_2Ph)_2$ gives $Re(NAr')_2(CHCMe_2Ph)(CH_2CMe_2Ph)$, a new compound that is similar to the neopentylidene species.[10] At this point the neophyl system has not been as extensively explored as the neopentyl system, but we expect the chemistry of neopentyl and neophyl-based compounds to be very similar. The structure of a complex such as $Re(NAr')_2(CH\text{-}t\text{-}Bu)(CH_2\text{-}t\text{-}Bu)$ is believed to be analogous to that of $ReO_2(CH\text{-}t\text{-}Bu)(CH_2\text{-}t\text{-}Bu)$[20] on the basis of NMR data, i.e., the imido ligands are inequivalent and the methylene protons in the neopentyl ligand are diastereotopic, viz.

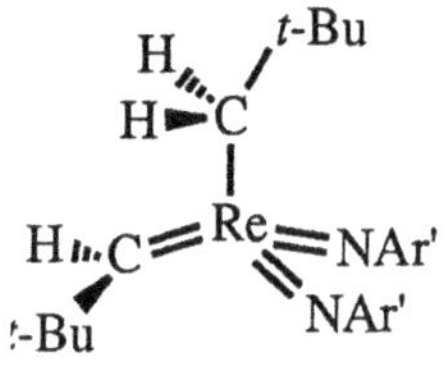

This structure is consistent with the proposed structural analogy between pseudotetrahedral bisimido complexes and metallocenes.[21,22]

Addition of excess HCl(g) to $Re(NAr')_2(CH\text{-}t\text{-}Bu)(CH_2\text{-}t\text{-}Bu)$ in dimethoxyethane yields $[Re(C\text{-}t\text{-}Bu)(CH\text{-}t\text{-}Bu)(Ar'NH_2)Cl_2]_2$ in 85% isolated yield (eq 10). We presume the mechanism to be essentially the same as that proposed for forming $[Re(C\text{-}t\text{-}Bu)(CH\text{-}t\text{-}Bu)(t\text{-}BuNH_2)Cl_2]_2$ (see above).[13] The structure of $[Re(C\text{-}t\text{-}Bu)(CH\text{-}t\text{-}Bu)(Ar'NH_2)Cl_2]_2$, as determined in an X-ray study,[10] is basically that shown in equation 10 in which the t-butyl group of the neopentylidene ligand points toward the

$$2\ Re(NAr')_2(CH\text{-}t\text{-}Bu)(CH_2\text{-}t\text{-}Bu) + 6\ HCl \xrightarrow[-\ 2\ Ar'NH_3Cl]{dme}$$

L = Ar'NH$_2$

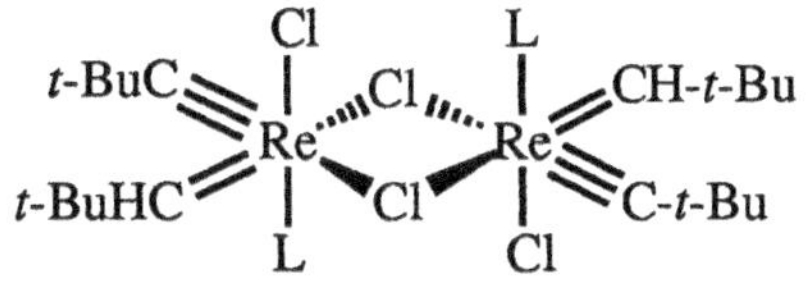

(10)

neopentylidyne ligand (the *syn* orientation). [Re(C-*t*-Bu)(CH-*t*-Bu)(Ar'NH$_2$)Cl$_2$]$_2$ does not react readily with olefins. [Re(C-*t*-Bu)(CH-*t*-Bu)(Ar'NH$_2$)Cl$_2$]$_2$ reacts with donor molecules to give pseudo six-coordinate adducts such as Re(C-*t*-Bu)(CH-*t*-Bu)(*t*-BuNH$_2$)$_2$Cl$_2$ (eq 11) or Re(C-*t*-Bu)(CH-*t*-Bu)(py)$_2$Cl$_2$, but these six-coordinate species also do not react readily with olefins.

$$0.5\ [Re(C\text{-}t\text{-}Bu)(CH\text{-}t\text{-}Bu)(Ar'NH_2)Cl_2]_2 \xrightarrow[-\ Ar'NH_2]{excess\ t\text{-}BuNH_2}$$

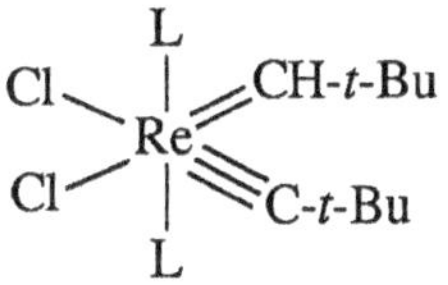

(11)

(L = *t*-BuNH$_2$)

The presence of the aniline in [Re(C-*t*-Bu)(CH-*t*-Bu)(Ar'NH$_2$)Cl$_2$]$_2$ complicates reactions involving some nucleophiles. Fortunately, aniline-free [Re(C-*t*-Bu)(CH-*t*-Bu)Cl$_2$]$_x$ can be prepared by first selectively hydrolyzing the imido ligands in Re(NAr')$_2$(CH-*t*-Bu)(CH$_2$-*t*-Bu) with water (on wet alumina as the catalyst; eq 12) and then treating ReO$_2$(CH-*t*-Bu)(CH$_2$-*t*-Bu) with HCl in dimethoxyethane (eq 13). Therefore [Re(C-*t*-Bu)(CH-*t*-Bu)Cl$_2$]$_x$ can be prepared from Re$_2$O$_7$ or [NH$_4$]ReO$_4$ in four high yield steps (eq 14), and is the precursor from which a compound containing virtually any alkoxide ligand can be prepared (see below).

$$Re(NAr')_2(CHR)(CH_2R) + 2\ H_2O \xrightarrow[-\ 2\ Ar'NH_2]{cat.} ReO_2(CHR)(CH_2R) \qquad (12)$$

(R = *t*-Bu)

$$ReO_2(CHR)(CH_2R) + 2\ HCl \xrightarrow[-\ 2\ H_2O]{DME} [Re(CR)(CHR)Cl_2]_x \qquad (13)$$

(R = *t*-Bu)

$$Re_2O_7 \longrightarrow Re(NAr')_2(py)Cl_3 \longrightarrow Re(NAr')_2(CHR)(CH_2R)$$

$$\longrightarrow ReO_2(CHR)(CH_2R) \longrightarrow [Re(CR)(CHR)Cl_2]_x \quad (R = t\text{-}Bu) \qquad (14)$$

$[Re(C\text{-}t\text{-}Bu)(CH\text{-}t\text{-}Bu)Cl_2]_x$ reacts with two equivalents of lithium *tert*-butoxide in tetrahydrofuran to yield previously reported $Re(C\text{-}t\text{-}Bu)(CH\text{-}t\text{-}Bu)(O\text{-}t\text{-}Bu)_2$[13] quantitatively, while addition of two equiv of $LiOCMe_2(CF_3)$ ($LiOR_{F3}$) or $KOCMe(CF_3)_2$ (KOR_{F6}) yields $Re(C\text{-}t\text{-}Bu)(CH\text{-}t\text{-}Bu)(OR_{F3})_2$ or $Re(C\text{-}t\text{-}Bu)(CH\text{-}t\text{-}Bu)(OR_{F6})_2$, respectively (eq 15). If only one equivalent of lithium alkoxide

$$1/x\ [Re(CR)(CHR)Cl_2]_x + 2\ MOR \xrightarrow[\substack{-2\ MCl \\ M = K\ or\ Li}]{THF} Re(CR)(CHR)(OR)_2 \qquad (15)$$

$(R = t\text{-}Bu)$

is added to $[Re(C\text{-}t\text{-}Bu)(CH\text{-}t\text{-}Bu)Cl_2]_x$ a 50% yield of $Re(C\text{-}t\text{-}Bu)(CH\text{-}t\text{-}Bu)(OR)_2$ is obtained. $Re(C\text{-}t\text{-}Bu)(CH\text{-}t\text{-}Bu)(OR_{F3})_2$, like $Re(C\text{-}t\text{-}Bu)(CH\text{-}t\text{-}Bu)(O\text{-}t\text{-}Bu)_2$, is a low-melting yellow solid that is extremely soluble in pentane. It can be obtained as yellow crystals from pentane at -40°, but these melt to an orange oil at room temperature. All three derivatives sublime readily (30-40 °C, 10^{-5} torr), but show some tendency to decompose when left in the solid state at room temperature for more than several hours. They are stable indefinitely in dilute solution (<0.01 M in C_6D_6) or as solids when stored at -40 °C.

When bisalkoxide complexes are first obtained from $[Re(C\text{-}t\text{-}Bu)(CH\text{-}t\text{-}Bu)Cl_2]_x$, exclusively the *syn* rotamer is observed (eq 16). When a solution of the *syn* rotamer is heated, a mixture of *syn* and *anti* rotamers is obtained, the ratio varying with the steric bulk and electronic nature of the ligands. The H_α resonance for the *anti* rotamer is found downfield of that for the *syn* rotamer and has a relatively high value for

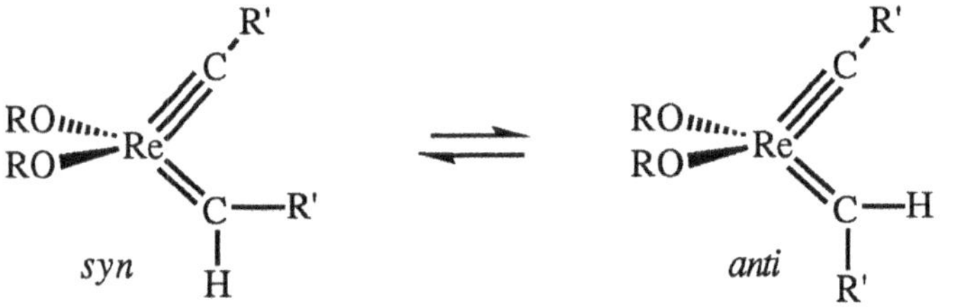

(16)

J_{CH} (~155-160 Hz) compared to that for the *syn* rotamer (J_{CH} = 120-125 Hz). (*Syn* and *anti* rotamers are well-known in $M(CHR')(NAr)(OR)_2$ complexes[23] and recently have been found to interconvert in a first order manner at rates that vary by five orders of magnitude, depending on the nature of OR.[24]) The *syn* and *anti* rotamers can be interconverted either thermally or (more rapidly) photochemically. "Crossover" experiments involving *syn*-$Re(C\text{-}t\text{-}Bu)(CH\text{-}t\text{-}Bu)(O\text{-}t\text{-}Bu)_2$ and *syn*-$Re(CCMe_2Ph)(CHCMe_2Ph)(O\text{-}t\text{-}Bu)_2$ suggest that alkylidene or alkylidyne ligands do not transfer from one metal to another under the conditions employed for interconversion of rotamers. On the other hand, an NMR spectrum of a mixture of $Re(C\text{-}t\text{-}Bu)(CH\text{-}t\text{-}Bu)(OR_{F3})_2$ and $Re(C\text{-}t\text{-}Bu)(CH\text{-}t\text{-}Bu)(O\text{-}t\text{-}Bu)_2$ showed that $Re(C\text{-}t\text{-}Bu)(CH\text{-}t\text{-}Bu)(OR_{F3})(O\text{-}t\text{-}Bu)$ was present within minutes at 25 °C as approximately 90% of the mixture. Therefore, O-*t*-Bu and OR_{F3} ligands exchange rapidly on the chemical time scale at room temperature in complexes of this type. The rates of rotamer equilibration in three $Re(C\text{-}t\text{-}Bu)(CH\text{-}t\text{-}Bu)(OR)_2$ compounds (OR = O-*t*-Bu, OR_{F3}, or OR_{F6}) were found to be first order with free energies of activation at 298K of 25.3, 28.0 and 30.3

kcal mol^{-1}, respectively, and entropies of activation -20, -15, and -16 e.u., respectively. Addition of THF (up to 10 equiv; free exchange is observed at room temperature) did not change the rate of interconversion of rotamers of Re(C-*t*-Bu)(CH-*t*-Bu)$(OR_{F3})_2$ at 113 °C.

Nitrogen or phosphorous base adducts of Re(C-*t*-Bu)(CH-*t*-Bu)$(OR)_2$ species can be prepared by adding excess base to a solution of the alkylidene complex at room temperature. For example, addition of PMe_3 yields five-coordinate monoadducts in which the phosphine ligand is firmly bound to the metal on the NMR time scale. The *syn* rotamer gives rise to a *syn* adduct and a given *syn*/*anti* mixture gives rise to the same mixture of *syn*/*anti* adducts. It should be noted that alkylidene ligand rotation in adducts ceases entirely in the temperature range where rotation was observed for the pseudotetrahedral species. Therefore, as was found in complexes of the type Mo(CH-*t*-Bu)(NAr)$(OR)_2$,[23] alkylidene ligands rotate more readily in pseudotetrahedral species than in higher coordinate species.[23,24] In both *syn* and *anti* rotamers the alkoxide ligands are inequivalent by NMR. One plausible structure is a trigonal bipyramid in which the alkylidyne and alkylidene ligands lie in the equatorial plane (eq 17). This structure is attractive on the basis of the recent crystallographic characterization of *syn* and *anti* adducts of M(CH-*t*-Bu)(NAr)$(OR)_2$ complexes.[23] However, the X-ray structure of

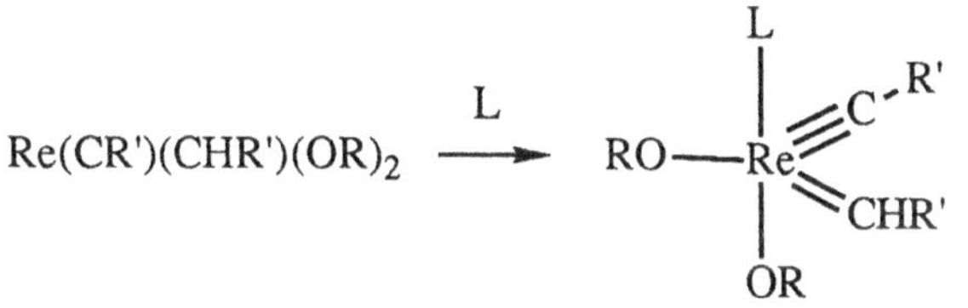

(17)

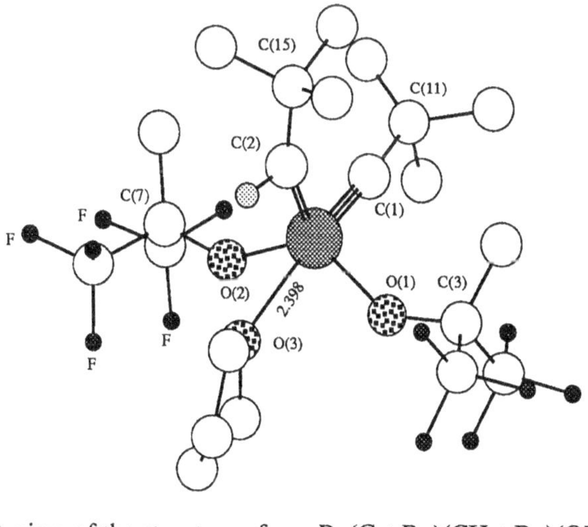

Figure 1. A view of the structure of *syn*-Re(C-*t*-Bu)(CH-*t*-Bu)(OR_{F6})(THF).

syn-Re(C-*t*-Bu)(CH-*t*-Bu)$(OR_{F6})_2$(THF)[10] shows it to be approximately a trigonal bipyramid in which the axial THF is bound *trans* to the neopentylidyne ligand (Figure 1). The neopentylidene Re=C distance (1.85 (1) Å) is slightly shorter than Re=C distances in

[Re(C-*t*-Bu)(CH-*t*-Bu)(Ar'NH$_2$)Cl$_2$]$_2$ (1.89 (1) Å)[10] and Re(C-*t*-Bu)(CH-*t*-Bu)(py)$_2$I$_2$ (1.873 (9) Å),[13] while the Re-C(2)-C(15) angle of 151 (1)° is more comparable to that in Re(C-*t*-Bu)(CH-*t*-Bu)(py)$_2$I$_2$ (150.3 (7)°) than that in [Re(C-*t*-Bu)(CH-*t*-Bu)(Ar'NH$_2$)Cl$_2$]$_2$ (140 (1)°). The Re≡C-C angle (175 (1)°) and Re-C(1) distance (1.75

It remains to be seen the extent to which the structure of adducts of various types may serve as models for the transition state in a metathesis reaction in which a weak adduct presumably is formed first between the metal and an incoming olefin. In complexes of the type M(CH-*t*-Bu)(NAr)(OR)$_2$ the most attractive theory at this time is that an olefin approaches the C/N/O face of the tetrahedral complex. The analogous approach in rhenium alkylidene/alkylidyne complexes would be on the C/C/O face, as shown in equation 17.

Table 1. Structural Data for Re(VII) Alkylidene/Alkylidyne Complexes.

Compound[a]	Re=C(Å)	Re≡C(Å)	Re=C-X(°)	Re≡C-X(°)	ref.
Re(C-*t*-Bu)(CH-*t*-Bu)(py)$_2$I$_2$	1.873 (9)	1.742 (9)	150.3 (7)	174.8 (7)	13
[Re(C-*t*-Bu)(CH-*t*-Bu)(Ar'NH$_2$)Cl$_2$]$_2$	1.89 (1)	1.76 (1)	140 (1)	167 (1)	10
Re(C-*t*-Bu)(CH-*t*-Bu)(OR$_{F6}$)$_2$(THF)	1.85 (1)	1.75 (1)	151 (1)	175 (1)	10
anti-Re(C-*t*-Bu)(CHFc)(OR$_{F6}$)$_2$	1.88 (1)	1.70 (1)	114.8 (7)	174.0 (8)	25
Re(C-*t*-Bu)(CHOEt)(OR$_{F6}$)$_2$(THF)$_2$	1.883 (9)	`1.713 (8)	129.9 (6)	177.7 (8)	25

[a] py = pyridine; Ar' = 2,6-C$_6$H$_3$Me$_2$; OR$_{F6}$ = OCMe(CF$_3$)$_2$; Fc = ferrocenyl; THF = tetrahydrofuran. All are *syn* rotamers unless otherwise noted.

METATHESIS REACTIONS INVOLVING TERMINAL OLEFINS

Reactions between ordinary terminal olefins and complexes of the type Re(CR')(CHR')(OR)$_2$ (R' = *t*-Bu or CMe$_2$Ph) in the absence of coordinating solvent (e.g., THF) are slow (hours) when OR = O-*t*-Bu and fast (seconds) when OR = OR$_{F6}$. For example, the reaction of one equivalent of 1-decene with a mixture of *anti* (H$_\alpha$ at 12.48 ppm) and *syn* (H$_\alpha$ at 11.05 ppm) Re(C-*t*-Bu)(CH-*t*-Bu)(OR$_{F6}$)$_2$ in benzene-d$_6$ produces new resonances for *anti* (H$_\alpha$ at 12.54 ppm; J_{HH} = 7 Hz) and *syn* (H$_\alpha$ at 11.19 ppm; J_{HH} = 5 Hz) rotamers of Re(C-*t*-Bu)[CH(CH$_2$)$_7$Me](OR$_{F6}$)$_2$ in equilibrium with *syn* and *anti*-Re(C-*t*-Bu)(CH-*t*-Bu)(OR$_{F6}$)$_2$ (Figure 2a). If several equivalents of 1-decene are added, most of the Re(C-*t*-Bu)(CH-*t*-Bu)(OR$_{F6}$)$_2$ is consumed and H$_\alpha$ resonances for *syn* and *anti*-Re(C-*t*-Bu)[CH(CH$_2$)$_7$Me](OR$_{F6}$)$_2$ appear. However, the new H$_\alpha$ resonances lose intensity, consistent with decomposition of *syn* and *anti*-Re(C-*t*-Bu)[CH(CH$_2$)$_7$Me](OR$_{F6}$)$_2$ in the presence of excess 1-decene, and the residual Re(C-*t*-Bu)(CH-*t*-Bu)(OR$_{F6}$)$_2$ also eventually all disappears. Terminal alkylidene complexes of this type are relatively stable in dilute solution when they are prepared from an internal olefin (see later), so their instability in the presence of excess 1-decene can be ascribed either to some adverse reaction involving 1-decene, or one involving ethylene that is formed by productive metathesis (see later section). Methyl acetate does not react readily with the sample of *syn* and *anti*-Re(C-*t*-Bu)[CH(CH$_2$)$_7$Me](OR$_{F6}$)$_2$ generated in situ, although H$_\alpha$ resonances shift downfield to varying degrees, consistent with coordination of the methyl acetate to the metal.

Addition of one equivalent of methyl-9-decenoate to a mixture of *syn* and *anti*-Re(C-*t*-Bu)[CH(CH$_2$)$_7$Me](OR$_{F6}$)$_2$ yields a mixture that contains methyl-9-decenoate, *t*-butylethylene, *anti* and *syn*-Re(C-*t*-Bu)[CH(CH$_2$)$_7$CO$_2$Me](OR$_{F6}$)$_2$, and *anti* and *syn*-

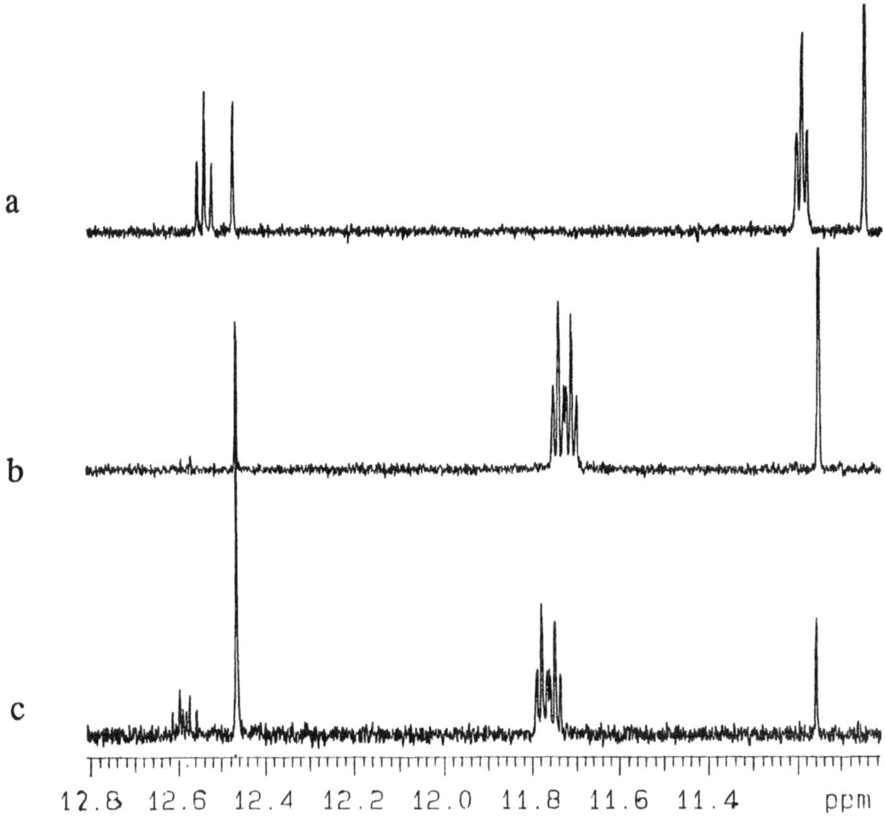

Figure 2. (a) Re(C-*t*-Bu)(CH-*t*-Bu)(OR_{F6})$_2$ (27 mM) in C_6D_6 plus 1 equiv of 1-decene.
(b) Re(C-*t*-Bu)(CH-*t*-Bu)(OR_{F6})$_2$ (34 mM) in C_6D_6 plus (i) 1-decene (0.7 equiv) followed by (ii) methyloleate (4.7 equiv); spectrum recorded after 15 min.
(c) Sample in (b) 14 hours later; H_α intensities had decreased by 30%.

Re(C-*t*-Bu)[CH$(CH_2)_7$Me]$(OR_{F6})_2$ (Figure 2).[25] The exact chemical shift for H_α in each of the four alkylidene complexes depends on conditions, as chemical shifts are sensitive to the presence of the ester group, as noted above. Addition of excess methyl-9-decenoate to Re(C-*t*-Bu)(CH-*t*-Bu)$(OR_{F6})_2$ also leads to loss of intensity for the H_α resonances in *syn* and *anti*-Re(C-*t*-Bu)[CH$(CH_2)_7CO_2$Me]$(OR_{F6})_2$, consistent with sample decomposition. Observation of H_α resonances for *syn* and *anti*-Re(C-*t*-Bu)[CH$(CH_2)_7CO_2$Me]$(OR_{F6})_2$ is important in establishing the identity of complexes formed in reactions involving methyl oleate (see later).

The reaction between *syn*-Re(C-*t*-Bu)(CH-*t*-Bu)$(OR_{F6})_2$ and 1.0 equiv of vinylferrocene in dichloromethane affords crystalline, red Re(C-*t*-Bu)(CHFc)$(OR_{F6})_2$ in modest yield (eq 18). Some benzene-insoluble FcCH=CHFc is formed, and yields of Re(C-*t*-Bu)(CHFe)(OR_{F6}) are not high, consistent with some sample decomposition,

Re(C-*t*-Bu)(CH-*t*-Bu)$(OR_{F6})_2$ + 1.0 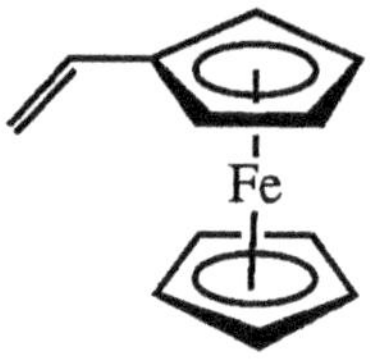 $\underset{\longleftarrow}{\xrightarrow{CH_2Cl_2}}$

Re(C-*t*-Bu)(CHFc)$(OR_{F6})_2$ + H_2C=CH(*t*-Bu) (18)

probably involving ethylene that is generated via productive metathesis. The Re(C-*t*-Bu)(CHFc)$(OR_{F6})_2$ isolated in this reaction is typically 85-90% *anti*. Photolysis of *anti*-Re(C-*t*-Bu)(CHFc)$(OR_{F6})_2$ with a medium pressure Hg lamp yields a mixture containing both *anti*-Re(C-*t*-Bu)(CHFc)$(OR_{F6})_2$ (δH_α = 13.15 ppm) and *syn*-Re(C-*t*-Bu)(CHFc)$(OR_{F6})_2$ (δH_α = 11.72 ppm). The reaction between *syn* and *anti* Re(C-*t*-Bu)(CHFc)$(OR_{F6})_2$ and *t*-butylethylene proceeds readily to yield *syn* and *anti*-Re(C-*t*-Bu)(CH-*t*-Bu)$(OR_{F6})_2$ and vinylferrocene, i.e., the reaction shown in equation 18 is reversible. An X-ray study of *anti*-Re(C-*t*-Bu)(CHFc)$(OR_{F6})_2$[25] shows the expected distorted tetrahedral core structure in which the Re≡C(2) bond distance is 1.88 (1) Å, the Re=C_α-C_β angle is relatively acute (114.8 (7)°), and the rhenium-carbon triple bond length is 1.70 (1) Å.

Simple terminal olefins such as propylene, 1-butene, and 1-decene react much more slowly with *syn*-Re(C-*t*-Bu)(CH-*t*-Bu)$(OR_{F6})_2$ in THF or 1,2-dimethoxyethane (DME) than in a noncoordinating solvent such as C_6D_6, and the new alkylidene complexes are formed in high yield.[25] The reaction is faster if only a few equivalents of THF or DME are present in a non-coordinating solvent. New alkylidene complexes can be prepared conveniently by treating *syn*-Re(C-*t*-Bu)(CH-*t*-Bu)$(OR_{F6})_2$ with a slight excess of the appropriate alkene in the presence of DME (~1-2 equiv) in benzene (eq 19; R = Me, Et, Ph). Addition of a stoichiometric amount of olefin limits productive metathesis to give

Re(C-*t*-Bu)(CH-*t*-Bu)$(OR_{F6})_2$ + CH_2=CHR + DME $\xrightarrow{C_6H_6}$

syn-Re(C-*t*-Bu)(CHR)$(OR_{F6})_2$(DME) + CH_2=CH-*t*-Bu (19)

ethylene, and the back reaction between Re(C-*t*-Bu)(CHR)(OR_{F6})$_2$(DME) complexes and *t*-butylethylene is relatively slow, possibly because DME is bound more strongly in the Re(CHR) complex than in the Re(CH-*t*-Bu) complex for steric reasons. The DME adducts have the added advantage of being solids which can be recrystallized from pentane.

syn-Re(C-*t*-Bu)(CH-*t*-Bu)(OR_{F6})$_2$ reacts cleanly with ethyl vinyl ether (1 to 5 equiv) in benzene or toluene in the presence of THF, or even in toluene at -80 °C or in neat THF at -10°C, to produce pale orange, crystalline *syn*-Re(C-*t*-Bu)(CHOEt)(OR_{F6})$_2$(THF)$_2$ and neohexene quantitatively (eq 20). No productive metathesis is observed. This reaction is dramatically faster (at least an order of magnitude) than an analogous reaction involving an ordinary terminal olefin such as 1-butene. We believe at this stage that the vinyl ether reacts more rapidly with the electrophilic metal center predominantly because it is more electron rich than an ordinary olefin, not because of some assistance by coordination of the ether oxygen to the metal.

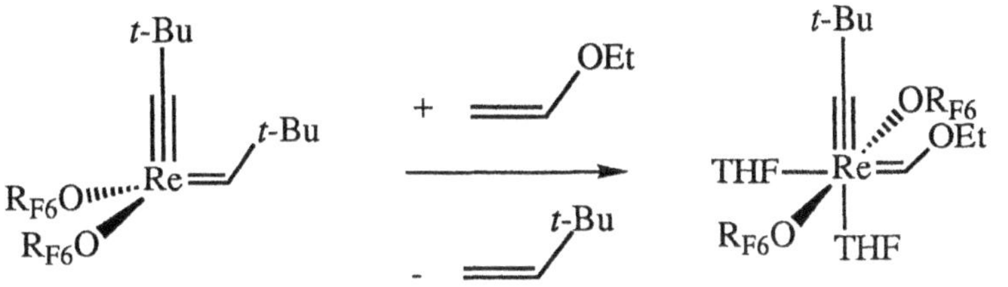

(20)

An X-ray study of *syn*-Re(C-*t*-Bu)(CHOEt)(OR_{F6})$_2$(THF)$_2$ (Figure 3; Table 1) showed it to be a distorted octahedron in which the Re≡C(1) bond length (1.713 (8) Å) is comparable to the Re≡C bond length found in *anti*-Re(C-*t*-Bu)(CHFc)(OR_{F6})$_2$ (1.70 (1) Å) and the Re=C(6) bond length (1.883 (9) Å) is similar to that found in other six-coordinate high oxidation state rhenium alkylidene complexes. The C(6)-O(1) bond (1.35 (1) Å) is approximately 0.09 Å shorter than the O(1)-C(7) bond, as one might expect, but it is significantly longer than that observed in a typical octahedral Fischer-carbene complex such as *cis*-[Mn(CO)$_5$](CO)$_4$Re[C(OMe)Me] (1.299 (8) Å).[26] This finding is consistent with very little or no π bonding between C(6) and O(1), in contrast to the C-O π bonding found in a typical Fischer-type carbene complex.[27,28]

Low temperature NMR studies of *syn*-Re(C-*t*-Bu)(CHOEt)(OR_{F6})$_2$(THF)$_2$ in toluene-d_8 show that the THF ligands are inequivalent and exchange readily at different rates with free THF; the THF that exchanges most readily we propose to be that more weakly bound *trans* to the neopentylidyne ligand (Figure 3). Re(C-*t*-Bu)(CHOEt)(OR_{F6})$_2$(THF)$_2$ (approximately 0.01M) is stable in C_6D_6 over a period of several hours in the presence of several equivalents of ethyl vinyl ether. The relative stability of the Re(CHOEt) complex in the presence of excess ethyl vinyl ether might be ascribed to slow productive metathesis to give ethylene, or to stabilization of the complex in the presence of an ether donor, or both. Base-free, four-coordinate *syn*-Re(C-*t*-Bu)(CHOEt)(OR_{F6})$_2$ can be prepared in situ, but solutions decompose when concentrated to give [Re(C-*t*-Bu)(OR_{F6})$_2$]$_2$[29] and EtOCH=CHOEt as the only identifiable product. Photolysis of the *syn* rotamer gives a mixture that contains both the *syn* and *anti* rotamers.

Phenyl vinyl sulfide reacts with *syn*-Re(C-*t*-Bu)(CH-*t*-Bu)(OR_{F6})$_2$ in tetrahydrofuran to afford *syn*-Re(C-*t*-Bu)(CHSPh)(OR_{F6})$_2$(THF)$_2$ in high yield as orange crystals (eq 21). NMR data suggest that the structure of *syn*-Re(C-*t*-Bu)(CHSPh)(OR_{F6})$_2$(THF)$_2$ is analogous to that of Re(C-*t*-Bu)(CHOEt)(OR_{F6})$_2$(THF)$_2$. Photolysis of *syn*-Re(C-*t*-Bu)(CHSPh)(OR_{F6})$_2$(THF)$_2$ yields a 1:1 mixture of *syn* (δH_α = 12.31, δC_α = 238.6 in

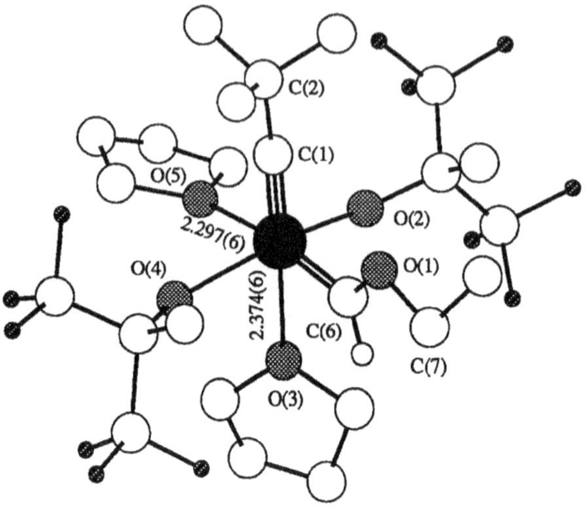

Figure 3. A view of the structure of *syn*-Re(C-*t*-Bu)(CHOEt)(OR_{F6})$_2$(THF)$_2$.

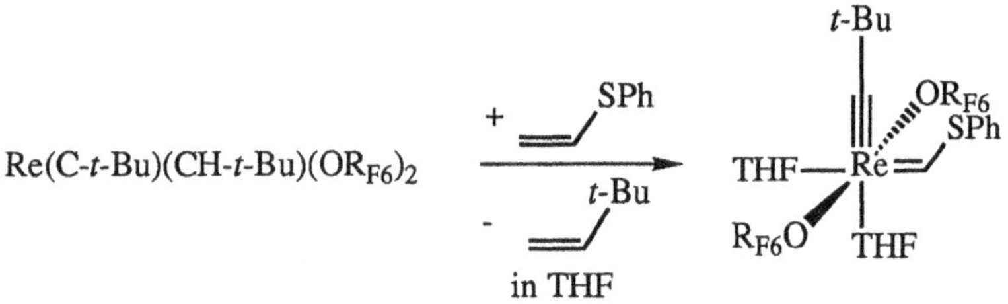

(21)

CD_2Cl_2, $J_{CH\alpha}$ = 143 Hz) and *anti* (δH_α = 12.50, δC_α = 224.3 in CD_2Cl_2, $J_{CH\alpha}$ = 184 Hz) rotamers. The 184 Hz CH coupling constant in the *anti* isomer is the largest yet observed in a d^0 alkylidene complex, and is consistent with some bonding of sulfur to the rhenium and consequent rehybridization of the CH to one having a higher p character. Re(C-*t*-Bu)(CHSPh)(OR_{F6})$_2$(THF)$_2$ also is formed in seconds upon adding phenyl vinyl sulfide to Re(C-*t*-Bu)(CHOEt)(OR_{F6})$_2$(THF)$_2$ in C_6D_6, a fact that suggests that degenerate metathesis (methylene exchange) is fast. Re(C-*t*-Bu)(CHSPh)(OR_{F6})$_2$(THF)$_2$ is relatively stable in the absence of phenyl vinyl sulfide, but it decomposes to a large extent over a period of one hour in the presence of excess phenyl vinyl sulfide, in contrast to the relative stability of Re(C-*t*-Bu)(CHOEt)(OR_{F6})$_2$(THF)$_2$ in the presence of ethyl vinyl ether.

1-Vinyl-2-pyrrolidinone reacts with *syn* or *anti*-Re(C-*t*-Bu)(CH-*t*-Bu)(OR_{F6})$_2$ in methylene chloride or other noncoordinating solvents to give orange, crystalline *anti*-Re(C-*t*-Bu)[CH(N(CH_2)$_3$CO)](OR_{F6})$_2$ (eq 22; $J_{CH\alpha}$ = 173 Hz). This compound shows no evidence for isomerization to a *syn* rotamer, is stable in the solid state in the absence of coordinating solvent, and shows a decreased carbonyl stretching frequency (1614 cm^{-1} in Nujol) in its IR spectrum relative to that in 1-vinyl-2-pyrrolidinone (1723 cm^{-1}). All three facts are consistent with coordination of the carbonyl oxygen to the metal and therefore stabilization of the *anti* rotamer. The two alkoxide ligands are equivalent in this complex; the structure shown in equation 22 is one idealized possibility.

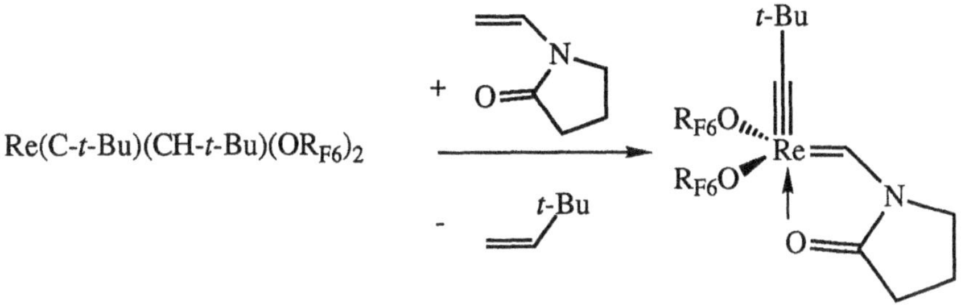

(22)

REACTIONS BETWEEN $Re(C\text{-}t\text{-}Bu)(CH\text{-}t\text{-}Bu)(OR_{F6})_2$ AND INTERNAL OLEFINS

syn-$Re(C\text{-}t\text{-}Bu)(CH\text{-}t\text{-}Bu)(OR_{F6})_2$ (~0.01 M) reacts over a period of several hours in C_6D_6 with *cis*-2-pentene to form a mixture of *syn* and *anti* ethylidene and propylidene species along with the expected amounts of 5,5-dimethyl-2-pentene and 2,2-dimethyl-3-hexene, the primary metathesis products, all quantitatively according to NMR integration versus an internal standard. In the presence of 10 equiv of *cis*-3-hexene *syn*-$Re(C\text{-}t\text{-}Bu)(CH\text{-}t\text{-}Bu)(OR_{F6})_2$ is converted completely into *syn* and *anti*-$Re(C\text{-}t\text{-}Bu)(CHEt)(OR_{F6})_2$ in six hours in C_6D_6 ([Re] = 22 mM, eq 23). The *syn* and *anti* propylidene complexes are stable at concentrations < 10 mM for days in C_6D_6 or CD_2Cl_2 in the presence of *cis*-3-hexene, but $Re(C\text{-}t\text{-}Bu)(CHEt)(OR_{F6})_2$ decomposes when solutions containing it (and *cis*-3-hexene and the initial metathesis products) are taken to dryness *in vacuo*. These findings suggest that the decomposition that results in reactions between *syn*-$Re(C\text{-}t\text{-}Bu)(CH\text{-}t\text{-}Bu)(OR_{F6})_2$ and excess *terminal* olefins noted earlier cannot be ascribed to an instability of the new terminal alkylidene complexes *per se*.

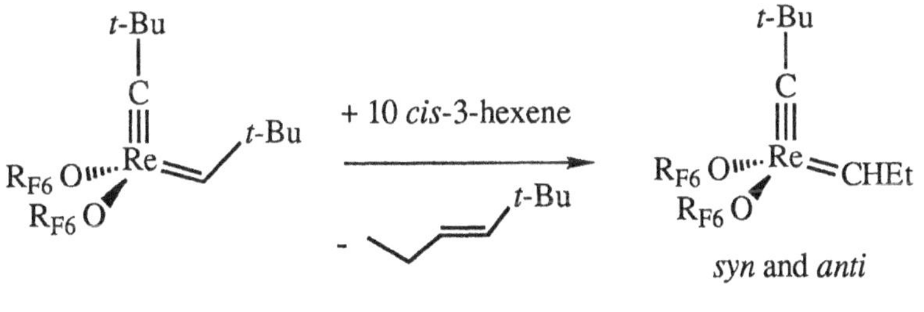

(23)

Addition of 100 equiv of *cis*-2-pentene to *syn*-$Re(C\text{-}t\text{-}Bu)(CH\text{-}t\text{-}Bu)(OR_{F6})_2$ in benzene yielded a 1:2:1 mixture of 2-butenes, 2-pentenes, and 3-hexenes in 2.5 h at 25 °C. When an additional 100 equiv of *cis*-2-pentene was added to this mixture, equilibrium was restored in less than 30 minutes. Observation of an induction period and an increase in the rate of metathesis with time is consistent with a large increase in the rate of metathesis by complexes containing less bulky propylidene and ethylidene ligands (versus a neopentylidene ligand), a finding that qualitatively has been observed in several other circumstances involving well-characterized metathesis catalysts of the type $M(CHR)(NAr)(OR')_2$.[30] This catalyst system is stable indefinitely at low concentrations (< 10 mM), and its absolute activity at room temperature is estimated to be approximately 250 equiv h^{-1} for metathesis of *cis*-2-pentene, a rate that is perhaps as much as two orders of magnitude slower than that observed for $W(NAr)(CH\text{-}t\text{-}Bu)(OR_{F6})_2$.[16]

Syn-Re(C-*t*-Bu)(CH-*t*-Bu)$(OR_{F6})_2$ reacts slowly with methyl oleate. After 12 h in C_6D_6 five equivalents of methyl oleate converted 40% of the initial *syn*-Re(C-*t*-Bu)(CH-*t*-Bu)$(OR_{F6})_2$ (ca. 0.01 M) to two new alkylidene complexes with H_α resonances that we now know are characteristic of those for *syn*-Re(C-*t*-Bu)[CH$(CH_2)_7$Me]$(OR_{F6})_2$ and *syn*-Re(C-*t*-Bu)[CH$(CH_2)_7CO_2$Me]$(OR_{F6})_2$ (Figure 2b). Methyl oleate reacts much more rapidly with *syn* and *anti*-Re(C-*t*-Bu)[CH$(CH_2)_7$Me]$(OR_{F6})_2$ than with *syn* and *anti*-Re(C-*t*-Bu)(CH-*t*-Bu)$(OR_{F6})_2$. Addition of 5 equivalents of methyl oleate to a sample of *syn* and *anti*-Re(C-*t*-Bu)[CH$(CH_2)_7$Me]$(OR_{F6})_2$ produces alkylidene H_α resonances for *syn*-Re(C-*t*-Bu)[CH$(CH_2)_7$Me]$(OR_{F6})_2$ and *syn*-Re(C-*t*-Bu)[CH$(CH_2)_7CO_2$Me]$(OR_{F6})_2$ initially (Figure 2b), and only with time (and concomitant decomposition) are the *anti* alkylidene complexes formed (Figure 2c).

Addition of 50 equiv of methyl oleate to Re(C-*t*-Bu)(CH-*t*-Bu)$(OR_{F6})_2$ in dichloromethane produced a 1:2:1 mixture of $Me(CH_2)_7CH{=}CH(CH_2)_7Me$, $Me(CH_2)_7CH{=}CH(CH_2)_7CO_2Me$, and $MeO_2C(CH_2)_7CH{=}CH(CH_2)_7CO_2Me$ in 12 h. After this mixture had stood for 24 hr another 50 equiv of methyl oleate could be brought to equilibrium in less than 7 hr, again demonstrating an increase in the rate of metathesis as the neopentylidene initiator is consumed. As we now expect, the initial rate of metathesis can be increased by first forming a more reactive alkylidene complex. For example, treatment of Re(C-*t*-Bu)(CH-*t*-Bu)$(OR_{F6})_2$ with 10 equiv of *cis*-3-hexene for several hours followed by adding 50 equiv of methyl oleate lead to equilibrium between $Me(CH_2)_7CH{=}CH(CH_2)_7Me$, $Me(CH_2)_7CH{=}CH(CH_2)_7CO_2Me$, and $MeO_2C(CH_2)_7CH{=}CH(CH_2)_7CO_2Me$ in 3 h. Addition of 100 equiv of additional methyl oleate to this remaining mixture led to an equilibrium mixture in six hours. Therefore the absolute rate for methyl oleate metathesis by this catalyst system appears to be approximately 17 equiv h^{-1} at room temperature, at least an order of magnitude slower than metathesis of *cis*-2-pentene.

REACTIONS INVOLVING ETHYLENE

Reactions between rhenium(VII) alkylidene alkylidyne complexes of the type Re(CR')(CHR')$(OR)_2$ (R' = *t*-Bu or CMe_2Ph) and ethylene rapidly lead to mixtures of organometallic species, at least one of which is often a complex analogous to complexes of the type $[Re(C\text{-}t\text{-}Bu)(OR)_2]_2$ that contain an unsupported Re=Re double bond.[29] Upon closer examination of reactions involving ethylene we found that ethylene adds across the C=Re≡C system in a 3+2 cycloaddition that we believe has no precedent.[31]

Ethylene reacts with *syn*-Re(C-*t*-Bu)(CH-*t*-Bu)(O-*t*-Bu)$_2$ below -20°C to give one isomer of a compound (**1a**) in low yield whose proton NMR spectra are analogous to those of structurally characterized trigonal bipyramidal tungstacyclobutane complexes such as W[CH(*t*-Bu)CH_2CH_2](NAr)[OCMe$(CF_3)_2$]$_2$ (Table 2).[30,32,33] Another compound (**2a**) is also formed below -20 °C (exclusively at 25°) in which the former neopentylidyne α carbon atom (here called the γ carbon atom; see Scheme) and the former neopentylidene α carbon atoms are *both* coupled to ^{13}C (when $^{13}CH_2{}^{13}CH_2$ is employed) by ca. 30 Hz. We propose that this maroon, pentane-soluble powder is a rhenacyclopentene complex (**2a**; see Scheme). As a solid at -40 °C **2a** is stable indefinitely, but preliminary studies suggest that in C_6D_6 at room temperature it decomposes to give some *syn*-Re(C-*t*-Bu)(CH-*t*-Bu)(O-*t*-Bu)$_2$ and ethylene, in addition to as yet unidentified products. Analogous reactions involving *syn*- or *anti*-Re(C-*t*-Bu)(CH-*t*-Bu)$(OR_{F3})_2$ yield a TBP rhenacyclobutane complex below -50 °C (**1b**) and an analogous rhenacyclopentene complex above ca. -30 °C (**2b**). Compound **2b** decomposes above ca. 5 °C to as yet uncharacterized products. Analogous reactions

Table 2. Pertinent NMR Data for Rhenacyclobutane Complexes and Rhenacyclopentene Complexes.[a]

Rhenacycle	δH_α	δH_β	$\delta H_{\alpha'}$	δC_α	J_{CC}[b]	δC_β	J_{CC}	$\delta C_{\alpha'}$	J_{CC}	δC_γ
Re(C-*t*-Bu)[CH_2CH_2CH(*t*-Bu)](O-*t*-Bu)$_2$ (**1a**)	7.10	-2.03 -2.40								
Re(C-*t*-Bu)[CH_2CH_2CH(*t*-Bu)](OR_{F3})$_2$ (**1b**)	6.55	-2.58 -2.77	5.37 5.50			-4.1	12	94.3	n.o.	286.2
Re(C-*t*-Bu)[CH_2CH_2CH(*t*-Bu)](OR_{F6})$_2$ (**1c**)	6.85	-2.36 -2.73	5.51 5.47	142.3	7	-4.9	13	94.5	n.o.	291.1
Re[C(*t*-Bu)CH_2CH_2CH(*t*-Bu)](O-*t*-Bu)$_2$ (**2a**)	9.49	0.63 1.89	3.13 4.15	60.1	31	57.0	39	83.5	29	209
Re[C(*t*-Bu)CH_2CH_2CH(*t*-Bu)](OR_{F3})$_2$ (**2b**)	12.07	0.30[c]	3.76[c]	77.8	30	42.3	37	85.0	29	214.8
Re[C(*t*-Bu)CH_2CH_2CH(SPh)](OR_{F6})$_2$	10.64	2.90 1.96	3.49 3.15	77.8	30	49.9	38	62.3	28	276
Re[C(*t*-Bu)CH_2CH_2CH(SPh)](OR_{F6})$_2$(PMe_2Ph)	12.0	3.57[d] 1.89	2.25[d] 0.12							
Re[C(*t*-Bu)CH_2CH_2CH(SPh)](OR_{F6})$_2$(PMe_3)	11.30	2.71 0.89	3.94 2.17	63.1		46.0	40	63.2		

[a] See reference 31; n.o. = not observable; see Scheme for labeling format. [b] Hz. [c] Other resonance obscured.
[d] α' and β assignments not confirmed.

involving *syn-* or *anti-*Re(C-*t*-Bu)(CH-*t*-Bu)(OR_{F6})$_2$ yield a TBP rhenacyclobutane complex (**1c**) at low temperatures that loses ethylene at 25° in solution or in the solid state to reform Re(C-*t*-Bu)(CH-*t*-Bu)(OR_{F6})$_2$ (along with other decomposition products) in a few minutes at 25 °C; no rhenacyclopentene complex of type **2** is observed. Re(C-*t*-Bu)(CHSPh)(OR_{F6})$_2$, on the other hand, reacts with ethylene to give a stable rhenacyclopentene complex (**2d**); no TBP rhenacyclobutane complex of type **1** is observed. TBP rhenacyclobutane complexes do not form readily upon cooling samples of rhenacyclopentene complexes to a temperature where the rhenacyclobutane complex is known to be stable. All compounds decompose in the presence of excess ethylene to as yet not fully characterized products.

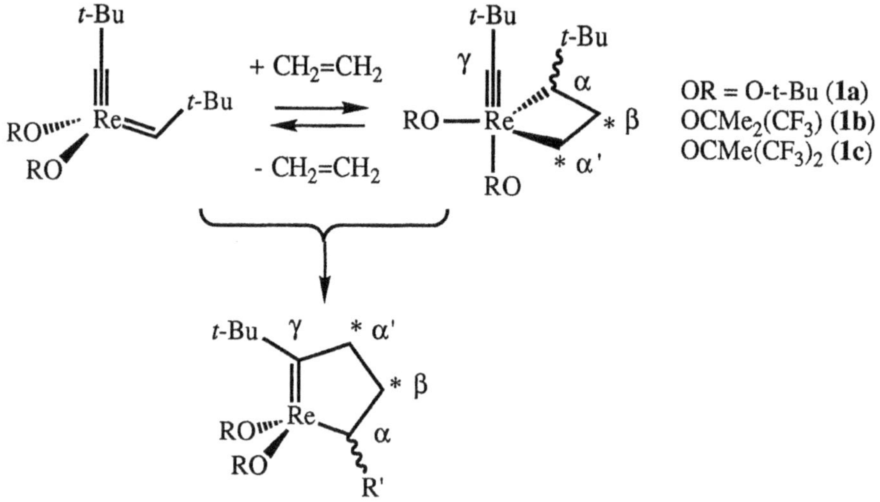

OR = O-*t*-Bu, R' = *t*-Bu (**2a**); OR = $OCMe_2(CF_3)$, R' = *t*-Bu (**2b**);
OR = $OCMe(CF_3)_2$, R' = SPh (**2d**); * = ^{13}C.

In contrast to **2a** and **2b, 2d** forms a relatively stable monoadduct upon addition of PMe_2Ph in which the essential features of the rhenacyclopentene ring are retained (Table 2). An X-ray study showed that Re[CH(SPh)CH_2CH_2C(*t*-Bu)](OR_{F6})$_2$(PMe_2Ph) is roughly a trigonal bipyramid that contains an axial alkoxide and an axial phosphine ligand (Figure 4). The Re-C(4) bond length (1.94 Å) is consistent with it being a double bond and the Re-C(1) bond length (2.14 Å) is consistent with it being a single bond, as one would expect if the ethylene has added across the C=Re≡C system. The ReC_5 ring is envelope-shaped, with C(2) being tipped significantly below the Re-C(1)-C(4) plane. An important feature of the structure is the strong Re-S interaction (Re-S = 2.36 Å). We have no reason to believe that the structures of **2a**, **2b**, and **2d** do not contain the essential features of the ReC_4 ring found in **3d**, judging from the similarity of the NMR data for all complexes of type **2** and the phosphine adducts of **2d** (Table 2). Coordination of sulfur could be part of the reason why adducts of **2d** are stable toward loss of ethylene.

Details concerning the mechanism of forming these Re(V) metallacyclopentene complexes are not yet available. Knowing the precise mechanism is extremely important because these results may have some relevance to the question concerning the mechanism of 3+2 cycloadditions involving olefins and oxo or imido ligands in Os(VIII) complexes.[34,35] Experiments are under way that should distinguish between the direct conversion of TBP rhenacyclobutane complexes (**1**) to rhenacyclopentene complexes (**2**), a competition between the parent alkylidyne alkylidene complex and ethylene to form **1** or **2**, and a more convoluted and less obvious mechanism involving proton migrations. What is certain is that the formation of rhenacyclopentene complexes is a fourth way in which d^0 metals in well-characterized alkylidene complexes now are known to be reduced; the more common are (i) rearrangement of metallacyclobutane complexes to olefin complexes, (ii) rearrangement of alkylidene complexes to olefins (for Ta), and (iii) bimolecular coupling of alkylidene ligands to give olefins.[30] This result also points out the limitation of alkylidyne ligands as ancillary ligands in metathesis reactions that involve ethylene or terminal olefins and Re(VII) alkylidyne/alkylidene complexes, at least those that contain neopentylidyne or neophylidyne ligands.

RING OPENING METATHESIS POLYMERIZATION

Norbornene is the prototypical monomer that is often used in order to test the efficiency of a catalyst in ROMP.[36,37] $Re(CCMe_2Ph)(CHCMe_2Ph)(O\text{-}t\text{-}Bu)_2$ reacts impractically slowly with norbornene. However, $Re(CCMe_2Ph)(CHCMe_2Ph)(OR_{F3})_2$ reacts with 2-25 equiv of norbornene (NBE) in C_6D_6 or CD_2Cl_2 at 25 °C to yield the new alkylidene complexes, *anti* and *syn*-$Re(CCMe_2Ph)[(CHC_5H_8CH)_xCHCMe_2Ph](OR_{F3})_2$, that have H_α resonances at 11.94 ppm and 10.58 ppm (ratio of approximately 1:6). The living oligomers are very sensitive to oxygen, but they are stable for days (at a conc < 10 mM) under an inert atmosphere. Only 90% of $Re(CCMe_2Ph)(CHCMe_2Ph)(OR_{F3})_2$ is consumed by 10 equiv of norbornene at room temperature in 1-2 h. From these and similar data we can calculate a value of $k_p/k_i = 5.0$ for norbornene, where k_p is the rate constant for propagation and k_i the rate constant for initiation.

The rate of norbornene polymerization is first order in catalyst and first order in NBE over greater than three half lives when the catalyst concentration is less than 10 mM (correlation coefficients > 0.995). At higher catalyst concentrations we suspect that bimolecular decomposition reactions that destroy the alkylidene ligand compete with polymerization. We have found that $k_p = 0.027$ (3) $M^{-1}s^{-1}$ at 23 °C. Since $k_p/k_i = 5$ (0.5), $k_i = 0.0050\ M^{-1}s^{-1}$. PolyNBE can be cleaved from the metal by adding a large excess of benzaldehyde (e.g., 50 equiv). GPC analysis showed the samples to be essentially monodisperse, indicative of a well-behaved living polymerization. By proton NMR we can say that the double bonds in the polymer are ~ 60% *cis*.

2,3-Dicarbomethoxynorbornadiene is polymerized by $Re(CCMe_2Ph)(CHCMe_2Ph)(OR_{F3})_2$ as an initiator, but considerably more slowly than NBE. Five equiv of monomer were not consumed completely in 6-7 h at room temperature (0.24 M in Re) in C_6D_6. Raising the temperature of the reaction mixture to 60 °C for 15 minutes led to consumption of all monomer, with no apparent decomposition of living alkylidene species. The polymers were cleaved off the metal by adding benzaldehyde and shown by gel permeation chromatography to have low polydispersities, although polydispersities >1.2 are not uncommon; a fine balance of catalyst concentration, reaction time, and temperature are required to obtain polymers with the lowest polydispersities. The proton NMR spectrum of a 100-mer suggests that the double bonds in the polymer are approximately 55% *trans*, in contrast to the 90 to 95% *trans* polymer prepared with $Mo(CH\text{-}t\text{-}Bu)(NAr)(O\text{-}t\text{-}Bu)_2$ as the initiator.[38]

5-Cyanonorbornene can be polymerized to give polymers with polydispersities of 1.1-1.6 employing $Re(CCMe_2Ph)(CHCMe_2Ph)(OR_{F3})_2$ as the initiator in dichloromethane. At a catalyst concentration of 7 mM, 100 equiv of monomer was consumed in two hours to give a polymer having a PDI = 1.15. However the molecular weight of the major fraction was not proportional to the number of equiv of the monomer. Polymers with broader molecular weight distributions (PDI = 1.4-1.6) were obtained when solutions were more concentrated in catalyst (> 10 mM).

Preliminary (unpublished) studies suggest that ROMP reactions initiated by Re(C-*t*-Bu)(CHPh)(OR_{F6})$_2$(DME) in toluene are the best behaved so far and will yield polymers with the lowest polydispersities. Norbornene and methyltetracyclododecene (MTD) have been polymerized smoothly and in high yield, and the polymers have been cleaved from the metal by treating the living polymers with benzaldehyde, a vinyl ether, or styrene. The k_p/k_i ratio appears to be 1 to 2, an extremely low value that is useful for preparing low polydispersity polymers. We expect to be able to demonstrate that reactions initiated by Re(C-*t*-Bu)(CHPh)(OR_{F6})$_2$(DME) are living, that block copolymers can be prepared routinely, and that functionality tolerance is high.

Discussion

Our finding that complexes that contain rhenium in its highest possible oxidation state are active for the metathesis of ordinary olefins is consistent with what is known about metathesis by well-characterized and well-defined catalysts that contain molybdenum or tungsten,[30] the other two metals that are active for metathesis by classical catalyst systems.[5] Not surprisingly, therefore, the chemistry of well-defined Mo, W, and Re complexes that give rise to alkylidene complexes is similar in many respects; alkylidene ligands are formed in α hydrogen abstraction reactions,[11] four-coordinate species are the most active and the least stable toward bimolecular decomposition,[4] and the activity of the catalysts can be controlled by varying the nature of the alkoxide ligands, the most active catalysts being those that contain the most electron-withdrawing ligands. Donor solvents such as THF or DME can dramatically alter the reactivity of rhenium catalysts, as one might expect on the basis of findings for phosphine adducts of Mo(VI)[23] and W(VI) alkylidene complexes[1,23,39] and studies involving ring-opening of cyclooctatetraene by W(CH-*t*-Bu)(NAr)(OR_{F6})$_2$.[40]

Alkylidyne complexes of rhenium in its highest possible oxidation state are not common. Examples are Re($CSiMe_3$)(CH_2SiMe_3)$_3$Cl, which has been isolated in ~10% yield from the reaction between Re(THF)$_2$$Cl_4$ and Me_3SiCH_2MgCl in THF,[41] ReCp*Br_3(C-*t*-Bu),[42] Re(C-*t*-Bu)(NAr)X_2 species (X = Cl, alkoxide, etc.),[43] and [ReH_2(C-n-Bu)(2-mercaptoquinolinide)(PPh_3)$_2$]$^+$.[44] The alkylidyne ligand plays an important role in the rhenium systems discussed here. First and foremost it allows one to synthesize neutral, four-coordinate rhenium alkylidyne analogs of Mo or W imido complexes of the type M(CH-*t*-Bu)(NAr)(OR)$_2$. One can argue that four-coordination is the most logical coordination number for metathesis catalysts, since such species can be reasonably stable in solution, yet can readily react with an olefin to yield a fluxional five-coordinate metallacyclobutane intermediate. Second, the Re≡C bond does not react readily with an olefin compared to the rate at which the Re=C bond reacts. Third, the alkylidyne ligand raises the barrier for interconversion of alkylidene rotamers compared to what it is in most complexes of the type M(CH-*t*-Bu)(N-2,6-C_6H_3-i-Pr_2)(OR)$_2$. The higher barrier might be ascribed to the relative unavailability of an orbital that lies in the C≡Re=C plane that can stabilize an alkylidene ligand that has rotated by 90°; that orbital can be made available in complexes of the type M(CH-*t*-Bu)(N-2,6-C_6H_3-i-Pr_2)(OR)$_2$ if the lone pair on the imido ligand is not effectively donated to the metal, but in alkylidyne complexes that orbital is involved in forming a covalent bond between Re and the alkylidyne carbon atom. The rate at which an alkylidene ligand can rotate relative to the rate at which it reacts with an incoming olefin could be important in determining the cis/trans selectivity in metathesis reactions involving ordinary olefins or in ROMP reactions.[24] There undoubtedly are other more subtle but important roles for the alkylidyne ligand in systems of this sort that we do not yet appreciate, some of which could differ from the roles played by the imido ligand in Mo and W systems.

Prior to this work no documented homogeneous rhenium metathesis catalyst had been reported, and no definitive evidence had been presented that implicated Re(VII) as the active species in heterogeneous catalyst systems. There appears to be no reason to expect that the oxidation state of the metal in the active site in heterogeneous Re-based metathesis catalysts will be different from that observed in these new homogeneous systems. An important question is what type of species might be present in active heterogeneous rhenium metathesis catalyst systems, and in particular whether alkylidyne complexes might not be formed in classical metathesis systems. An intriguing recent finding is that Re(VII) alkylidyne complexes can be formed from lower oxidation state species from acetylenes, even internal ones.[44] With a slight stretch of the imagination one could imagine that alkylidyne ligands could be formed also from olefins and rhenium in a low oxidation state. As an alternative one must consider the possibility that at least in heterogeneous systems, cationic rhenium centers analogous to hypothetical $[Re(CHR)(O)(OR')_2]^+$ might be the most active. However, we have found so far that homogeneous species of the type $[Re(CHR)(NR')(OR'')_2]^+$ are transformed into inactive amido alkylidyne complexes, $[Re(CR)(NHR')(OR'')_2]^+$, and that complexes of the type $Re(CHR)(NR')(OR'')_3$ are inactive.[45]

We have always been somewhat surprised by the stability of the alkylidyne ligand in the presence of olefins. It is now becoming clear that "high oxidation state" chemistry of this type is being pushed to its limit of viability, since we now know that ethylene *will* react with the alkylidyne ligand and reduce the metal. Interestingly this process *appears* to be reversible under some conditions and therefore conceivably could be a method of *forming* alkylidyne/alkylidene complexes from a Re(V) metallacyclopentene intermediate, which in turn could be formed from an even lower oxidation state species. The mechanism of this "3+2 cycloaddition" across the C≡Re=C unit and its relevance, if any, to addition of an olefin to osmium(VIII) containing an OsO_2, OsO(NR), or $Os(NR)_2$ unit has yet to be determined. It also remains to be determined whether related 3+2 cycloaddition reactions, e.g., across a C≡Re=N unit, are possible.

Several important differences between Re and Mo or W have been revealed in this study. First, metathesis activity appears to be significantly lower for four-coordinate Re catalysts compared to Mo or W catalysts. Lower activity does not appear to be linked to significantly different rates of reactivity of *syn* versus *anti* rotamers, as is the case in certain circumstances for $Mo(CH\text{-}t\text{-}Bu)(NAr)(OR)_2$ catalysts.[24] Activity also appears to be attenuated more dramatically for rhenium complexes upon changing from hexfluoro-*t*-butoxide to *t*-butoxide ligands. For example, $Re(C\text{-}t\text{-}Bu)(CH\text{-}t\text{-}Bu)(O\text{-}t\text{-}Bu)_2$ does not react with norbornene. At this stage it is not known whether fundamental differences between W(VI) and Re(VII) are the most plausible reason for the difference in reactivity, or whether the presence of a M≡C bond instead of an M=N bond is at least as important a factor in determining reactivity.

Ever since the discovery of "d^0" alkylidene complexes,[11] we have been looking for complexes that contain a heteroatom directly bound to the alkylidene α carbon atom, a situation that appears to be necessary to ensure the stability of many "low oxidation state" carbene complexes.[11,27,28] We have now prepared several examples. So far we can say that such species do not appear to have structures or reactivities that clearly set them apart from their hydrocarbon analogs, with the notable exception of possible bonding of the heteroatom to the metal in an *anti* rotamer of the Re=CHSPh complex. It is interesting to note that we have not yet found any interpretable reactions of complexes of the type

$Mo(CH\text{-}t\text{-}Bu)(NAr)(OR)_2$ with vinylethers or vinylsulfides, so perhaps rhenium complexes are inherently more tolerant of functional groups directly attached to the alkylidene α carbon atom, a proposal that would be consistent with the observed greater tolerance of classical heterogeneous Re catalysts toward functional groups.[5]

Metathesis of methyl oleate (an example of an olefin with a "remote" functionality) and related natural products has been a high priority for more than two decades.[5,46-51] To our knowledge there are no other reports in the literature in which a metal-alkylidene complex derived from methyl oleate in a catalytic metathesis reaction has been observed. The rhenium catalyst disclosed in this work appears to be relatively stable and long-lived, in part because the ester coordinates to, but does not react with, the metal center. As a consequence the rate of metathesis in the presence of an ester group drops approximately an order of magnitude compared to what it might be for an analogous hydrocarbon in the absence of coordinating solvents. Therefore a continued search for catalysts for oleate metathesis with rates that approach those for metathesis of ordinary olefins seems futile. Since the alkylidene moiety is unlikely ever to be indefinitely stable to the ester functionality, the challenge will be to design a catalyst system that will maximize the difference in reactivity of the internal olefin versus the ester and that will be long lived. High turnovers in oleate metathesis ultimately will be limited by impurities that destroy the alkylidene functionality in a catalyst relatively efficiently.

Conclusions

There will be circumstances other than those I have described here in which "high oxidation state" organometallic chemistry of rhenium in a catalytic reaction will be viable, although it is becoming clear that the balance necessary to achieve this feat is more difficult to maintain as one moves to the right in the transition metal series, and that some of the d^0 rhenium chemistry in fact may look like chemistry of d^0 osmium species. On this basis it would seem unlikely that the principles that have been used to prepare Re(VII) alkylidyne and alkylidene complexes (α hydrogen migration reactions) can be extended further (to technetium, or especially osmium or ruthenium), at least in a routine fashion.

References

(1) Schrock, R. R., Rocklage, S. M., Wengrovius, J. H., Rupprecht, G., and Fellmann, J. (1980), J. Molec. Catal. 8, 73.
(2) Kress, J. R. M., Russell, M. J. M., Wesolek, M. G., and Osborn, J. A. (1980), J. Chem. Soc., Chem. Commun., 431.
(3) Kress, J., Wesolek, M., Le Ny, J.-P., and Osborn, J. A. (1981), J. Chem. Soc., Chem. Commun., 1039.
(4) Schrock, R. R. (1986), J. Organometal. Chem. 300, 249.
(5) Ivin, K. J. (1983) Olefin Metathesis, Academic Press, New York.
(6) Dragutan, V., Balaban, A. T., and Dimonie, M.(1985) Olefin Metathesis and Ring-Opening Polymerization of Cyclo-Olefins, 2nd ed., Wiley, New York.
(7) Beattie, I. R. and Jones, P. J. (1979), Inorg. Chem. 18, 2318.
(8) Mertis, K. and Wilkinson, G. (1976), J. Chem. Soc., Dalton Trans., 1488.
(9) Edwards, D. S. and Schrock, R. R. (1982), J. Am. Chem. Soc. 104, 6806.
(10) Toreki, R. and Schrock, R. R. (1992), J. Am. Chem. Soc. 114, 3367.
(11) Schrock, R. R. (1986) 'Alkylidene Complexes of the Earlier Transition Metals', in

P. R. Braterman (ed.), Reactions of Coordinated Ligands, Plenum, New York, p. 221.
(12) Nugent, W. A. (1983), Inorg. Chem. 22, 965.
(13) Edwards, D. S., Biondi, L. V., Ziller, J. W., Churchill, M. R., and Schrock, R. R. (1983), Organometallics 2, 1505.
(14) Horton, A. D., Schrock, R. R., and Freudenberger, J. H. (1987), Organometallics 6, 893.
(15) Horton, A. D. and Schrock, R. R. (1988), Polyhedron 7, 1841.
(16) Schrock, R. R., DePue, R., Feldman, J., Schaverien, C. J., Dewan, J. C., and Liu, A. H. (1988), J. Am. Chem. Soc. 110, 1423.
(17) Schrock, R. R., DePue, R. T., Feldman, J., Yap, K. B., Yang, D. C., Davis, W. M., Park, L. Y., DiMare, M., Schofield, M., Anhaus, J., Walborsky, E., Evitt, E., Krüger, C., and Betz, P. (1990), Organometallics 9, 2262.
(18) Fox, H. H., Yap, K. B., Robbins, J., Cai, S., and Schrock, R. R. (1992), Inorg. Chem. 31, 2287.
(19) Schrock, R. R., Murdzek, J. S., Bazan, G. C., Robbins, J., DiMare, M., and O'Regan, M. (1990), J. Am. Chem. Soc. 112, 3875.
(20) Cai, S., Hoffman, D. M., and Wierda, D. A. (1988), J. Chem. Soc., Chem. Commun. 1489.
(21) Williams, D. S., Schofield, M. H., Anhaus, J. T., Crowe, W. E., and Schrock, R. R. (1990), J. Am. Chem. Soc. 112, 6728.
(22) Weinstock, I. A., Schrock, R. R., Williams, D. S., and Crowe, W. E. (1991), Organometallics 9, 1.
(23) Schrock, R. R., Crowe, W. E., Bazan, G. C., DiMare, M., O'Regan, M. B., and Schofield, M. H. (1991), Organometallics 10, 1832.
(24) Oskam, J. H. and Schrock, R. R. (1992), J. Am. Chem. Soc. 114, 7588.
(25) Toreki, R., Vaughan, G. A., Schrock, R. R., and Davis, W. M. (1992), J. Am. Chem. Soc. 114, in press.
(26) Casey, C. P., Cyr, C. R., Anderson, R. L., and Marten, D. F. (1975), J. Am. Chem. Soc., 97, 3053.
(27) Dötz, K. H., Fischer, H., Hofmann, P., Kreissl, F. R., Schubert, U., and Weiss, K. (1983), Transition Metal Carbene Complexes, Verlag Chemie, Weinheim.
(28) Dötz, K. H. (1986) Carbene Complexes of Groups VIA, VIIA and VIII, Plenum, New York, Vol. 1.
(29) Toreki, R., Schrock, R. R., and Vale, M. G. (1991), J. Am. Chem. Soc. 113, 3610.
(30) Feldman, J. and Schrock, R. R.(1991), Prog. Inorg. Chem. 39, 1.
(31) Vaughan, G. A., Toreki, R., Schrock, R. R., and Davis, W. M. (1992), J. Am. Chem. Soc., submitted.
(32) Feldman, J., Davis, W. M., Thomas, J. K., and Schrock, R. R. (1990), Organometallics 9, 2535.
(33) Feldman, J., Davis, W. M., and Schrock, R. R. (1989), Organometallics 8, 2266.
(34) Finn, M. G. and Sharpless, K. B. (1991), J. Am. Chem. Soc. 113, 113.
(35) Woodard, S. S., Finn, M. G., and Sharpless, K. B. (1991), J. Am. Chem. Soc. 113, 106.
(36) Grubbs, R. H. and Tumas, W. (1989), Science 243, 907.
(37) Schrock, R. R. (1990), Acc. Chem. Res. 23, 158.
(38) Bazan, G., Khosravi, E., Schrock, R. R., Feast, W. J., Gibson, V. C., O'Regan, M. B., Thomas, J. K., and Davis, W. M. (1990), J. Am. Chem. Soc. 112, 8378.
(39) Wu, Z., Wheeler, D. R., and Grubbs, R. H. (1992), J. Am. Chem. Soc. 114, 146.

(40) Klavetter, F. L. and Grubbs, R. H. (1989), J. Am. Chem. Soc. 110, 7807.
(41) Savage, P. D., Wilkinson, G., Motevalli, M., and Hursthouse, M. B. (1987), Polyhedron 6, 1599.
(42) Herrmann, W. A., Felixberger, J. K., Anwander, R., Herdtweck, E., Kiprof, P., and Riede, J. (1990), Organometallics 9, 1434.
(43) Schrock, R. R., Weinstock, I. A., Horton, A. D., Liu, A. H., and Schofield, M. H. (1988), J. Am. Chem. Soc. 110, 2686.
(44) Leeaphon, M., Fanwick, P. E., and Walton, R. A. (1992), J. Am. Chem. Soc. 114, 1890.
(45) Schofield, M. H., Schrock, R. R., and Park, L. Y. (1991), Organometallics 10, 1844.
(46) Mol, J. C. (1982), J. Mol. Catal. 15, 35.
(47) Mol, J. C. (1991), Journal of Molecular Catalysis 65, 145.
(48) Couturier, J. L., Paillet, C., Leconte, M., Basset, J.-M., and Weiss, K. (1992), Angew. Chem. Int. Ed. Eng. 31, 628.
(49) Murdzek, J. S., (1988), Ph.D. Thesis, Massachusetts Institute of Technology.
(50) Quignard, F., Leconte, M. B., and Basset, J.-M. (1985) J. Chem. Soc., Chem. Commun. 1816.
(51) Schaverien, C. J., Dewan, J. C., and Schrock, R. R. (1986), J. Am. Chem. Soc. 108, 2771.

PROTONATION REACTIONS OF ALKYLIDYNE(CARBABORANE) COMPLEXES OF THE GROUP 6 METALS

Stephen A. Brew, Nicholas Carr and F. Gordon A. Stone*
Department of Chemistry,
Baylor University,
Waco, Texas 76798-7348
U.S.A.

ABSTRACT. Upon protonation, salts of the anionic complexes [*closo*-1,2-R'_2-3-(≡CR)-3,3-$(CO)_2$-3,1,2-$MC_2B_9H_9$]$^-$ (**1**) and [*closo*-1,6-R'_2-4-(≡CR)-4,4-$(CO)_2$-4,1,6-$MC_2B_{10}H_{10}$]$^-$ (**2**) (M = W or Mo, R = alkyl, aryl or alkynyl; R' = H or Me) afford a remarkable variety of unusual mono- or di-metal complexes, resulting from the juxtaposition of M≡CR and B–H groups on the surface of the respective metallacarbaborane cages. The nature of the products isolated depends on many factors including:

(*i*) whether one or one-half of a molar equivalent of acid is used,
(*ii*) whether the anion of the acid employed has ligating properties, *e.g.* HBF_4 *versus* HCl or HI,
(*iii*) whether Lewis bases (L = CO, PPh_3, $CNBu^t$, $Ph_2P(CH_2)_nPPh_2$ or PhC≡CPh) or donor molecules [M(≡CR)$(CO)_2$(η-C_5H_5)] are present,
(*iv*) whether the carbaborane cages contain CH or CMe vertices, and
(*v*) whether the substituent R in the ≡CR fragment contains a functional group.

Novel features of the chemistry reported include non-spectator behaviour of the carbaborane cages, chemically induced polytopal rearrangements of the cage vertices which are both remarkably facile and partially reversible, and acid-induced cage degradations of the species with MC_2B_{10} frameworks.

F. R. Kreißl (ed.), Transition Metal Carbyne Complexes, 23–37.

1. INTRODUCTION.

Salts [NEt_4^+, PPh_4^+, *etc.*] of the anionic complexes **1** and **2** are versatile reagents for preparing complexes with metal–metal bonds, since low-valent metal-ligand fragments readily add to the M≡C groups.[1]

[Y] 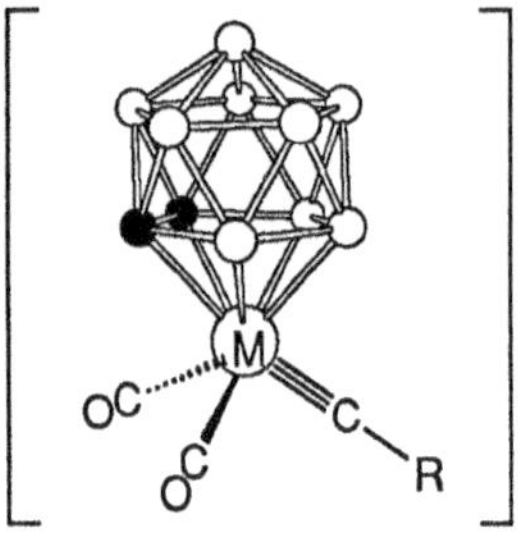

○ BH

[Y] 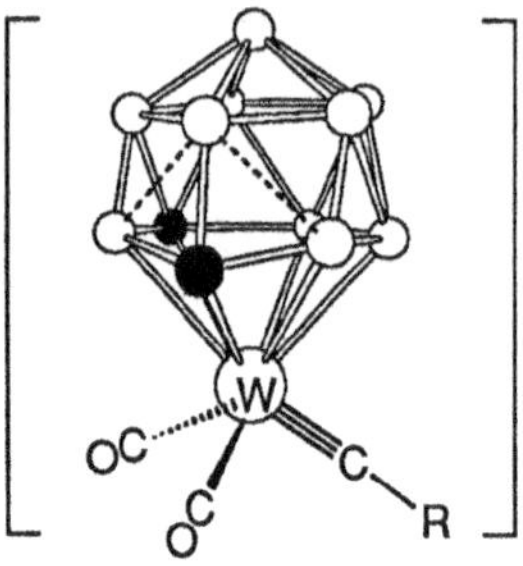

	M	R	●
1a	W	C_6H_4Me-4	CMe
1b	W	Me	CMe
1c	W	C_6H_4OMe-2	CMe
1d	W	$C_6H_4CH_2OMe$-2	CMe
1e	W	C_6H_4Me-4	CH
1f	W	Me	CH
1g	W	$C{\equiv}CBu^t$	CH
1h	Mo	C_6H_4Me-4	CMe

	R
2a	C_6H_4Me-4
2b	Me

● CMe

Y = NEt_4, PPh_4, $AsPh_4$, NMe_3Ph, $N(PPh_3)_2$, *etc.*

An added dimension of their chemistry, which is the subject of this paper, involves protonation reactions in the presence of substrate molecules. New B–C, P–C, C–C, and metal–metal bond forming processes have been identified as a consequence of the formation of mono- and di-metal species having unusual structures. The pathways followed in the new reactions are strongly influenced by the various factors listed in the Abstract, and these will be addressed during this presentation.

A number of these reactions involve the neutral alkylidyne(cyclopentadienyl)metal complexes **3**, which are the isolobally related fore-runners of the metallacarbaborane salts **1** and **2**. For a comparison of the protonation reactivity of the compounds **3** and related species, the reader is referred to reference 2.

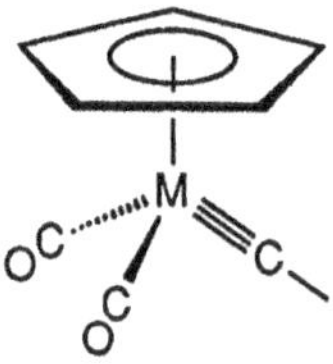

	M	R
3a	W	C_6H_4Me-4
3b	W	Me
3c	W	$C{\equiv}CBu^t$
3d	Mo	C_6H_4Me-4

2. PROTONATIONS WITH $HBF_4 \cdot Et_2O$ IN THE PRESENCE OF LEWIS BASES.

Protonation reactivity of the complexes **1** and **2** is comprehensively expanded by the addition of Lewis base substrate molecules to solutions containing these alkylidyne-metal species, prior to treatment with acid. Thus, when the alkylidyne(carbaborane)tungsten complexes of type **1** are protonated with $HBF_4 \cdot Et_2O$ in the presence of donor molecules (L), the species [*closo*-1,2-Me_2-8-(CH_2R)-3,3-$(CO)_2$-3,3-$(L)_2$-3,1,2-$WC_2B_9H_8$] **4** are obtained.[3] The metal atom in each of the products is ligated on one side by two CO molecules and by two of the 2e donor molecules L, whilst maintaining its position as a vertex of a 3,1,2-WC_2B_9 carbaborane cage. However, one boron atom adjacent to the tungsten, and in a β-position with respect to the cage carbon vertices, now bears a CH_2R substituent. The products **4** are presumed to derive from an intermediate [*closo*-1,2-Me_2-3,3-$(CO)_2$-3-{C(H)R}-3,1,2-$WC_2B_9H_9$], the alkylidene ligand of which inserts into the B–H bond of this neighbouring vertex; the β site apparently being activated by the cage CMe groups. The coordinatively unsaturated metal centre captures two donor molecules (L) from the mixture to yield as products, species with $M(CO)_2(L)_2$ groups. If no donor molecules are added to solutions of the reagent **1a** prior to treatment with acid, the only product isolated is the tetra-carbonyl metal complex **4a**, formed in low yield by scavenging of CO from the solution.

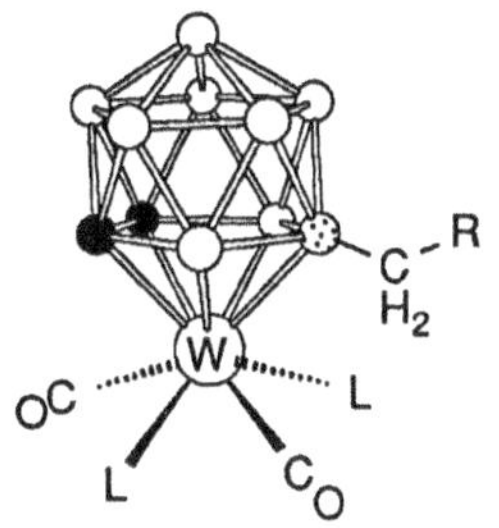

	R	L
4a	C_6H_4Me-4	CO
4b	C_6H_4Me-4	PhC≡CPh
4c	C_6H_4OMe-2	$PHPh_2$
4d	C_6H_4Me-4	PPh_3

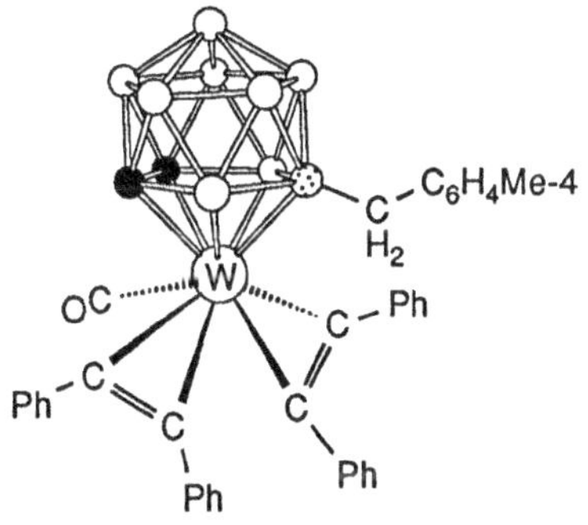

5

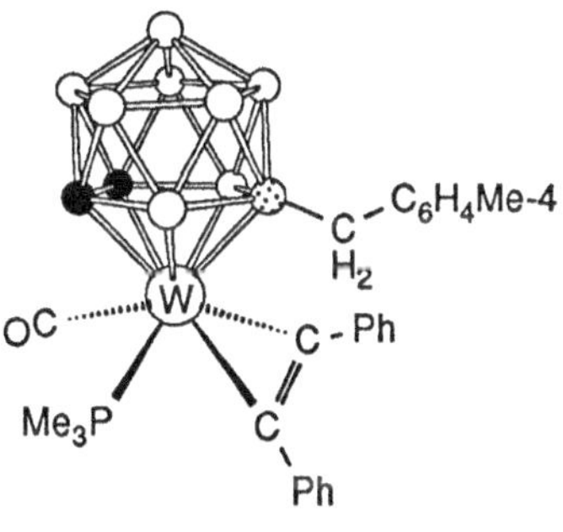

6

Whilst solutions of complex **4a** are reasonably stable at room-temperature, the remaining species **4** undergo a variety of transformations.[3] Dichloromethane solutions of **4b** release CO if warmed above -20 °C, cleanly affording complex **5**. Treatment of **4b** with PMe_3, at low temperature, yielded the mono-alkyne complex **6**. Complexes **4b**, **5** and **6** thus represent a series in which diphenylacetylene acts, respectively, as a formal two-, three- and four-electron donor to a metal centre.

Upon protonating salts **1a** and **1c** in the presence of mono-dentate phosphines, an interesting balance of reactivity was observed. When using the secondary phosphine $PHPh_2$ the expected bis-phosphine product **4c** was obtained from protonation of **1c**.[3b] Simply replacing this phosphine in the reaction mixture with the tertiary phosphine PPh_3 afforded complex **7b**, in which only one substrate molecule is present; CO having been scavenged from the solution. A similar product **7c** was isolated from the reaction between **1d**, PPh_3 and $HBF_4{\cdot}Et_2O$. In contrast, the smaller C_6H_4Me-4 carbyne substituent present in **1a**, resulted in the latter yielding a mixture of **4d** and **7a** when protonated in the presence of PPh_3.[3a] This reactivity trend is attributed primarily to steric requirements within the molecules.

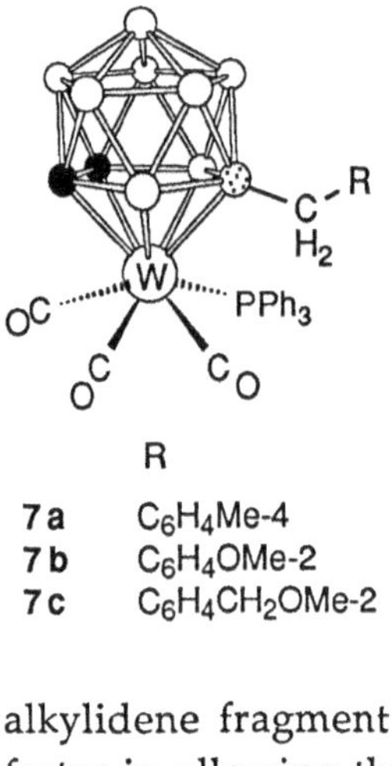

	R
7a	C_6H_4Me-4
7b	C_6H_4OMe-2
7c	$C_6H_4CH_2OMe$-2

Despite these subtle differences in reactivity, all the compounds **4** to **7** share an identical 3,1,2-WC_2B_9 icosahedral cage framework, with the carbyne-derived CH_2R fragment attached to the β-boron atom (with respect to the C atoms of the cage). Migration of the alkylidene fragment from the metal to an adjacent cage vertex, is the key factor in allowing the synthesis of this type of species. Such a process is not open to the alkylidyne(cyclopentadienyl)metal species **3**. Moreover, the carbaborane cage is a fully three dimensional structure, which offers more varied reaction possibilities than its "two dimensional" cyclopentadienyl counterpart.

This feature is well demonstrated by syntheses of the complexes [*closo*-1,2-Me_2-8,9-($CH_2C_6H_4CH_2$-2)-3,3,3,3-$(CO)_4$-3,1,2-$WC_2B_9H_7$] **8** and [*closo*-1,2-Me_2-8,9-($CH_2C_6H_4CH_2$-2)-3-(CO)-3,3-(η-$PhC_2Ph)_2$-3,1,2-$WC_2B_9H_7$] **9**.[4] These products are formed *via* protonation of the reagent **1d**, in which tungsten bears the $\equiv CC_6H_4CH_2OMe$-2 ligand, in the presence of CO or PhC≡CPh respectively. In these compounds, the metal-ligating groups are exactly as

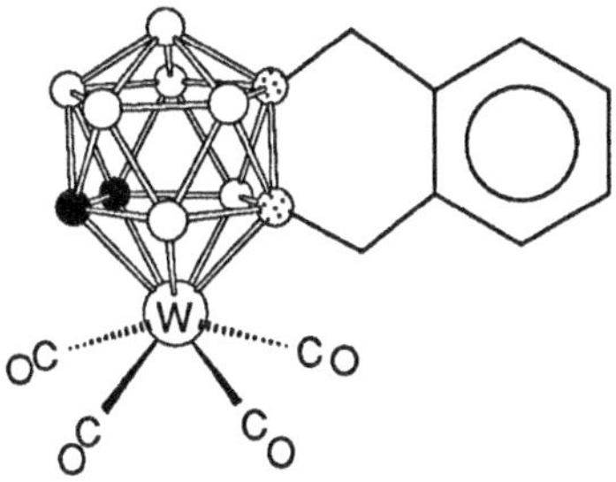

8

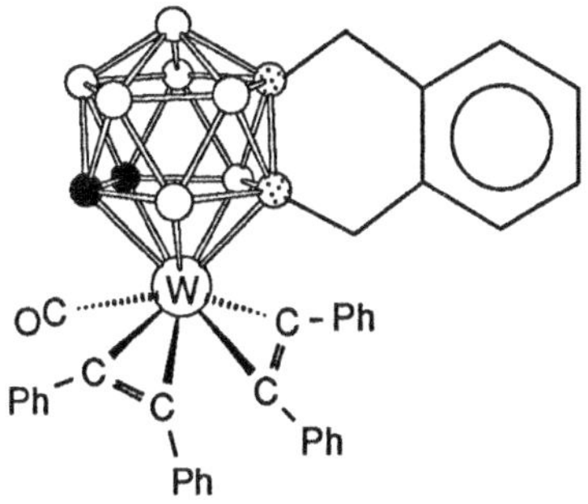

9

those of the complexes discussed above, but the "upper" pentagonal ring of the carbaborane cage is exo-polyhedrally linked to the lower. This arises from elimination of MeOH between the *ortho*-CH_2OMe group and a B–H vertex. Thus a second exopolyhedral B–C bond is formed between the cage and the aryl group derived from the alkylidyne ligand of the precursor. It is interesting to note, however, that when a single molecule of the bulkier PPh_3 ligand is present as in complex **7c**, the elimination step does not occur. This is apparently due to conformational restraints imposed by the steric requirements of these large groups.

Similar experiments with the 13-vertex cage system present in the salts **2**, were a natural extension of the above studies. Thus CO saturated solutions of **2a**, upon treatment with $HBF_4{\cdot}Et_2O$, afford mixtures of the tetra-carbonyl tungsten complexes **10** and **11**.[5] The former compound mutates slowly into the latter in solution by extrusion of a BCH_2R fragment, the precise nature of which is presently unknown. This process is an extremely novel degradation of the docosahedral cage system in which one boron vertex is expelled. The product **11** contains the icosahedral core structure which is fundamental in boron chemistry, with a 2,1,7-WC_2B_9 arrangement of atoms. The stability of

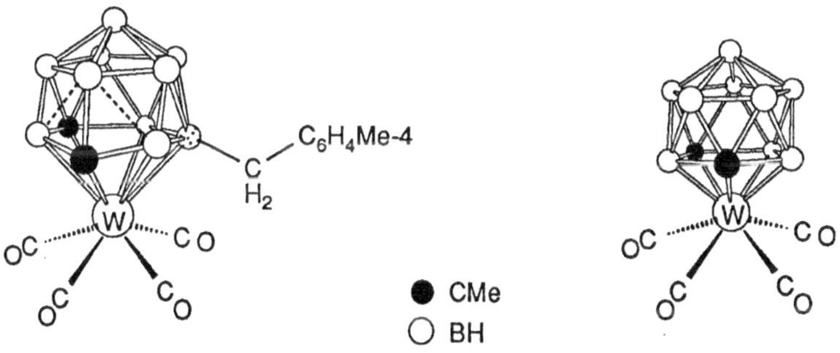

10

11

this structural unit is apparently a sufficient driving force for the degradation to occur at room temperature. However, it is interesting to note that in the absence of carbon monoxide, treatment of **2a** with $HBF_4{\cdot}Et_2O$ affords only **11** (*ca.* 40% yield) by scavenging of CO ligands from solution. The implications of this observation for the mechanism of cage degradation are presently unclear. Nevertheless, it is evident from the above, that products resulting from protonations of the 13-vertex metalla-carbaborane carbyne salts **2** are characterised by initial retention of the docosahedral structure bearing a CH_2R substituent, which subsequently loses this group along with the boron atom cage vertex to which it is attached.

Given that the majority of products from the above protonation reactions contain two Lewis base substrate molecules ligating the metal, it was of interest to study such reactions in the presence of bidentate species. When treated with $HBF_4{\cdot}Et_2O$ in the presence of bis(diphenylphosphino)methane (dppm), the alkylidyne(carbaborane) salts of type **1** yield two types of product, depending on the cage-carbon atom substituents, the particular phosphine and the substituent on the alkylidyne-carbon atom. Hence the ylid zwitterion complexes **12** and the chelated species **13** are produced as follows.[6] Protonation of **1a** with $HBF_4{\cdot}Et_2O$ in the presence of dppm yields [*closo*-1,2-Me_2-3,3-$(CO)_2$-3,3-{$P(Ph)_2CH_2P(Ph)_2C(H)(C_6H_4Me\text{-}4)$}-3,1,2-$WC_2B_9H_9$] **12a**, whilst similar treatment of **1a**, **1e** or **1f** in the presence of dppe (bis-diphenyl-phosphino-ethane) gives **12b**, **12c** or **12d**, respectively.

By contrast, protonation of **1b** in the presence of dppe or dmpe (bis-dimethylphosphino-ethane) gives only the chelate complexes [*closo*-1,2-Me_2-8-Et-3,3-$(CO)_2$-3,3-{$PR'_2(CH_2)_2PR'_2$}-3,1,2-$WC_2B_9H_8$] **13a** or **13b** respectively. In

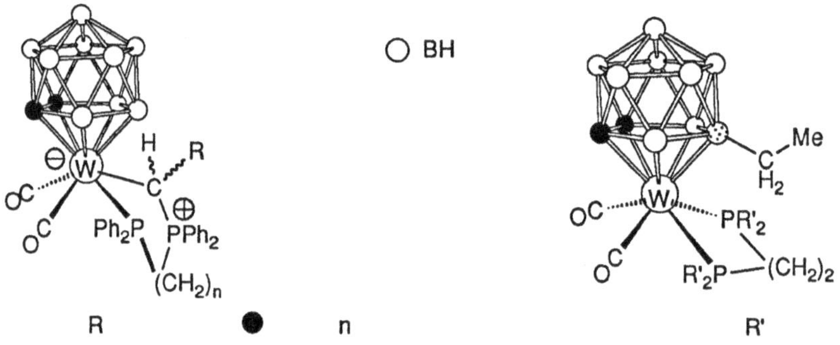

	R	●	n
12a	$C_6H_4Me\text{-}4$	CMe	1
12b	$C_6H_4Me\text{-}4$	CMe	2
12c	$C_6H_4Me\text{-}4$	CH	2
12d	Me	CH	2

	R'
13a	Ph
13b	Me

● CMe ⊙ B

these compounds the metalla-carbaborane cage displays its more familiar involvement in the protonation reactivity of the reagents **1**, by adopting the CH_2R substituent discussed earlier. The coordination sphere of the metal vertex is thus open to chelation by the bis-phosphine ligands.

It appears from these results that formation of ylid structures is favoured starting from those reagents **1** containing W≡CC_6H_4Me-4 groups. However, whether CH or CMe fragments are present in the C_2B_9 cages also influences the nature of the product. Thus, formation of the ylid complex **12d** *versus* the chelated molecule **13a** reflects an apparent activation of the B(β)–H bond by methyl substitution of the cage carbon vertices.

3. PROTONATIONS WITH HI

Treatment of dichloromethane solutions of **1a** with aqueous HI yields, within seconds, the salt [Y][*closo*-1,8-Me_2-11-($CH_2C_6H_4$Me-4)-2-I-2,2,2-$(CO)_3$-2,1,8-$WC_2B_9H_8$] **14a**.[3b,7] Complex **15** resulted from similar treatment of **1d** with HI, and shows the exo-polyhedral aryl substituent seen earlier. In both of these products an icosahedral WC_2B_9 cage bears one halide and three carbonyl ligands on the metal vertex and a CH_2R substituent on an adjacent boron vertex, as in the $HBF_4\cdot Et_2O$ reactions described above. However, the most striking feature of the anions of **14** and **15** is that the carbon-atom vertices of the icosahedron are no longer adjacent, so that there is now only one metal–carbon connectivity. Thus the metallacarbaborane species have undergone a remarkably facile polytopal rearrangement to give a 2,1,8-WC_2B_9 core structure, rather than the 3,1,2-WC_2B_9 cage system existing in the

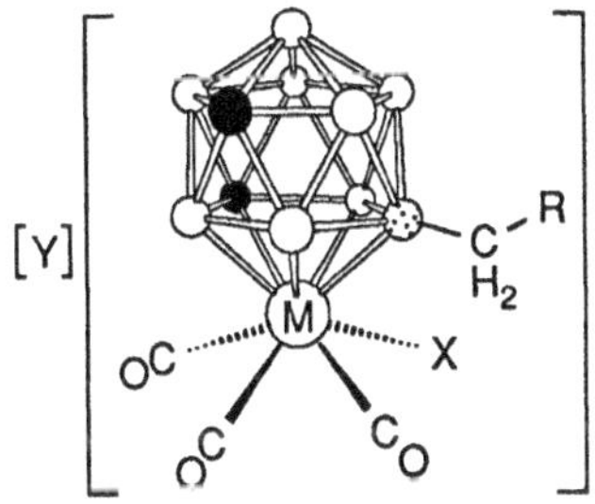

	M	X
14a	W	I
14b	W	Cl
14c	Mo	I

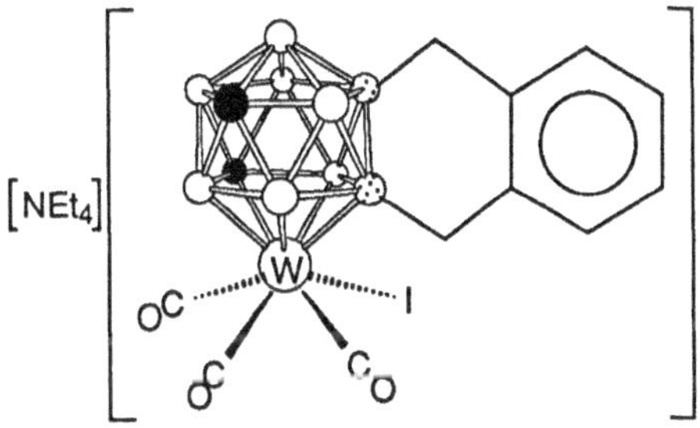

15

R = C_6H_4Me-4 Y = NEt_4, PPh_4, $AsPh_4$, NMe_3Ph, $N(PPh_3)_2$, *etc.*

precursors and found in the products of the $HBF_4{\cdot}Et_2O$ reactions. This rearrangement may be regarded as a 120 ° rotation of one C–B–B triangular face of the WC_2B_9 cage.

To test the ease with which this process may occur, a dichloromethane solution of **1a** was cooled to -78 °C prior to treatment with HI. Under these conditions complex **14a** was again formed, the reaction being complete within 10 minutes.[7] Although polytopal isomerisations of metallacarbaborane cages are common, elevated temperatures are generally required. Complex, **14a** is formed at a significantly lower temperature than any previously reported for the rearrangement of a carbametallaborane polyhedron.

Interestingly, treatment of the molybdenum-alkylidyne salt **1h** (Y = NEt_4) with HI allows the discrete observation of both the protonation and the rearrangement processes. The first-formed product is [NEt_4][*closo*-1,2-Me_2-8-($CH_2C_6H_4Me$-4)-3-I-3,3,3-$(CO)_3$-3,1,2-$MoC_2B_9H_8$] **16**, which retains the cage topology of its precursor. However, **16** isomerises quantitatively into **14c** at ambient temperatures within 18 hours.[8]

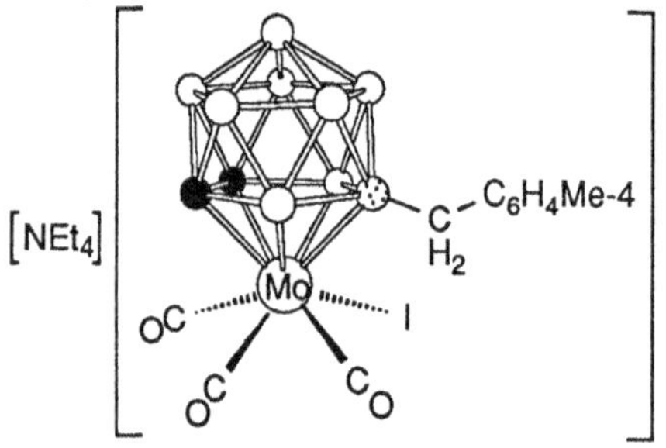

16

The divorce of the rearrangement process from the protonation step is further emphasised by the reaction of tetracarbonyl complex **4a** with NEt_4Cl. This reaction affords **14b** (Y = NEt_4) in high yield, either by stirring in THF at ambient temperature for 5 days or by ultra-violet irradiation of solutions for *ca.* 3 hours (Scheme 1). Remarkably, the process is partially reversible.[7] Treatment of a CO saturated solution of **14b** with an excess of $AgBF_4$ affords a 1:1 mixture of the polytopal isomers [*closo*-1,2-Me_2-8-($CH_2C_6H_4Me$-4)-3,3,3,3-$(CO)_4$-3,1,2-$WC_2B_9H_8$] **4a** and [*closo*-1,8-Me_2-11-($CH_2C_6H_4Me$-4)-2,2,2,2-$(CO)_4$-2,1,8-$WC_2B_9H_8$] **17**. However, if only 1 molar equivalent of $AgBF_4$ is used, complex **17**, having the 2,1,8-WC_2B_9 cage, is the only isomer obtained.

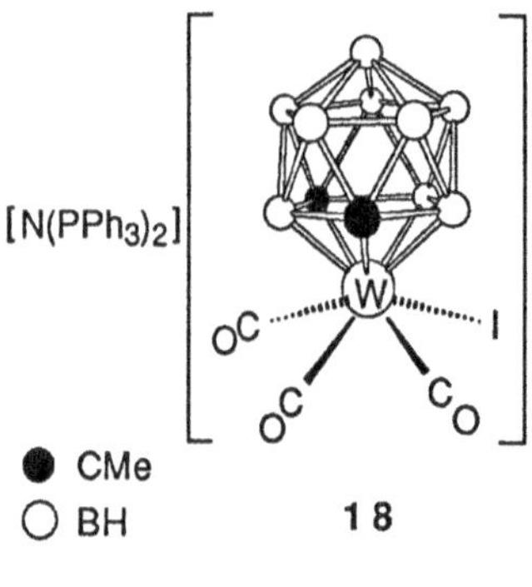

18

By now the reader will be aware that protonation reactions of the docosahedral (13-vertex) salts **2** and the icosahedral (12-vertex) complexes **1** follow broadly similar routes with respect to the metal vertex, but frequently differ in the form of the polyhedra found in the products. Protonations of the reagents **2** with

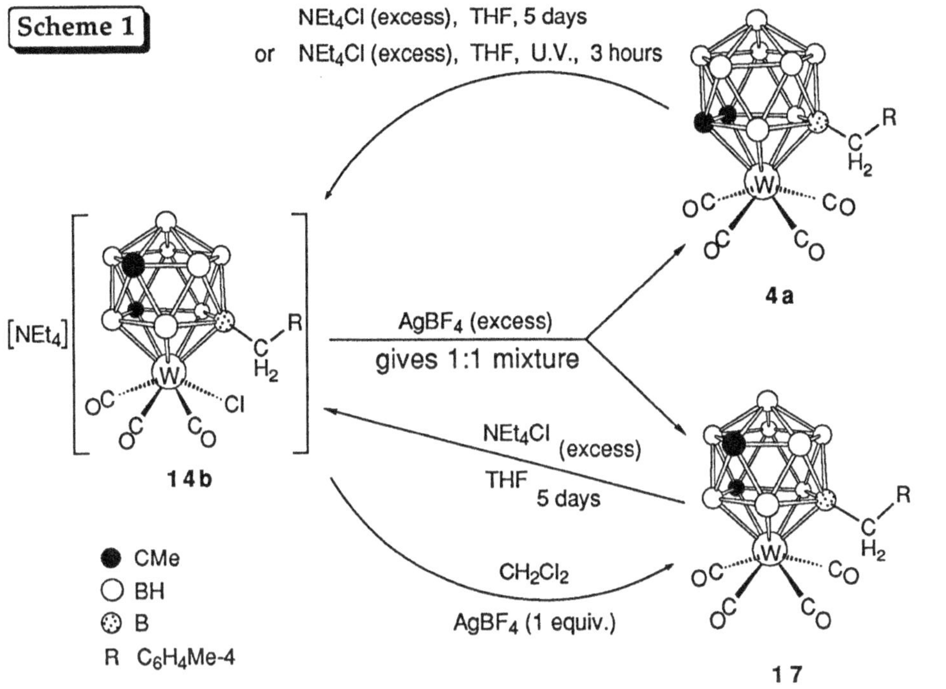

HX are no exception. Thus treatment of **2a** [Y = $N(PPh_3)_2$] with HI afforded $[N(PPh_3)_2]$[*closo*-1,7-Me_2-2-I-2,2,2-$(CO)_3$-2,1,7-$WC_2B_9H_9$] **18**.[7] The product **18** contains an icosahedral carbaborane cage with the 2,1,7-WC_2B_9 arrangement of the core atoms observed earlier in the product **11** from protonation of the salt **2a** with $HBF_4{\cdot}Et_2O$. The metal vertex of this cage bears the halide and three carbonyl ligands just as observed in the species **14a**. Thus the effects of HI and $HBF_4{\cdot}Et_2O$ on the 13-vertex framework are identical, a feature distinguishing the behaviour of salts **1** from salts **2**.

4. FORMATION OF ALKYLIDYNE DIMETAL COMPOUNDS.

Two different methodologies are available for the proton-mediated formation of dimetal complexes from the compounds **1** or **2**. In the first, mixtures containing one of the reagents **1** or **2** together with one of the neutral alkylidyne(cyclopentadienyl)metal complexes **3** are protonated. Protonation of the salts **1** or **2** generates an electronically and coordinatively unsaturated alkylidene-metal species, which can bind a molecule of type **3** to generate a metal–metal bond. This process is formally similar to the protonation of the salts **1** or **2** in the presence of alkynes, described in Section 2, since the latter are isolobal with the compounds **3**. The second method depends on the fact that if the alkylidyne-metal complexes **1 - 3** are treated with half a molar equivalent of acid the electronically and coordinatively unsaturated alkylidene-metal species so formed is present in a mixture with unreacted starting reagent, able to complex with this intermediate. Such a process also generates a metal–metal bond but the product is now a salt.

4.1. PROTONATION OF MIXTURES OF TWO ALKYLIDYNE-METAL COMPLEXES.

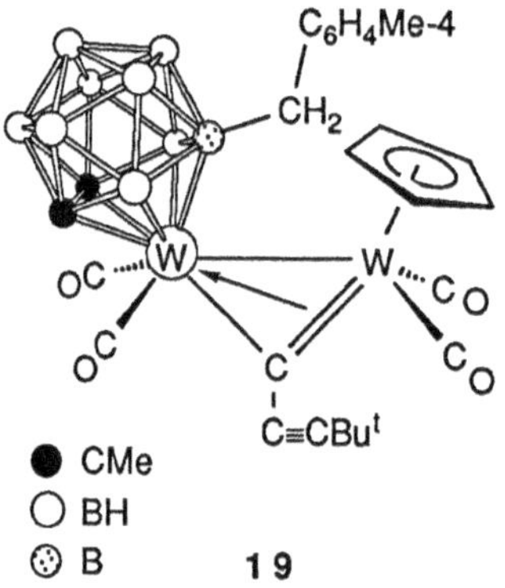

19

Treatment of an equimolar mixture of the (alkynyl-alkylidyne)metal complex **3c** and the alkylidyne(carbaborane)metal salt **1a**, with $HBF_4 \cdot Et_2O$, yields compound **19**.[9] In this product the icosahedral cage from the precursor **1a** adopts the form familiar from protonation reactions described earlier. Thus, atom B(β) bears a CH_2R substituent whilst the tungsten vertex retains its ligated carbonyl ligands. The coordination sphere of the metal is completed by a molecule of the neutral species **3c**, bound in an analogous manner to a 4e-donor alkyne molecule.

When a THF solution of **19** is heated at reflux temperature for 2 hours carbon monoxide is liberated and complex **20a**, with a three-centre-two-electron (3c-2e) B–H⟶W bridge bond, is formed, thereby maintaining the valence electron count at the metal centre bearing the cyclopentadienyl group. Compound **20a** is structurally related to complexes **20b** and **20c**, which are obtained directly by protonating mixtures of **1a** and **3b**, or **1a** and **3d**, respectively. In these syntheses,[10] intermediates analogous to the tetra-carbonyl dimetal complex **19** are clearly implicated, but none were detected.

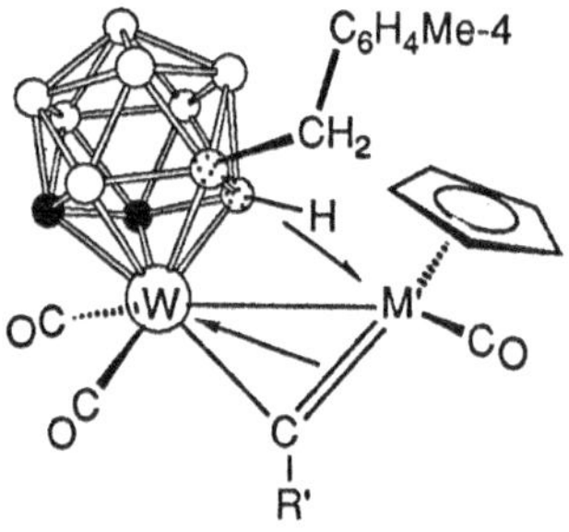

	M'	R'
20a	W	C≡CBu^t
20b	W	Me
20c	Mo	C_6H_4Me-4

● CMe ○ BH ⊙ B

When similar protonation studies were conducted using mixtures of the docosahedral salts **2** and the neutral alkylidyne(cyclopentadienyl)metal complexes **3**, subtly different species were produced. Thus, two types of compound were identified, differing in both the topology and geometry of the cage structures, as exemplified by compounds **21** and **22**.[10b] The combinations of reagents leading to these products, which contain 12- and 13-vertex frameworks, respectively, are detailed in Scheme 2.

This shows that the type of product isolated is dependent on the nature of the cyclopentadienyl(alkylidyne)metal reagent used. Thus protonation of an equimolar mixture of the tungsten compounds **2a** and **3a** affords **21a** as the only product, whereas when the cyclopentadienyl-molybdenum complex **3d** is employed with **2a**, a mixture of B_9 and B_{10} products results. On the other hand mixtures of **2a** with **3b**, which bears the ≡CMe group, yield only complex **22a** upon protonation.[10b] The overall form of the dimetal species **21** and **22** is related to the species **20**, however, unlike the latter compound no CH_2R boron cage substituent is observed in the products from **2a**. The compounds

Scheme 2

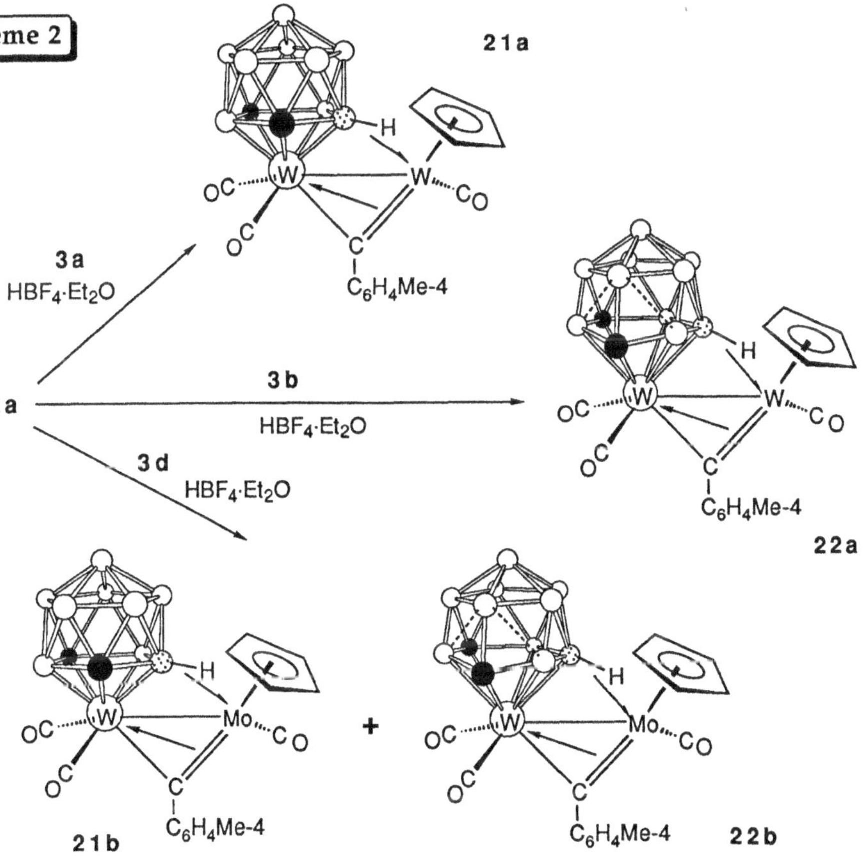

21 are formed by ejection of a boron vertex along with an alkylidyne fragment, in a manner familiar from Section 2. On the other hand, formation of the species **22** involves the loss only of the alkylidene fragment. This is an unusual feature of the protonations of these particular combinations of reagents, as the less stable 13-vertex cage structure is retained in the products **22**.

The reasons for these variations in reactivity patterns are presently unclear, however, unpredictable and subtle differences in reactivity are a feature of the compounds **20** - **22**, and analogues reported elsewhere. This is perhaps not surprising as they constitute an assembly of reactive ligands around an unsaturated dimetalla-cyclopropene core, each group exerting its own specific electronic and steric influence. This has previously been observed to result in coupling of a variety of ligands with the bridging alkylidyne group and/or the carbaborane cage, leading to B–C and/or C–C bond-forming reactions.[1] Such coupling may be proton-mediated, as when mixtures of the alkylidyne-metal complexes **1e** and **3c**, or **1g** and **3a** are treated with $HBF_4{\cdot}Et_2O$.[11] Compound **23** is the product no matter which combination of precursors is used. Thus the acid-derived proton reaches its preferred site on the tolyl-bearing carbon atom. A probable pathway, in which the proton may be transferred to the tolyl-bearing carbon *via* a μ-hydrido(μ-alkyne)-ditungsten intermediate has been proposed.[11] By such a process the carbene ligand, initially formed upon protonation, is thus prevented from inserting into a B–H bond of the icosahedral cage to form a BCH_2R group of the type described earlier.

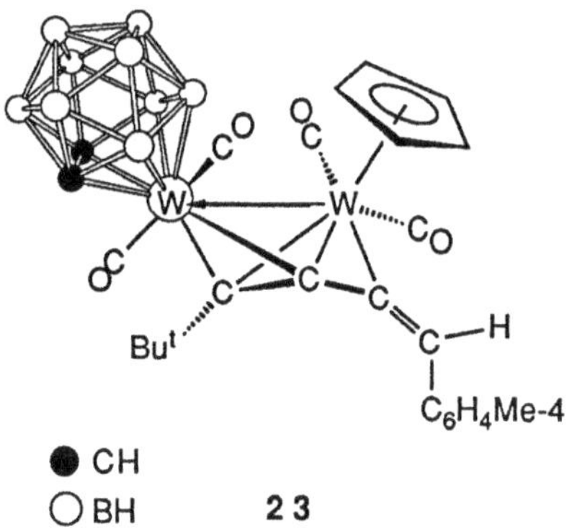

23

4.2. PROTONATIONS INVOLVING HALF MOLAR EQUIVALENTS OF $HBF_4{\cdot}Et_2O$.

Dimetal complexes may equally well be obtained from the reagents **1**, **2** or **3** alone, by treatment with 0.5 molar equivalents of acid. With this stoicheiometry a mixture is effectively created upon protonation, containing alkylidyne- and alkylidene-metal species in equal proportions. The validity of this methodology was first demonstrated in 1985 during protonation studies of the reagents **3a** and **3b**.[2c] In practice, it was found that 0.4 mol equivalents of $HBF_4{\cdot}Et_2O$ produced the best results by ensuring that the alkylidyne-metal species remained in excess. Thus, from **3a** and **3b** the ditungsten complexes **24** were isolated. The reagent **1e** containing the $WC_2B_9H_{11}$ cage, follows a similar pathway leading to the anionic species **25**, upon treatment with half an equivalent of $HBF_4{\cdot}Et_2O$.[12]

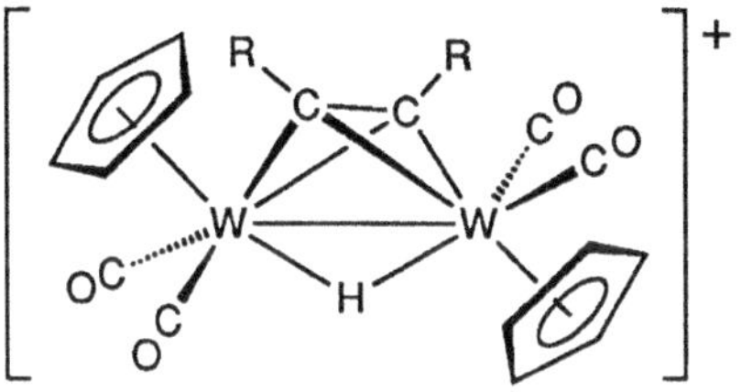

	R
24a	C_6H_4Me-4
24b	Me

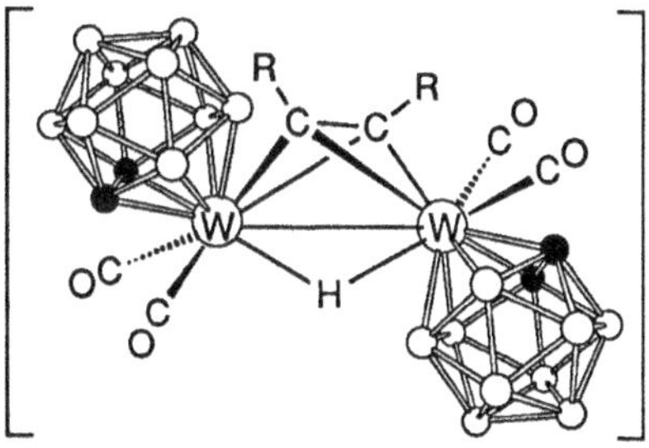

● CH
○ BH

25

R = C_6H_4Me-4

Contrasting with these results are protonation studies of the $WC_2B_9H_9Me_2$ cage systems of **1a** and **1b** or the MC_2B_{10} polyhedra of the salts **2**, using 0.4 mol equivalents of $HBF_4{\cdot}Et_2O$. Although dimetal species are again formed, interactions of the metallacarbaborane cages with the rest of the cluster are not limited to the metals alone. Thus, treatment of **1a** with 0.4 mol equivalents of $HBF_4{\cdot}Et_2O$ affords the μ-σ,η^2 ketenyl species **26**.[13] In both cages the β-boron atom exhibits "non-innocent" behaviour, being involved either in a B–H⟶M interaction or BCH_2R group formation. Furthermore, the polyhedron bearing the latter group exhibits a pronounced distortion arising from the formal deficiency of one electron-pair. This is known as a

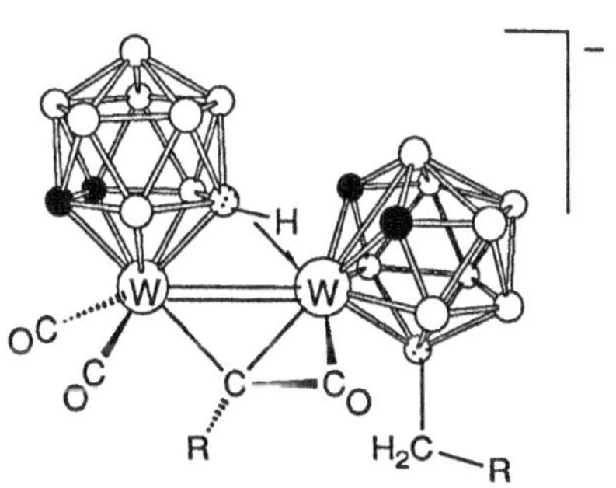

26

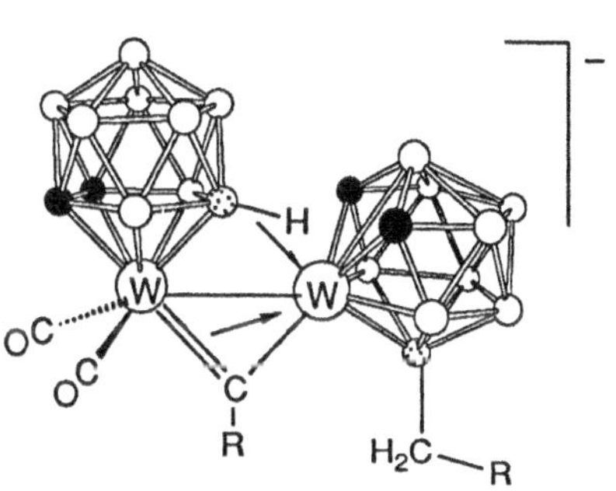

27

R = C_6H_4Me-4

● CMe ○ BH ◌ B

hyper-closo cage structure, which has no connectivity between the two carbon vertices. The two metallacarbaborane cluster fragments are linked by the aforementioned B–H—M interaction and by a formal double bond between the metal vertices. This bond is supported by the bridging C(R)C(O) ketenyl group. However, upon heating solutions of this salt, CO is expelled from the ketenyl ligand to form the μ-alkylidyne bridged compound **27**.

Protonations of the docosahedral salts **2** in the mole ratio 2:1 (carbyne : acid) yield the products **28**.[5] The X-ray determined structure of the anion of **28b** shows two 13-vertex cages linked both by a metal-metal double bond and by a B–H—M 3c - 2e interaction. The second cage again bears a BCH_2R fragment, whilst the metal–metal bond is spanned by the μ-σ,η^2 ketenyl ligand. However, the already strained 13-vertex cage system is apparently unable to adopt a *hyper-closo* configuration and a third carbonyl ligand is retained to donate the required electron pair to the cluster. This necessitates a transfer of the ketenyl ligand such that in the solid state it becomes σ-bonded to the metal vertex of the cage bearing the BCH_2R substituent, and η^2-bonded to the metal vertex of the cage involved in the B–H—M interaction. However, NMR and IR spectroscopy firmly indicate that solutions of the salts **28** in organic solvents adopt the alternative μ-η^1 bonding mode in which the ketenyl oxygen atom carries a formal positive charge.

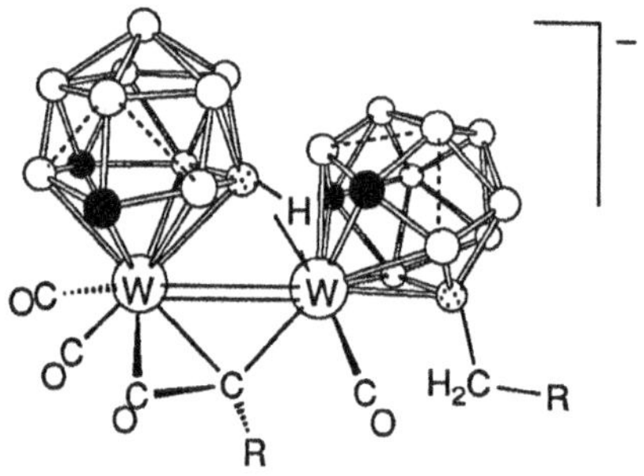

	R
28a	C_6H_4Me-4
28b	Me

● CMe ○ BH ⊙ B

5. CONCLUSION.

The above summary of recent results indicates the remarkable scope, within the as-yet-undiscovered chemistry of the reagents **1** and **2**, for the formation of unusual structures and the observation of new chemical processes. The future potential of these molecules stems from the union of alkylidyne-metal and metallacarbaborane complexes which they represent, and appears to offer unlimited opportunity for exploration.

Acknowledgement: We thank the Robert A. Welch Foundation for support under grant AA-1201.

References:

1. F. G. A. Stone, *Adv. Organomet. Chem.*, **31**, 53 (1991).
2. (a) A. Mayr, M. F. Asaro, M. A. Kjelsberg, K. S. Lee and D. van Engen, *Organometallics*, **6**, 432 (1987); A. Mayr, M. A. Kjelsberg, K. S. Lee, M. F. Asaro and T. C. Hsieh, *Organometallics*, **6**, 2610 (1987); (b) M. Bottrill, M. Green, A. G. Orpen, D. R. Saunders and I. D. Williams, *J. Chem. Soc., Dalton Trans.*, 511 (1989); (c) J. A. K. Howard, J. C. Jeffery, J. C. V. Laurie, I. Moore, F. G. A. Stone and A. Stringer, *Inorg. Chim. Acta.*, **100**, 23 (1985); (d) F. R. Kreissl, W. J. Sieber, M. Wolfgruber, and J. Riede, *Angew. Chem., Int. Ed. Engl.*, **23**, 640 (1984); F. R. Kreissl, W. J. Sieber, H. Keller, J. Riede, and M. Wolfgruber, *J. Organomet. Chem.*, **320**, 83 (1987); (e) K. E. Garrett, J. B. Sheridan, D. B. Pourreau, W. C. Feng, G. L. Geoffroy, D. L. Staley, and A. L. Rheingold, *J. Am. Chem. Soc.*, **111**, 8383 (1989).
3. (a) S. A. Brew, D. D. Devore, P. D. Jenkins, M. U. Pilotti, and F. G. A. Stone, *J. Chem. Soc., Dalton Trans*, 393 (1992); (b) J. C. Jeffery, S. Li, D. W. I. Sams, and F. G. A. Stone, *J. Chem. Soc., Dalton Trans*, 877 (1992).
4. J. C. Jeffery, S. Li, and F. G. A. Stone, *Organometallics*, **11**, 1902 (1992).
5. N. Carr and F. G. A. Stone, unpublished results.
6. S. A. Brew, P. D. Jenkins, J. C. Jeffery, and F. G. A. Stone, *J. Chem. Soc., Dalton Trans*, 401 (1992).
7. S. A. Brew, N. Carr, J. C. Jeffery, M. U. Pilotti, and F. G. A. Stone, *J. Am. Chem. Soc.*, **114**, 2203 (1992).
8. S. Li and F. G. A. Stone, unpublished results.
9. G. C. Bruce and F. G. A. Stone, *Polyhedron*, in press.
10. (a) S. A. Brew and F. G. A. Stone, *J. Chem. Soc., Dalton Trans*, 867 (1992); (b) S. A. Brew, N. Carr, M. D. Mortimer, and F. G. A. Stone, *J. Chem. Soc., Dalton Trans*, 811 (1991).
11. G. C. Bruce, D. F. Mullica, E. L. Sappenfield, and F. G. A. Stone, *J. Chem. Soc., Dalton Trans*, in press.
12. A. P. James and F. G. A. Stone, *J. Organomet. Chem.*, **310**, 47 (1986). See also reference 49.
13. N. Carr, D. F. Mullica, E. L. Sappenfield, and F. G. A. Stone, *Organometallics*, in press.

CYCLOMETALATED ARYLOXY(CHLORO)NEOPENTYLIDENE - TUNGSTEN COMPLEXES. SYNTHESES FROM NEOPENTYLIDYNE COMPLEXES AND CATALYTIC PROPERTIES IN OLEFIN METATHESIS

J.-L. COUTURIER, M. LECONTE, and J.-M. BASSET*
Institut de Recherches sur la Catalyse and Ecole Supérieure de Chimie Industrielle de Lyon, CNRS
2, Avenue Albert Einstein
69626 Villeurbanne Cedex
France

ABSTRACT. Cyclometalated aryloxy(chloro)neopentylidene-tungsten complexes can be synthesized starting from $WCl_4(OAr)_2$ (OAr = 2,6-disubstituted phenoxide), but also starting from the neopentylidyne complex $W(CCMe_3)Cl_3(dme)$ (by reaction with LiOAr). Some of these cyclometalated neopentylidenes are probably among the most active and stereoselective one-component metathesis catalysts. In particular, they are fairly active in the metathesis of an olefinic ester such as ethyl oleate and they have been successfully used in the metathesis of olefinic sulfides.

1. Introduction

The design of versatile, highly active, and well-defined catalysts remains one of the main objectives of the research in olefin metathesis [1-3]. This is especially true for the application of metathesis to acyclic or cyclic olefins bearing functional groups which probably constitutes one of the most promising uses of this reaction. In fact, metathesis offers many interesting possibilities for the synthesis of valuable organic products or polymers that are often difficult to obtain by other methods [4-6].

In the past few years, it was shown that some neopentylidene-tungsten(VI) complexes with alkoxide [7-10], imido [9, 10], or aryloxide [11-13] ligands were efficient homogeneous metathesis catalysts. Aryloxides proved to be particularly useful ligands since their steric and electronic properties can be varied in a wide range by changing the nature, the number and the position of the substituents on the phenoxide. Using these ligands, it has been possible to control the activity and, in some cases, the stereoselectivity of the metathesis reaction [11-19].

A general method for the synthesis of aryloxy(chloro)neopentylidene-tungsten complexes involves the reaction of $WCl_4(OAr)_2$ (OAr = 2,6-disubstituted phenoxide) [20] with 1 equiv or 1.5 equiv of $MgNp_2$(dioxane) ($Np=CH_2C(CH_3)_3$) [11, 21] (Scheme 1). The reaction proceeds via a double (or triple) alkylation of tungsten followed by an α-H abstraction and elimination of neopentane [22]. In the particular case of the 2,6-diphenylphenoxide complexes, the activation of an *ortho* C-H bond in one of the phenyl substituents of the aryloxide ligands leads to the elimination of HCl with the formation of a cyclometalated compound (Scheme 2) [23, 24].

F. R. Kreißl (ed.), Transition Metal Carbyne Complexes, 39–50.

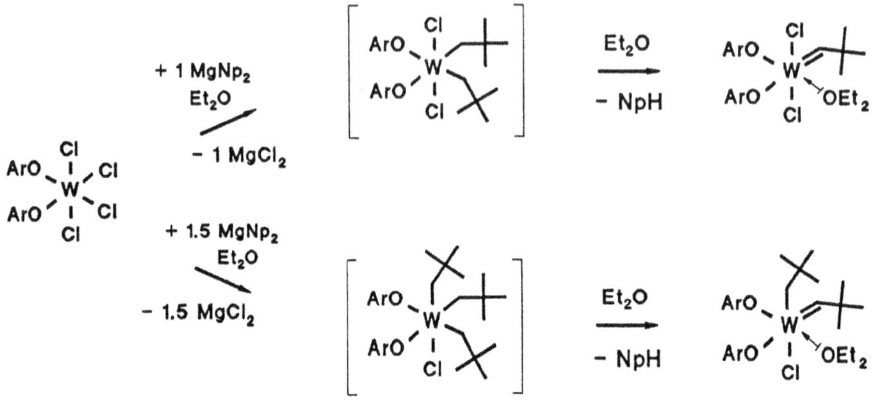

Scheme 1.

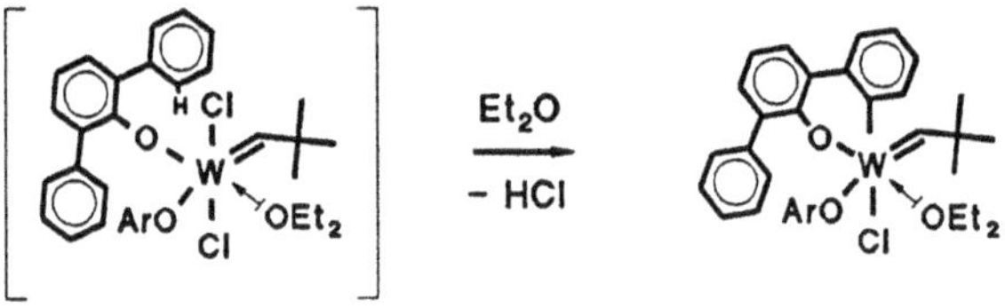

Scheme 2.

Interestingly, this type of cyclometalated complex can be obtained via an other route starting from the Schrock's neopentylidyne complex $W(CCMe_3)Cl_3(dme)$ (dme = dimethoxyethane) [25, 26]. The purpose of this paper is to report the main features of this new way of synthesis and to give some examples of the catalytic properties of these cyclometalated aryloxy(chloro)neopentylidene-tungsten complexes in olefin metathesis.

2. Results and Discussion

2.1. SYNTHESIS OF CYCLOMETALATED NEOPENTYLIDENES STARTING FROM A NEOPENTYLIDYNE COMPLEX

Reaction of $W(CCMe_3)Cl_3(dme)$ [25, 26] with 2 equiv of LiOAr (OAr = O-2,6-$C_6H_3Ph_2$, O-2,4,6-$C_6H_2Ph_3$, or O-2,6-C_6H_3(*t*-Bu)$_2$) in diethyl ether leads to orange-brown solid compounds **1-3** which likely results from the sequences of reactions depicted in Scheme 3. Complexes **1** to **3** were characterized by elemental analysis, mass spectroscopy, and ^{1}H and ^{13}C NMR. Some characteristic ^{1}H and ^{13}C NMR data for **1**, **2**, and **3** are given in Table 1.

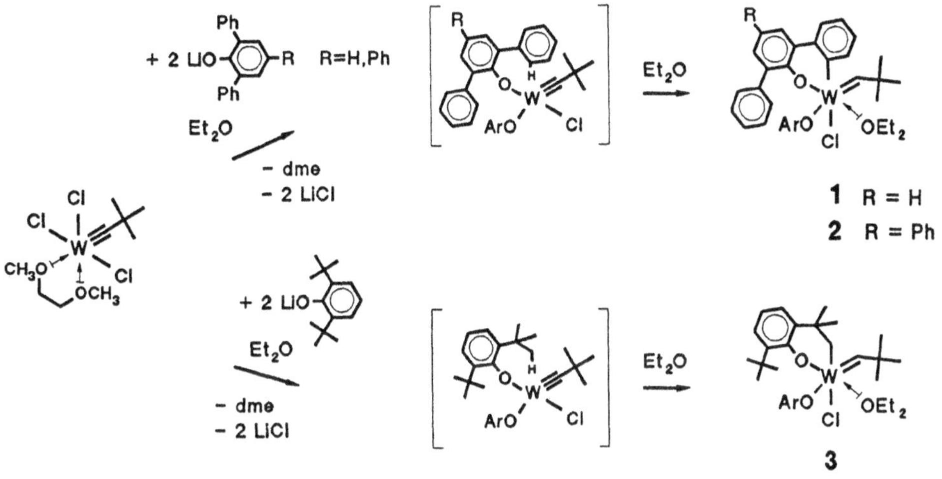

Scheme 3.

TABLE 1. Characteristic ^{1}H and ^{13}C NMR data for **1**, **2**, and **3** (C_6D_6, 25°C)

	δ (ppm/TMS)			
	1		**2**	**3**
^{1}H				
W(=C<u>H</u>CMe$_3$)	10.11	J_{WH}=14Hz	10.20	9.68[a]
W(=CHC<u>Me</u>$_3$)	0.57		0.65	-[b]
^{13}C				
W(=<u>C</u>HCMe$_3$)	296.67	J_{CH}=130.5Hz J_{CW}=166Hz	296.85	283[a]
W(-C)[c]	183.15	J_{CW}=115Hz	183.55	73.2[a]
W(=CH<u>C</u>Me$_3$)	45.24		45.29	-[b]
W(=CHC<u>Me</u>$_3$)	31.82		31.89	-[b]

[a]Values in agreement with those previously found by R. R. Schrock [27]. [b]Not attributed due to the presence of various *t*-Bu groups in the complex. [c]Cyclometalated carbon.

The key step of the synthesis is the intramolecular activation of the C-H bond of the phenyl (or *t*-Bu) substituent on the d^0 metal which leads to the addition of an hydrogen atom to the carbyne to form a carbene and a stable cyclometalated structure [27-31]. The role of the weakly coordinated ether is probably crucial in the reaction path where a pentacoordinated W(VI) with a possible agostic C-H bond should be the precursor of the carbene via a four center transition state as it is now currently admitted in electrophilic activation of C-H bonds [32].

Steric effects likely play an important role in the process of cyclometalation since it was found that reaction of $W(CCMe_3)Cl_3(dme)$ with 2 equiv of LiO-2-C_6H_4Ph or with 3 equiv of LiO-2,6-C_6H_3(*i*-Pr)$_2$ in Et_2O does not lead to cyclometalated neopentylidene complexes but to the new neopentylidynes $W(CCMe_3)Cl(O$-2-$C_6H_4Ph)_2(OEt_2)$ [33] and $W(CCMe_3)$(O-2,6-C_6H_3(*i*-Pr)$_2$)$_3$ [27] respectively.

2.2. CATALYTIC PROPERTIES OF CYCLOMETALATED ARYLOXY(CHLORO)-NEOPENTYLIDENE-TUNGSTEN COMPLEXES IN OLEFIN METATHESIS

2.2.1. *Activity and stereoselectivity of **1**, **2**, and **3** in the metathesis of cis- or trans-2-pentene.* The catalytic properties of **1**, **2**, and **3** were tested in the metathesis of *cis*- and *trans*-2-pentene and appeared very promising. For example, with **1** or **2**, the metathesis equilibrium of 500 equiv of 2-pentene is reached in 1 min at 25°C with initial turnover rates higher than 300 min^{-1} (an initial turnover rate of 800 min^{-1} was found with **1** by using 1000 equiv of 2-pentene) (Table 2). There is a drastic effect of the nature of the aryloxide ligand on both the rate and the stereochemistry of the metathesis of *cis*-2-pentene. Complexes **1** and **2**, with two phenyl groups in *ortho* positions on the phenoxides, appear to be both highly active and stereoselective whereas **3**, with two *tert*-butyl groups in *ortho* positions, shows a very poor activity and a very poor stereoselectivity.

TABLE 2. Activity and stereoselectivity of **1**, **2**, and **3** in the metathesis of *cis*-2-pentene[a]

	t^b(min)	$r_0{}^c$(min^{-1})	$[t/c\ C_4]_0{}^d$	$[t/c\ C_6]_0{}^d$
1	1	ca. 300	0.0	0.0
2	1	ca. 300	0.15	0.30
3	600	ca. 1	0.65	1.10

[a]Experimental conditions: catalyst: 0.02 mmol; *cis*-2-pentene/catalyst molar ratio = 500; reaction temperature = 25°C; solvent C_6H_5Cl (5 mL). [b]Reaction time necessary to reach the metathesis equilibrium (2-butene/2-pentene/3-hexene = 1/2/1). [c]Initial turnover rate for the conversion of *cis*-2-pentene. [d]*trans*/*cis*-2-butene (or *trans*/*cis*-3-hexene) ratio obtained at 0% conversion.

The initial values for *trans*/*cis* ratios of 2-butene (or 3-hexene) were accurately determined by plotting $t/c\ C_4$ (or $t/c\ C_6$) as a function of $t/c\ C_5$ and extrapolating at $t/c\ C_5 = 0$ (Figure 1). Interestingly, the high stereoselectivities obtained with **1** in the metathesis of *cis*-2-pentene or *trans*-2-pentene (in the metathesis of *trans*-2-pentene, a value of 0.004 was found for both *cis*/*trans* C_4 and *cis*/*trans* C_6 at 0% conversion) are maintained even at high conversions (Figure 2). When the equilibrium of "productive" metathesis is nearly reached (roughly 25% yield of 2-butene), there is a dramatic change of the stereochemistry of the products due to secondary metathetical isomerization reactions that progressively lead to the ultimate equilibrium *trans*/*cis* ratios (*trans*/*cis* $C_4 \approx 3$ and *trans*/*cis* $C_6 \approx 6$).

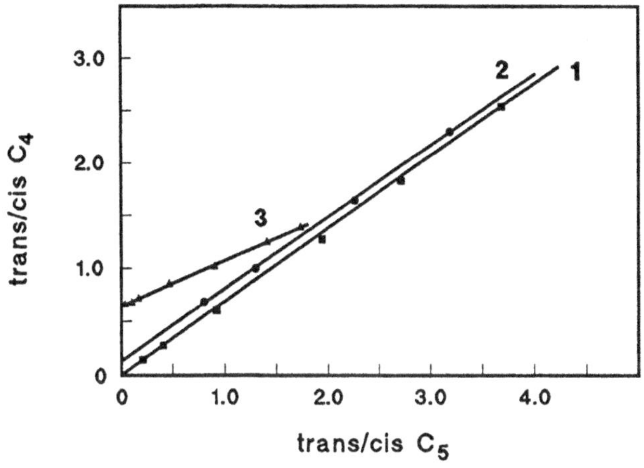

Figure 1. *Trans*/*cis*-2-butene vs. *trans*/*cis*-2-pentene in the metathesis of *cis*-2-pentene with **1**, **2**, and **3**.

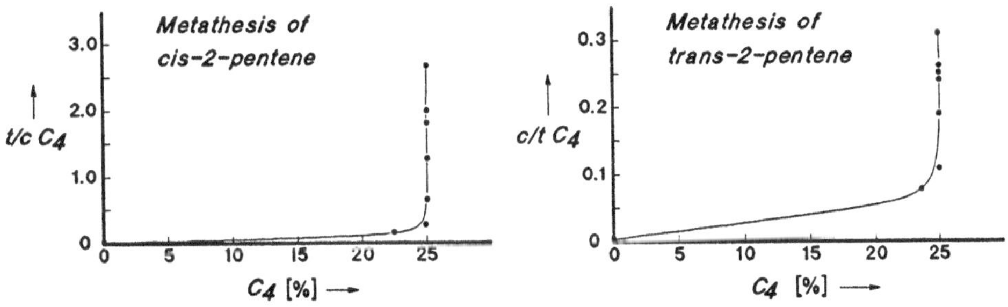

Figure 2. *Trans*/*cis*-2-butene and *cis*/*trans*-2-butene vs. 2-butene yield in the metathesis of *cis*-2-pentene and *trans*-2-pentene with **1**.

To our knowledge, such high stereoselectivities, which are kept almost up to thermodynamic metathesis equilibrium, have never been reported in metathesis of 2-pentene with any highly active tungsten-based catalyst. The retention of configuration of the starting olefin is easily explained on the basis of the favored configuration of the tungstacyclobutane intermediate with two 1,3 alkyl groups in equatorial-equatorial configuration (Scheme 4) [34]. One can assume that these favored configurations are particularly stabilized due to the presence of bulky ligands on the tungsten and/or due to the rigidity generated by the cyclometalation. Nevertheless, it is obvious that cyclometalation alone is not sufficient to induce a high stereoselectivity since **3** proves to be poorly stereoselective. As a consequence, possible steric (and/or electronic) effects of the aryloxide ligands have to be invoked.

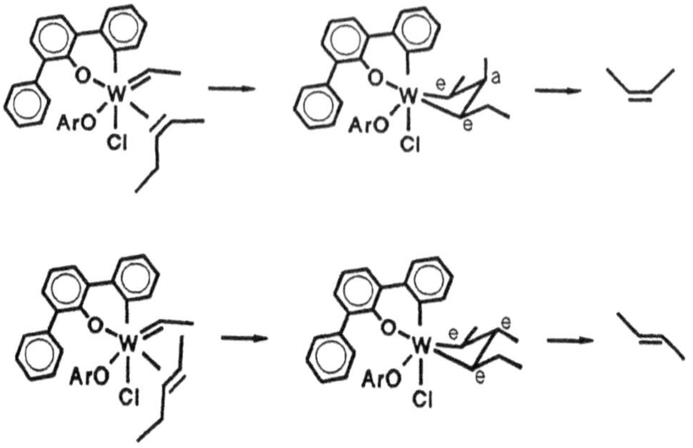

Scheme 4.

Comparison of the metathesis rates obtained with **1**, **2**, and **3** demonstrates the drastic effect of the nature of the aryloxide ligand on the activity of the catalyst. Electron-withdrawing groups on the phenoxide, such as phenyls in *ortho* or *para* position, make the tungsten center more electrophilic and, probably, more reactive towards olefins. On the other hand, electron-donating groups in *ortho* position such as *tert*-butyls, by increasing the electron density on the tungsten, strongly decrease the metathesis activity. Such a promoting effect of electron-withdrawing groups was also reported in the case of alkoxides ligands [35].

2.2.2. *Application of cyclometalated neopentylidene complexes to the metathesis of olefinic esters.* Very promising results were obtained in the metathesis of an olefin bearing an ester group such as ethyl oleate (ethyl-9-octadecenoate) (Scheme 5)

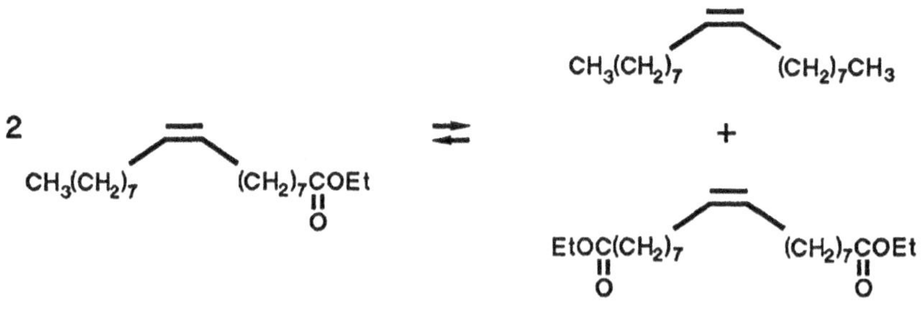

Scheme 5.

For example, **1** can convert selectively, in 60 min at 25°C, ca. 50% of 500 equiv of ethyl oleate into 9-octadecene and diethyl-9-octadecenedioate (with an initial turnover rate for the conversion of ethyl oleate higher than 800 h^{-1}) (Figure 3). To our knowledge, this is the highest activity reported so far with homogeneous catalysts in the metathesis of this type of substrate [4, 9].

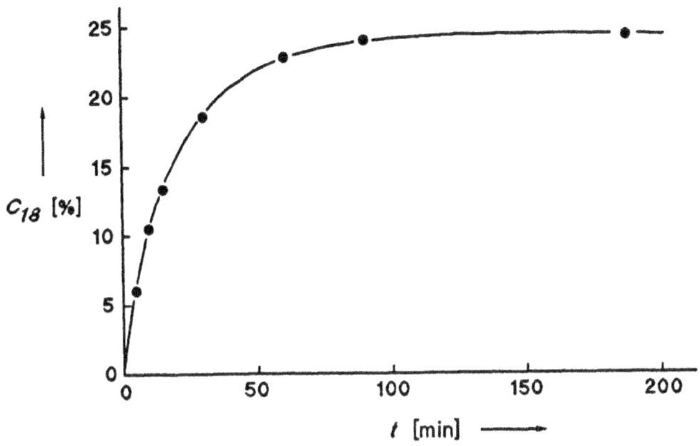

Figure 3. Yield of 9-octadecene vs. reaction time in the metathesis of ethyl oleate with **1** (ethyl oleate/**1** molar ratio = 500, T = 25°C, solvent: chlorobenzene (5 mL)).

2.2.3. *Application of cyclometalated neopentylidene complexes to the metathesis of sulfur-containing olefins.* To date, the metathesis reaction has been applied to olefins containing oxygen, nitrogen, halogens, or silicon atoms, but, to our knowledge, there is no successful metathesis reaction involving sulfur-containing acyclic or unstrained cyclic olefins reported in the literature [36, 37]. This lack of data is probably due to a poisoning effect of the functional groups containing sulfur atoms, as it was previously reported in the case of some classical heterogeneous [38] or homogeneous [39] catalytic systems. However, this specific application of olefin metathesis could present a particular interest by enlarging the field of thiochemistry and offering the possibility to synthesize known or new olefinic compounds containing sulfur, via a novel and simple route.

We have found that metathesis of olefinic sulfides can be successfully achieved by using, as homogeneous catalyst, the aryloxy(chloro)neopentylidene-tungsten complex **1**. Thus, the self-metathesis of allyl methyl sulfide (**4**) and of 5-alkylthiocyclooctenes (**5**) (or their co-metathesis with various acyclic or cyclic olefins without functional groups) can lead to a family of new olefins, dienes, or unsaturated polymers containing one or more thioether groups.

In the presence of **1**, the self-metathesis of allyl methyl sulfide (**4**), or its co-metathesis with 2-butene (**6a**) or 2-pentene (**6b**), actually occurs and leads to the expected products (Scheme 6) [40].

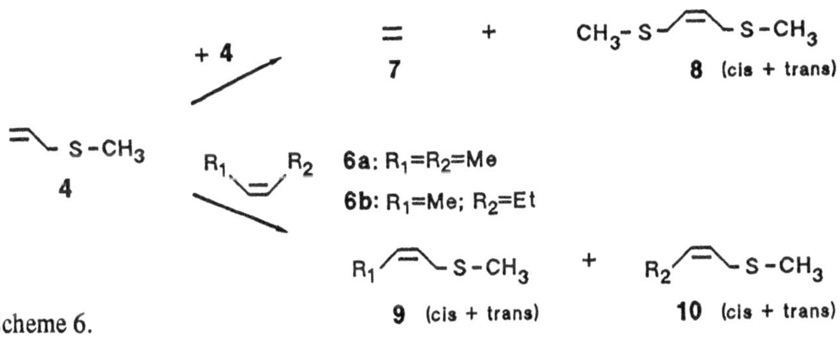

Scheme 6.

The turnover rates (Table 3) are lower than those obtained with **1** in the metathesis of 2-pentene or ethyl oleate, suggesting a reversible coordination of the sulfur compound to the metallocarbene leading to a partial deactivation of the catalyst. Nevertheless, the reaction is highly selective (only the expected metathesis products are detected) and the conversion of **4** can reach a high value when an excess of the co-reactant olefin is used.

TABLE 3. Self-metathesis of allyl methyl sulfide (**4**) and its co-metathesis with 2-butene (**6a**) or 2-pentene (**6b**) catalyzed by **1**[a]

4/1[b]	6a/1[b]	6b/1[b]	T[c](°C)	t[d](h)	Conversion of 4 (%)	Yields[e] (%)			
						7	8	9	10
25	-	-	80	15	40	20	20	-	-
25	100	-	20	15	95	-	-	95	-
20	-	20	80	15	60	5	5	30	20

[a]Typically, the metathesis reactions were carried out by using 0.025 mmol of **1** and 5 mL of C_6H_5Cl as solvent. [b]Initial molar ratios. [c]Reaction temperature. [d]Reaction time. [e]Yields based on initial **4**.

Ring-opening metathesis polymerization of various 5-alkylthiocyclooctenes (**5a-e**) [41] leads to the corresponding sulfur polymers **11** (Scheme 7), with turnover rates higher than those observed in the metathesis of allyl methyl sulfide (Table 4). The results of Table 4 show that there is a strong effect of the nature of the thioalkyl substituent on the rate of polymerization: the higher the steric crowding of the alkyl substituent, the higher the rate of polymerization; this confirms the hypothesis that steric hindrance around the sulfur atom could decrease the inbiting effect induced by a possible coordination of the sulfur to the tungsten. The turnover rates can be increased by increasing reaction temperature and/or by performing the polymerization in the absence of a solvent.

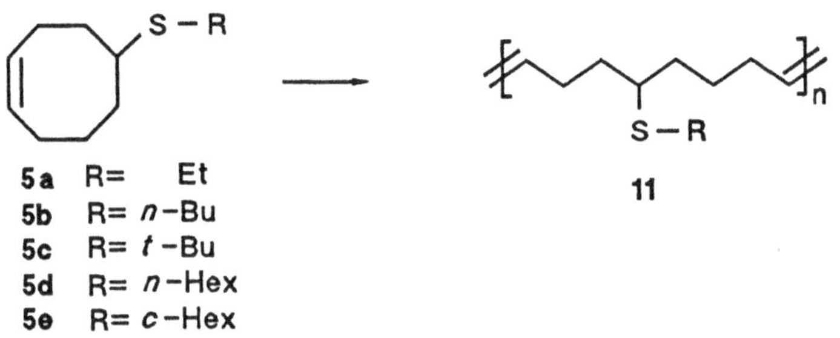

Scheme 7.

TABLE 4. Ring-opening metathesis polymerizations of 5-alkylthiocyclooctenes (**5a-e**) catalyzed by **1**[a]

	R	5/1[b]	T[c] (°C)	t[d] (min)	Conversion of **5** (%)
5a	Et	100	20	200	97
5b	*n*-Bu	100	20	120	97
5c	*t*-Bu	100	20	10	99
		100	80	5	99
		500	80	30	96
		100[e]	20	2	(65)[f]
5d	*n*-Hex	100	20	100	99
5e	*c*-Hex	100	20	30	99

[a]Typically, the metathesis reactions were carried out by using 0.025 mmol of **1** and 5 mL of C_6H_5Cl as solvent. [b]Initial molar ratios. [c]Reaction temperature. [d]Reaction time. [e]Polymerization performed without solvent. [f]Yield of isolated polymer (M_W = 140 100, M_N = 39 700, M_W/M_N = 3.5, T_g = -26°C).

Reaction of *n*-butylthiocyclooctene (**5b**) with ethylene (**7**) in the presence of **1** leads to the sulfur α,ω-diene **12** (Scheme 8). A decrease of the initial **5b/1** molar ratio leads to an increase of the final conversion of **5b**, while an increase of the initial **7/5b** molar ratio gives, as expected, a higher yield of **12** (Table 5). High reaction temperatures drastically enhance the turnover rates.

In summary, we have shown a new property of the neutral cyclometalated neopentylidene complex of tungsten **1** in metathesis of sulfur-containing olefins. The reason for which such carbene complex is active with such functionalized olefins is not fully elucidated yet. The unexpected activity observed might be due to the high rigidity of the cyclometalated structure associated with the rather high steric crowding at the tungsten which prevents strong coordination of the thioether moiety to the metal (due to the cyclometalation, the change of coordination number of tungsten from hexacoordination to pentacoordination is also prevented). This hypothesis is effectively corroborated by the steric effect at the sulfur atom which seems to play a similar role.

Scheme 8.

TABLE 5. Co-metathesis of *n*-butylthiocyclooctene (**5b**) with ethylene (**7**) catalyzed by (**1**)[a]

5b/1[b]	**7/1**[b]	T[c] (°C)	t[d] (h)	Conversion of **5b** (%)	Yield[e] of **12** (%)
200	200	20	8	82	39
100	200	20	8	92	59
50	200	20	8	98	72
50	200	80	0.25	93	61
25	200	20	6	99	75

[a]Typically, the metathesis reactions were carried out by using 0.025 mmol of **1** and 10 mL of C_6H_5Cl as solvent. [b]Initial molar ratios. [c]Reaction temperature. [d]Reaction time. [e]Yields based on initial **5b**.

3. Conclusion

The synthesis of a cyclometalated aryloxy(chloro)neopentylidene-tungsten complex can be achieved by an interesting intramolecular C-H activation step on a d^0 carbyne complex (Schrock's carbyne). The cyclometalated structure corresponds to a 6-coordinated tungsten with one mole of coordinated ether. This cyclometalated neopentylidene complex is probably quite rigid. Nevertheless it is, to our knowledge, one of the most active and stereoselective "one-component" metathesis catalysts reported to date. Besides its specific catalytic properties with simple acyclic olefins, the catalyst is also active in metathesis of sulfur-containing olefins such as thioethers. The reason why such catalyst is active for olefins with a thioether functionality is not clear. It is likely that the thioether is weakly and reversibly coordinated to the tungsten and competes both with the diethyl ether and with the olefinic double bond.

4. Acknowledgements

We thank the Société Nationale Elf-Aquitaine and the Commission of the European Communities for support, and Dr. Karin Weiss (University of Bayreuth) for help in the elaboration of the synthesis route to the cyclometalated compound **1** from the neopentylidyne complex.

5. References and notes

1. Ivin, K. J. (1983), Olefin Metathesis, Academic Press, London.
2. Dragutan, V., Balaban, A. T., and Dimonie, M. (1985), Olefin Metathesis and Ring-Opening Polymerization of Cyclo-Olefins, Wiley-Interscience, New-York.
3. Ivin, K. J. (1990), in Y. Imamoglu, B. Zumreoglu-Karan, and A. J. Amass (eds.), Olefin Metathesis and Polymerization Catalysts, Kluwer Academic Publishers, Dordrecht, pp. 1-43.
4. Mol, J. C. (1991), J. Mol. Catal. 65, 145-162.
5. Grubbs, R. H. and Tumas, W. (1989), Science 243, 907-915.
6. Schrock, R. R. (1990), Acc. Chem. Res. 23, 158-165.
7. Kress, J., Wesolek, M., and Osborn, J. A. (1982), J. Chem. Soc., Chem. Commun., 514-516.
8. Kress, J, Aguero, A., and Osborn, J. A. (1986), J. Mol. Catal. 36, 1-12.
9. Schaverien, C. J., Dewan, J. C., and Schrock, R. R. (1986), J. Am. Chem. Soc. 108, 2771-2773.
10. Johnson, L. K., Virgil, S. C., Grubbs, R. H., and Ziller, J. W. (1990), J. Am. Chem. Soc. 112, 5384-5385.
11. Quignard, F., Leconte, M., and Basset, J.-M. (1985), J. Chem. Soc., Chem. Commun., 1816-1817.
12. Basset, J.-M., Leconte, M., Ollivier, J., and Quignard, F. (1985), Fr. Pat. 85 01930 to SNEA; ibid. (1987), U.S. Pat. 4 654 462 to SNEA.
13. Quignard, F., Leconte, M., and Basset, J.-M. (1986), J. Mol. Catal. 36, 13-29.
14. Dodd, H. T. and Rutt, K. J. (1982), J. Mol. Catal. 15, 103-110
15. Dodd, H. T. and Rutt, K. J. (1985), J. Mol. Catal. 28, 33-36.
16. Basset, J.-M., Leconte, M., Ollivier, J., and Quignard, F. (1983), Fr. Pat. 83 09876 to SNEA; ibid. (1985), U.S. Pat. 4 550 216 to SNEA.
17. Quignard, F., Leconte, M., and Basset, J.-M. (1985), J. Mol. Catal. 28, 27-32.
18. Basset, J.-M., Leconte, M., Ollivier, J., and Quignard, F. (1986), Fr. Pat. 86 11978 to SNEA; ibid. (1989), U.S. Pat. 4 861 848 to SNEA.
19. Boutarfa, D., Paillet, C., Leconte, M., and Basset, J.-M. (1991), J. Mol. Catal. 69, 157-169.
20. Quignard, F., Leconte, M., Basset, J.-M., Hsu, L.-Y., Alexander, J. J., and Shore, S. G. (1987), Inorg. Chem. 26, 4272-4277.
21. The alkylating agent $MgNp_2$ can be replaced by MgNpCl (using 2 or 3 equiv).
22. A molecule of diethyl ether (the solvent of the reaction) remains coordinated to the tungsten center but can be easily replaced by another ether or a Lewis base such as diisopropyl ether, tetrahydrofurane, triphenylphosphine or pyridine.
23. Basset, J.-M., Boutarfa, D., Custodero, E., Leconte, M., and Paillet, C. (1990), in Y. Imamoglu, B. Zumreoglu-Karan, and A. J. Amass (eds.), Olefin Metathesis and Polymerization Catalysts, Kluwer Academic Publishers, Dordrecht, pp. 45-88.
24. Couturier, J.-L., Paillet, C., Leconte, M., Basset, J.-M., and Weiss, K. (1992), Angew. Chem., Int. Ed. Engl. 31, 628.
25. Schrock, R. R., Clark, D. N., Sancho, J., Wengrovius, J. H., Rocklage, S. M., and Pedersen, S. F. (1982), Organometallics 1, 1645.
26. Schrock, R. R., Sancho, J, and Pedersen, S. F. (1989), Inorg. Synth. 26, 44-51.
27. Churchill, M. R., Ziller, J. W., Freudenberger, J. H., and Schrock, R. R. (1984), Organometallics 3, 1554-1562.

28. Schrock, R. R., DePue, R. T., Feldman, J., Yap, K, B., Yang, D. C., Davis, W. M., Park, L., DiMare, M., Schofield, M., Anhaus, J., Walborsky, E., Evitt, E., Krüger, C., and Betz, P. (1990), Organometallics 9, 2262-2275.
29. Kerschner, J. L., Fanwick, P. E., and Rothwell, I. P. (1987), J. Am. Chem. Soc. 109, 5840-5842.
30. Kerschner, J. L., Rothwell, I. P., Huffman, J. C., and Streib, W. E. (1988), Organometallics 7, 1871-1873.
31. Kerschner, J. L., Fanwick, P. E., Rothwell, I. P., and Huffman, J. C. (1989), Organometallics 8, 1431-1438.
32. Thompson, M. E., Baxter, S. M., Bulls, A. R., Burger, B. J., Nolan, M. C., Santarsiero, B. D., Schaefer, W. P., and Bercaw, J. E. (1987), J. Am. Chem. Soc. 109, 203-219.
33. ^{13}C NMR: $\delta(W\equiv\underline{C}CMe_3) = 300.8$; $\delta(W\equiv C\underline{C}Me_3) = 48.8$; $\delta(W\equiv CC\underline{Me_3}) = 33.2$.
34. Leconte, M. and Basset, J.-M. (1979), J. Am. Chem. Soc. 101, 7296-7302.
35. Schrock, R. R., DePue, R. T., Feldman, J., Schaverien, C. J., Dewan, J. C., and Liu, A. H. (1988), J. Am. Chem. Soc. 110, 1423-1435.
36. Concerning acyclic olefins, it was only reported, in one sentence and without experimental data, that allyl pentyl sulfide could be metathesized to a small extent with the catalyst Re_2O_7/Al_2O_3-$(CH_3)_4Sn$: Mol, J. C. (1982), J. Mol. Catal. 15, 35-45.
37. Very recently, two examples were reported concerning the ring-opening metathesis polymerization of *highly strained cyclic olefins* (norbornenes and oxanorbornenes) substituted with functional groups containing sulfur atoms: a) Cummins, C. C., Schrock, R. R., and Cohen, R. E. (1992), Chem. Mater. 4, 27-30; b) Burns, E. G. and Grubbs, R. H. (1991), Poster presented at the 9th International Symposium on Olefin Metathesis and Polymerization, Collegeville (PA).
38. Pennella, F. (1981), J. Catal. 69, 206-208.
39. Nishiguchi, T., Fukuzumi, K, and Sugisaki, K. (1981), J. Catal. 70, 24-31.
40. For clarity, the olefins without sulfur produced in co-metathesis reactions are not represented in the schemes.
41. The various 5-alkylthiocyclooctenes **3a-e** are easily synthesized by radical addition of the corresponding thiols to cyclooctadiene: Griesbaum, K. (1970), Angew. Chem., Int. Ed. Engl. 9, 273-287.

REACTION OF SOME ALKYLIDYNE COMPLEXES OF TUNGSTEN WITH INORGANIC OXIDES: A GENERAL ROUTE TOWARDS ACTIVE SUPPORTED W BASED METATHESIS CATALYSTS?

R. BUFFON, M. LECONTE, A. CHOPLIN*, J.-M. BASSET
Institut de Recherches sur la Catalyse
2, Avenue A. Einstein
69 626 Villeurbanne Cedex
France

ABSTRACT. The reaction between Schrock-type W alkylidyne complexes and inorganic oxides leads to several surface complexes, at least one of them bearing an alkylidene ligand. In the case of $W(CCMe_3)Np_3/Nb_2O_5$, the high catalytic activity seems to be more related to the reaction of a reduced W- species with the olefin via a π-allyl mechanism than to the presence of the alkylidene ligand itself.

1. Introduction

In the field of olefin metathesis, the discovery of the metallocarbene mechanism by Chauvin has lead, in homogeneous catalysis, to the synthesis of a number of active metathesis catalysts; most of them contain a d° alkylidene centers (Re, Mo or W). Although the related heterogeneous catalysts are highly efficient, the nature of the active sites is still a matter of debate. We have selected carbyne complexes of W, i.e. $W(CCMe_3)(CH_2CMe_3)_3$ (1) and $W(CCMe_3)Cl_3(dme)$ (2) and inorganic oxides presenting surface Bronsted acidity, i.e. silica, alumina, niobia, silica-alumina; we expected to obtain surface alkylidene W complexes by protonation of the carbyne, as already proposed by K. Weiss.[1] The catalytic activities for olefin metathesis and chemical reactivity of these grafted complexes were then studied in order to get some insight into the coordination sphere of W in the active sites.

2. Results

2.1. CATALYTIC ACTIVITY FOR THE METATHESIS OF CIS-2-PENTENE

In solution, (1) and (2) present low or no catalytic activity for cis-2-pentene metathesis. However, when these two complexes are supported on inorganic oxides, they become very efficient catalysts. The activities are function of the support (nature and pretreatment temperature [T]= K).(Fig.1) The best performance is observed with SiO_2-Al_2O_3; however, only poor selectivities are obtained due to double bond migration, followed by cross metathesis reactions.

F. R. Kreißl (ed.), Transition Metal Carbyne Complexes, 51–53.

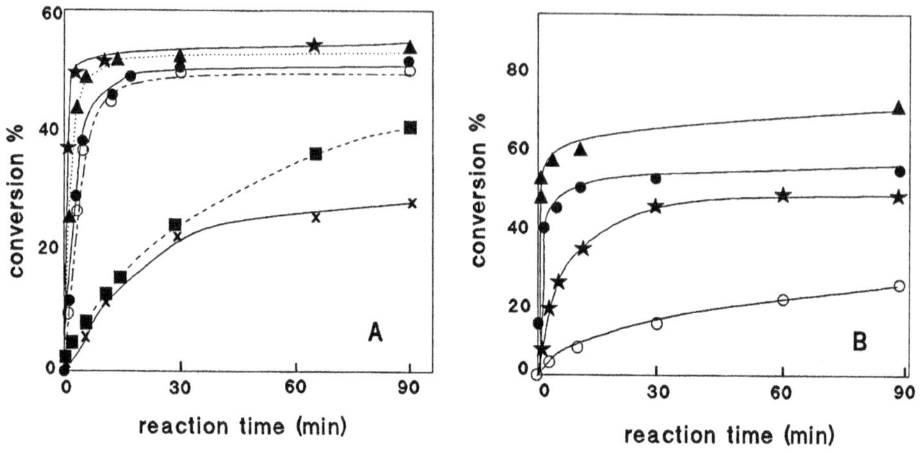

Figure: Catalytic activity for metathesis of cis-2-pentene with supported $W(CCMe_3)Np_3$ (A) and $W(CCMe_3)Cl(dme)$ (B). [W]/olefin: 1:1000; room temperature.
★Nb_2O_5[553]; ▲SiO_2-Al_2O_3[773]; ● Al_2O_3[673]; ○Nb_2O_5[393]; ■ SiO_2[473]; x SiO_2[773]

2.2. REACTIVITY OF (1) AND (2) TOWARDS INORGANIC OXIDES

As niobia appeared to be among the best supports, we have studied in more details the reaction between the functional groups present on its surface and the complexes **1** and **2**.

2.2.1. *W(CCMe₃)Np₃/niobia.* Impregnation of Nb_2O_5[T] by **(1)** produces between 1 and 2 mole of neopentane/ mole $[W]_S$, depending on T. These solids react with dimethylketone, with release of 2,2,4-trimethyl-2-pentene, suggesting the presence of surface neopentylidene W complexes; this is further confirmed by the formation of neohexene (although in small amounts) in presence of either 1-pentene, vinyltrimethylsilane or allylbromide. In the presence of trans-2,5-dimethyl-3-hexene, however, no exchange reaction of the alkylidene moieties occurs, but 3-methyl-1-butene (major product) as well as isobutene and propene are produced. Formation of these olefins can only be explained by the presence of a surface-carbene resulting from the reaction between a W surface complex and the starting olefin via a π-allyl mechanism.[2] This implies that some reduced W species is initially present on the surface.

2.2.2. *W(CCMe₃)Cl₃(dme)/niobia.* Impregnation of Nb_2O_5 [T] with a solution of **(2)** leads to the release of up to 0.5 mole HCl/ mole $[W]_S$, depending on T. These solids react with acetone in a Wittig type fashion. Metathetical exchange with a neopentylidene ligand occurs with allylbromide, while trans-2,5-dimethyl-3-hexene hardly reacts. These results suggest that in the case of **(2)**, a neopentylidene complex is present on the surface, which may actually be the active site, while no reactive reduced W species is present.

For comparison, similar studies were performed on SiO_2[473]. (See table below.)

TABLE . Some characteristics of supported complexes (1) and (2)

complexe	$W(\equiv CMe_3)Np_3$		$W(\equiv CCMe_3)Cl_3(dme)$	
support	SiO_2[473]	Nb_2O_5[573]	SiO_2[473]	Nb_2O_5[573]
initial turnover rate (min^{-1})[f]	15	370	0	60
gas evolved (mole/mole[W]s)	1.3[a]	1.4[a]	1.5[b]	0.2[b]
Wittig (mole/mole $[W]_s$)[c]	0.5	0.3	0.9	0.8
3-Me-1-butene (mole/mole $[W]_s$)[d]	0.03	0.3	traces	0.01
neohexene (mole/mole $[W]_s$)[e]	0.15	0.1	traces	0.4

[a]:neopentane;[b]: HCl; [c]: reaction with acetone;[d]: from reaction with trans-2,5-dimethyl-3-hexene; [e]: from reaction with allylbromide; [f]: [W]/cis-2-pentene: 1/1000, room temperature

3. Discussion and Conclusion

The amounts of NpH or HCl released in the gas phase indicate that the reaction of (1) and (2) with the surface of inorganic oxides such as SiO_2 and Nb_2O_5 leads to a mixture of surface complexes. The use of different chemical reactions for characterization revealed that the high catalytic activity is not correlated only to high concentration of neopentylidene W surface complexes. In the case of (1), reduced species (W^{IV} or less) seem to be involved; these could result from reductive elimination of dineopentyl (indeed detected during the impregnation step). When such reduction is not possible, as it is the case for (2)/ support, activity seems to be correlated to a neopentylidene surface complex, whose reactivity towards olefins is apparently dependent on electronic factors.

At least two types of reactions are involved in the process of grafting at the surface: addition of the surface hydroxyl groups to the carbynic bond and electrophilic cleavage of W-C or W-Cl bonds. Combination of these two reactions is possible, especially on the most hydroxylated surfaces. Finally, reductive elimination is possible for all species bearing at least two neopentyl ligands.

References

1. Weiss, K. and Lössel, G. (1989) 'Heterogeneous, metathesis-active Schrock-type carbene complexes by reaction of carbyne Tungsten(VI) complexes with silica gel', Angew. Chem. Int. Ed. Engl. 28, 62-64.

2. Buffon, R., Choplin, A., Leconte, M., Basset, J.-M., Touroude, R. and Herrmann, W. A., (1992) 'Surface organometallic chemistry of rhenium: attempts to characterize a surface carbene in metathesis of olefins with the catalyst CH_3ReO_3/Nb_2O_5', J. Mol. Catal. 72, L7-L10.

ON THE ROUTE FROM STOICHIOMETRIC TO CATALYTIC REACTIONS OF CARBYNE COMPLEXES. Part XX (1)

K.Weiss*, R.Goller**, M.Denzner**
G.Lößel*** and J.Ködel*

*Laboratorium für Anorganische Chemie der Universität Bayreuth, Postfach 101251 D-8580 Bayreuth F.R.G.
**Rehau Ag + Co, D-8673 Rehau F.R.G.
*** Wacker Chemie, D-8263 Burghausen F.R.G.

ABSTRACT: Stoichiometric reactions of the Schrock type carbyne tungsten(VI) complex Cl_3(dme)WCtBu with heteroallenes or heteroalkenes (isocyanates, carbodiimides, isothicyanates, imines and nitroso compounds) yield metathesis like products. Some of the products give further reactions like insertions or ring closures. With Cl_3(dme)WCtBu catalytic metatheses of differently substituted imines or carbodiimides occur. Consequently also linear or cyclic alkenes give catalytic metatheses with the Cl_3(dme)WCtBu. Polycyclic alkenes, like norbornene, give ringopening metathesis polymerisation (ROMP) not only with Cl_3(dme)WCtBu, but also with Np_3WCtBu and $(tBuO)_3$WCtBu. All 3 carbyne complexes are active catalysts for the polymerisation of 1-alkynes.
Heterogeneous, bimetallic metatheses catalysts are formed by reaction of Fischer type carbyne tungsten(O) or molybdenum(O) complexes with the surface chromium(II) atoms on silicagel of the reduced Phillips catalyst. Schrock type carbyne tungsten(VI) or molybenum(VI) complexes react with surface OH groups of silicagel or zeolithes to form surface bounded carbene tungsten(VI) complexes. The heterogeneous carbene tungsten(VI) complexes catalyse the metathesis of 1-alkenes up to 1000 turnovers per minute.

INTRODUCTION

The first transition metal complexes with a metal carbon triple bond was synthesized by Fischer and Kreis 1973 (2). They called the complexes with low valent transition metals and electrophilic carbyne carbons "carbyne complexes".
In 1975 Schrock published the synthesis of the first high valent (d°) transition metal complex with a nucleophilic

F. R. Kreißl (ed.), Transition Metal Carbyne Complexes, 55–66.

carbyne carbon atom. He called this type of complexes "alkylidyne complexes"(3). The third type of carbyne complexes with less polar metal carbon triple bonds was discovered by Roper in New Zealand (4).

In this paper we report of our work on stoichiometric and catalytic reactions of Fischer and Schrock type carbyne complexes. In the following text the term carbyne will be used also for alykidyne complexes.
When we started with our work on carbyne complexes, transition metal carbene complexes were well known synthons for organic syntheses and established catalysts for alkene metatheses reactions (5). Carbyne complexes just started to gain interest in this field.
In 1976 Fischer had tested the ROMP reactions of cycloalkenes with carbyne tungsten(O) complexes as catalysts (6). With addition of Lewis acids as cocatalysts the Fischer type carbyne complexes were active in cycloalkene metathesis polymerisation. Fischer type carbyne complexes are also active catalysts for alkyne polymerisations, as found by Katz in 1984 (7). The catalytic reactions of Schrock type carbyne tungsten(VI) or molybdenum(VI) complexes were focussed on alkyne metatheses reactions (8).

RESULTS AND DISCUSSION

Encouraged by our results on stoichiometric and catalytic metathesis reactions of carbodiimides and imines with Fischer type carbene tungsten(0) complexes (9) we started 1984 with metathesis like reactions of the Schrock type carbyne tungsten(VI) complex Cl_3(dme)WCtBu with heteroallenes (isocyanates, carbodiimides, isothiocyanates) and with heteroalkenes (imines and nitroso compounds).

Scheme 1 gives a postulated reaction pathway for all these metathesis like reactions. The heteroallenes or the heteroalkenes are supposed to react in a 2 + 2 cycloaddition reaction with Cl_3(dme)WCtBu to form metalla cyclobutene derivatives. Electrocyclic ringopening reactions of these (not isolated) metallacycles yield tungsten imido or oxo complexes with sigma bonded vinyl, iminyl, ketenyl or keteniminyl ligands.

Reaction products of this type, formed by reaction of isocyanates or carbodiimides with Cl_3(dme)WCtBu insert a further molecule of the heteroallene into the W-C bond.

Scheme 1: Reactions of Heteroalkenes and Heteroallenes with $Cl_3(dme)WCtBu$

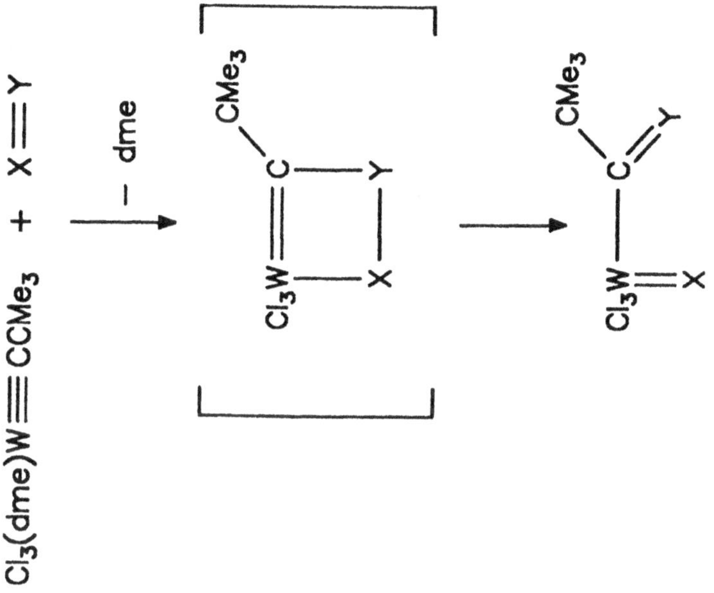

X=Y
RN=CHR
O=NR
RN=C=O
RN=C=NR

By these insertions a chelating ligand is formed (scheme 2). The structures of the reaction products were analysed by IR, NMR, mass spectra and for the reaction product of cycloexylisocyanate by X ray structure (10, 11). The chelating ligand is cleaved off the metal fragment with methanol and gives malonic acid derivatives.

Scheme 2: Insertion Reaction of a further Isocyanate

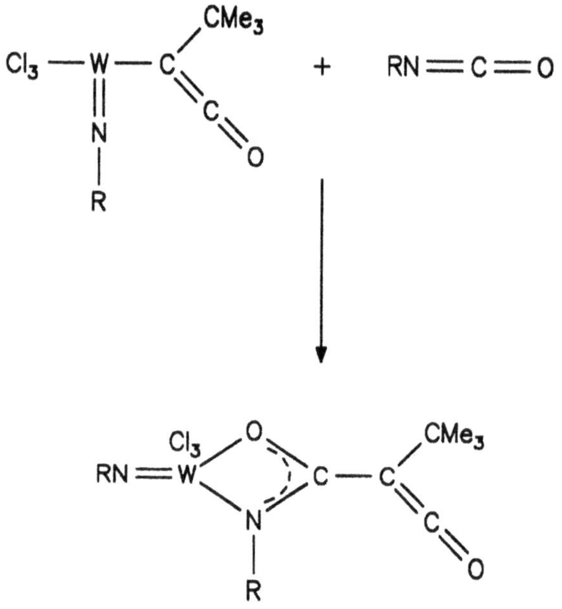

The reactions of alkylisothiocyanates with Cl_3(dme)WCtBu yields polymeric reaction products.The structure of the products was analysed by solid state NMR, by IR and a X ray analysis of a tetrameric derivative (10).
The postulated reaction pathway forms via a metallacycle and a ringopening reaction .the tungsten complex on which an imido and a thioketenylligand is coordinated. A proton shift to the nitrogen of the imido ligand forms a Mannich base. Ring closure reactions of the base and the thioketanylligand yields the metalla bicycles. The monomers polymerise by bridging chloro ligands.

Scheme 3: Reaction of nBu-N=C=S with Cl_3(dme)WCtBu

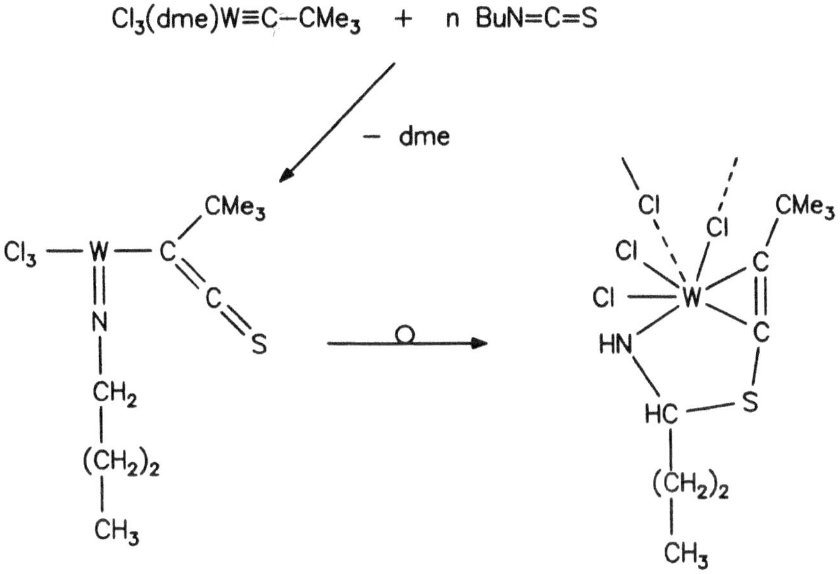

Cl_3(dme)WCtBu gives not only stoichiometric but also catalytic metathesis reactions with differently substituted carbodiimides or imines (10).

Following the catalytic metathesis of carbodiimides and isothiocyanates with Cl_3(dme)WCtBu, we tested the metatheses of alkenes with this complex. Linear 1-alkenes and monocyclic alkenes (scheme 4) give catalytic metathesis with the Schrock type carbyne complex in CH_2Cl_2 at 20°C (12). At 76°C and in 1,2 dichloroethane as solvent the turnover frequence enhance (Table 1) (13).

Table 1

Metathesis of Alkenes with the Carbyne-Tungsten-(VI)-Complex $Cl_3(dme)WCCMe_3$ (A) and the Heterogeneous Catalysts $SiO_2/Np_2WCHtBu$ (B) and $SiO_2/Cr/Cl(CO)_4WCPh$ (C)

Alkene	Cat.	[Alkene] / [W]	Temp.(°C)	Solvent	Conversion (%) 24 h	Activity per hour
1-Hexene	A	72	25	CH_2Cl_2	86	492
5-Methyl-1-hexene	A	72	25	CH_2Cl_2	81	73
4,4-Methyl-1-pentene	A	72	25	CH_2Cl_2	22	7
1-Octene	A	500	25	CH_2Cl_2	31	18
	A	500	25	$C_2H_4Cl_2$	38	30
	A	500	76	$C_2H_4Cl_2$	92	334
	B	10000	122	-	80	59520
	C	5000	122	-	85	15300
4-Perfluor-isopropyl-1-butene	A	100	25	CH_2Cl_2	20	5
	A	500	76	$C_2H_4Cl_2$	63	44
	B	1000	76	-	94	1240
	C	1000	76	-	97	1570

Molar ratio (Mol Alkene/Mol W), Reaction-temperature (°C), Metathesis activity (Mol Alkene/Mol W per hour).

Scheme 4: ROMP of Cyclopentene with Cl_3(dme)WCtBu

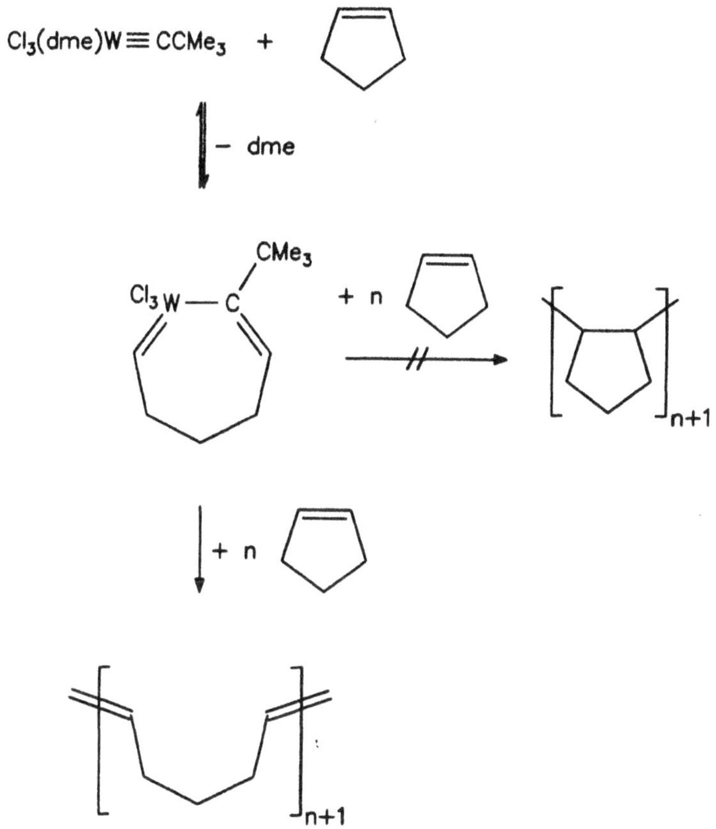

The monocyclic alkenes like cyclopentene, cyclooctene or 1,5 cyclooctadiene form polyalkenameres with high molar masses if Cl_3(dme)WCtBu is used as homogeneous catalyst. The heterogeneous catalyst produced polymers with lower molar masses (Table 2).

Table 2

ROMP of 1,5-Cyclooctadiene (COD) and 1,5,9-Cyclododecatriene (CDT) with Carbyne-Tungsten-(VI)-Catalysts in CH_2Cl_2 at 0-25 °C

Catalysts: A= Cl_3(dme)$WCCMe_3$ B= SiO_2/ $Np_2WCHCMe_3$

Alkene	Cat.	[Alkene]/[W]	Oligomer Yield (%)	Polymer Yield (%)	Polymer trans (%)	M_w $*10^{-3}$ (PS)	M_n $*10^{-3}$ (PS)	D
1,5-COD	A	500	10.9	79.0	77.5	198.1	112.0	1.8
	B	500	2.2	75.7	40.8	49.9	29.7	1.6
	A	2000	9.7	78.5	32.3	262.6	148.1	1.8
	B	2000	2.1	89.9	48.4	56.9	33.2	1.7
1,5,9-CDT	A	500	8.5	79.2	78.4	224.5	90.8	2.5
	B	500	9.8	65.1	66.6	3.8	0.9	3.8

Molar ratio (Mol Alkene/Mol W), Metathesis reactivity (Mol Alkene/Mol W per h).

For the ROMP reaction of polycyclic alkenes, like norbornene or dicyclopentadiene, not only Cl_3(dme)WCtBu, but also Np_3WC-tBu and $(tBuO)_3$WCtBu are active metathesis catalysts in contrast to the ROMP reactions of monocyclic alkenes (Table 3) (13).

Table 3

ROMP of Polycyclic Alkenes with Carbyne-Tungsten-(VI)-Complexes at 25°C in CH_2Cl_2 (3)

Catalysts: A= Cl_3(dme)WCCMe$_3$, D= (tBuO)$_3$WCCMe$_3$, E= Np$_3$WCCMe$_3$

Alkenes	Cat.	[Alkene] / [W]	Conversion (%)	Reaction-time(h)	Activity (per hour)	(M_w) (*10^{-3})
Dicyclopenta-diene	A	500	96	24	106	143
5-Ethylidene-2-norbornene	A	500	100	1	980	24
2-Norbornene	D	1000	100	5	842	166
5-Vinyl-2-nor-bornene	D	500	51	24	132	45
7-t-Butoxy-2,5-norbornadiene	D	200	25	24	21	24
2,5-Norborna-diene	E	1000	100	24	645	1
5-Methylidene-2-norbornene	E	500	36	24	156	-

Molar ratio (Mol Alkene/Mol W), Metathesis reactivity (Mol Alkene/Mol W per h), Molecular Weight (GPC, PS-Standard, in THF, not completely dissolved).

Many alkene metathesis catalysts are active catalysts for alkyne polymerisation. We tested the polymerisation of 1-alkynes with all 3 Schrock type carbyne complexes. Cl_3(dme)WCtBu, Np$_3$WCtBu and (tBuO)$_3$WCtBu are active catalysts for 1-alkyne polymerisation. In addition Cl_3(dme)WCtBu) catalyse the metathesis of internal alkynes and (tBuO)$_3$WCtBu gives alkyne metathesis as shown by Schrock (18).

Heterogeneous, bimetallic metathesis catalysts are formed by reactions of Fischer type carbyne tungsten or molybdenum complexes with the reduced Phillips catalyst, a suface chromium(II) compound on silica (14).(scheme 5). The bimetallic surface compounds can result from 2 + 1 cycloaddition reactions. Similar reactions are well known by the work of Stone (15).

Scheme 5: Formation of heterogeneous,bimetallic Metathesis Catalysts by Reaction of Fischer type Carbyne Complexes with Reduced Phillips Catalyst

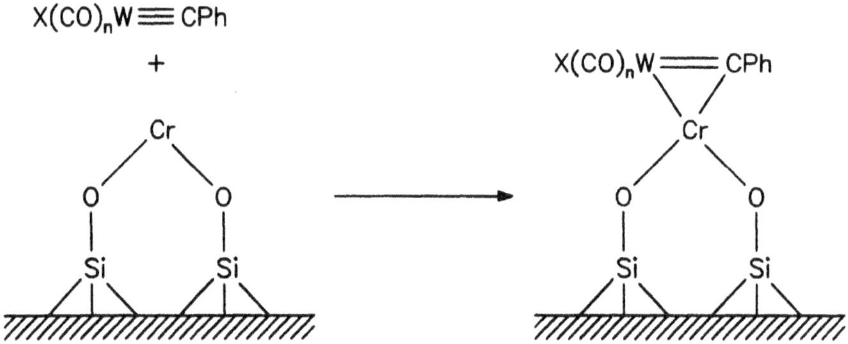

X = Cl, Br, I; n = 4   X = Cp; n = 2

By reaction with the Fischer type carbyne complexes loose the surface chromium(II) atoms of the reduced Phillips their polymerisations activity for 1-alkenes (16). The surface chromium(II) atoms enhance the metathesis activity of the bimetallic catalysts (Table 1). In contrast to the original Fischer type carbyne complexes are the bimetallic catalysts stable at room temperature. They can be stored for a year at 25°C without changing their metathesis activity (14).

The Schrock type carbyne tungsten or molybdenum complexes do not need surface chromium(II) atoms to activate their metathesis activity. 1989 we were able to show that the activity of these high valent carbyne complexes is drastically enhanced by reactions with surface OH groups of silicagel or zeolithes (17). By reactions with the surface OH groups heterogeneous Schrock type carbene complexes are formed. The formation of carbene ligands was proved by

Wittig like reactions with ketons (19) (Scheme 6) The heterogeneous carbene complex $SiO_2/Np_2WC(H)tBu$ is one of the most active heterogeneous metathesis catalyst (Table 1).

Scheme 6: Formation of Surface Carbene Tungsten(VI)complexes

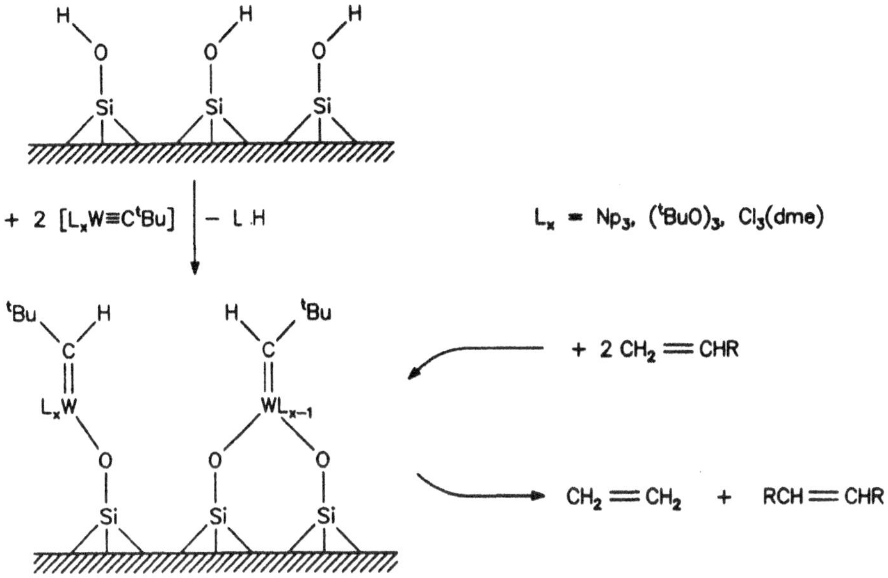

CONCLUSION

Schrock and Fischer type carbyne tungsten or molybdenum complexes are very interesting catalysts for alkene metathesis or alkyne polymerisation reactions. Within the first reaction steps they form carbene complexes and on these carbene complexes further metathesis or polymerisation occur.

ACKNOWLEDGEMENT

The authors thank Deutsche Forschungsgemeinschaft and Stiftung Volkswagenwerk for financial support.

REFERENCES

1. Part XX: Investigations of Polymerisation and Metathesis Reactions
Part XIX: K.Weiss, G.Lößel und M.Denzner in Y Immamoglu Metathesis and Polymerisation Catalysts,Kluwer Acad. Publ. Dordrecht, 521 (1990)
2. E.O.Fischer, G.Kreis, C.G.Kreiter, J.Müller, G.Huttner and H.Lorenz, Angew. Chem. 85 (1973) 618, Int. Ed. Eng. 12 (1973)564
3. L.J.Guggenberger and R.R. Schrock, J.Am.Chem.Soc. 97 (1975) 2935
4. G.R.Clark, C.M.Cochrane, K.Marsden, W.R.Roper and L.J. Wright, J. Organomet. Chem. 315 (1986) 211
5. K.Dötz, Transition Metal Carbene Complexes, Verlag Chemie, Weinheim 1983
6. E.O.Fischer and R.Wagner, J.Organomet.Chem. 116 (1970) C21
7. T.J.Katz, T.H.Shin, Y.C.Ying and V.I.W.Stuart, J.Am. Chem.Soc. 106 (1984) 2659
8. J.H.Wengrovius, J.Sancho and R.R.Schrock, J. Am. Chem. Soc.102 (1981) 3932
9. K.Weiss and P.Kindl, Angew.Chem. 96 (1984) 616 Int. Ed. Engl. 23 (1984) 629
10. R.Goller, Thesis University Bayreuth 1988
11. K.Weiss, U.Schubert and R.R.Schrock, Organometallics 5 (1986) 397
12. K.Weiss, Angew. Chem. 98 (1986) 360, Int. Ed.Engl. 25 (1986) 359.
13. J.Ködel, Thesis University Bayreuth 1993
14. K.Weiss and M.Denzner, J.Organomet. Chem. 355 (1988) 273
15. F.G.Stone, Inorg.Chim. Acta 50 (1981) 33
16. K.Weiss and H.L.Krauss, J.Catal. 88 (1984) 424
17. K.Weiss and G.Lößel, Angew. Chem. 101 (1989) 75 Int. Ed. Engl. 28 (1989) 62
18. R.R.Schrock, J.Organomet. Chem. 300 (1986) 249
19. G.Lößel, Thesis University Bayreuth 1990

YLIDE NICKEL CATALYSIS: PROGRESS IN ACETYLENE POLYMERIZATION

K.A. OSTOJA STARZEWSKI
Bayer AG, Zentrale Forschung
Wissenschaftliches Hauptlaboratorium
5090 Leverkusen, Deutschland

Ylids possess an outstanding ligand potential for catalytic applications of transition metal chemistry, - and especially for polymerization catalysis. [1, 2]

Table 1. Examples of unsaturated substrates in ylid-steered polymerization reactions

acetylene
ethylene
ethylene + α-olefins
ethylene + carbonmonoxide
butadiene
acrylic ester
methacrylic ester

Photoelectron spectroscopic studies place this class of isolable energy rich compounds R_3PCXY on the very top of an energy scale of ligands without net charge. (Tab. 2)

Table 2. First ionization potentials of representative phosphorus ylides (n_C) and related phosphines (n_P) [3]

$Me_3PCHCHCHMe$	6.02 eV
Me_3PCHPh	6.19 eV
Ph_3PCH_2	6.62 eV
Me_3PCH_2	6.81 eV
$Ph_3PC(CN)_2$	7.63 eV
Ph_3P	7.80 eV
Me_3P	8.60 eV

When two ylids (R_3PCXY and $R_3PCR'CR''O$) are reacted with Ni(O) complexes, polymerization catalysts form. Their ESCA spectra show low Ni core electron binding energies, which fall in the range of zerovalent nickel complexes. (Tab. 3)

F. R. Kreißl (ed.), Transition Metal Carbyne Complexes, 67–69.

Table 3. Ni $2p_{3/2}$ binding energies (rel. C 1s = 284.6 eV)

$Ni(COD)_2$	853.3 eV	[4]
$NiPh(Ph_2PCHCMeO)(Ph_3PCH_2)$	853.4 eV	
$Ni(PPh_3)_3$	854.0 eV	[4]
$NiPh(Ph_2PCHCMeO)(Ph_3P)$	854.4 eV	

An x-ray structure analysis shows a square planar nickel complex with an extremely long Ni-O bond of 1.95 A. (Tab. 4)

Table 4. Bond lengths in (PCCONi) metallocycles

$NiPh(Ph_2PCHCMeO)(i\text{-}Pr_3PCH_2)$	NiO: 1.951 A	CO: 1.302 A	[5]
$NiPh(Ph_2PCHCPhO)(PPh_3)$	NiO: 1.914 A	CO: 1.313 A	[2b]
$Ni(Ph_2PCHCPhO)_2$	NiO: 1.885 A	CO: 1.318 A	[2a]

Ylid nickel catalysts [Ni(O)/R_3PCR'CR"O/R_3PCXY] not only show superior performance in the polymerization of acetylene as compared to phosphine nickel catalysts [Ni(O)/ R_3PCR'CR"O/R_3P],- in highly polar solvents the normalized polymerization activity (mol acetylene/mol Ni h bar) probably exceeds that of all known nickel systems. (Fig. 1)

However, catalyst activity alone is not sufficient to make this chemistry useful, unless all of the technologically unattractive properties of classical polyacetylene (URPAC) can be overcome. (Tab. 5)

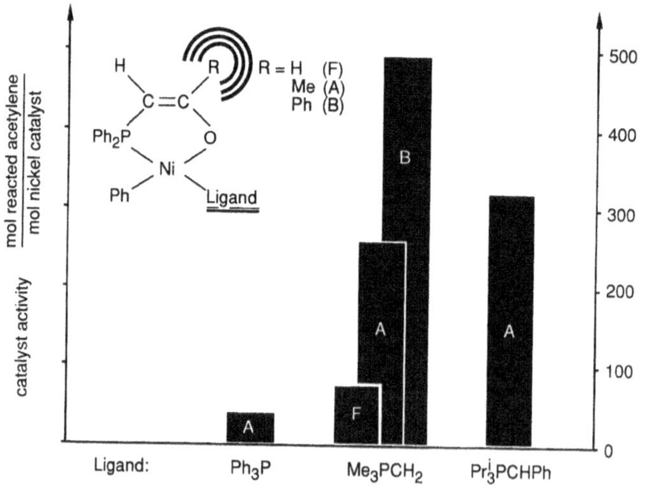

Figure 1. Catalyst activity in acetylene polymerization

Table 5. Evaluation of URPAC properties

+	a promising candidate for high tech applications	
-	insufficient synthetic control of polymer architecture	
-	insoluble	i.e. not processable
-	infusable	
-	unstable	

The breakthrough comes from ylid nickel catalysis: Soluble polyacetylenes in highly polar and thus stabilizing polymer matrices are accessible. The distribution of conjugation lengths is ligand-controlled (selectivity!). (Fig. 2) The novel matrix polyacetylenes (MATPAC) are readily processible, e.g. by injection molding, film casting, spin coating or fiber spinning, and - where necessary - the products may be oriented by drawing.

Ylid nickel catalysis thus fulfills the key prerequisites needed to develop PAC into an "advanced material".

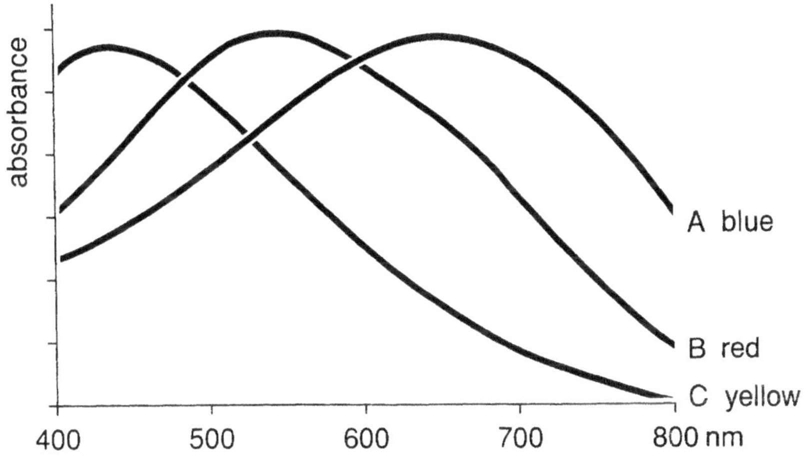

Figure 2. UV-vis spectra of colored PANPAC/DMF solutions, obtained with different ylide nickel catalysts. (PANPAC: polyacrylonitrile-polyacetylene)

[1] K.A. Ostoja Starzewski, G.M. Bayer, Angew. Chem. Int. Ed. Engl. (1991) 30, 961; and references therein. [2] a. U. Klabunde, R. Mulhaupt, T. Herskovitz, A.H. Janowicz, J. Calabrese, S.D. Ittel, J. Polym. Sci. Polym. Chem. (1987) 25, 1989; b. W. Keim, F.H. Kowaldt, R. Goddard, C. Krüger, Angew. Chem. Int. Ed. Engl. (1978) 17, 466; and references therein. [3] K.A. Ostoja Starzewski et al., Phosphorus Sulfur (1983) 18, 448; Inorg. Chem. (1979) 18, 3307; J. Amer. Chem. Soc. (1976) 98, 8486. [4] C.A. Tolman, W.M. Riggs, W.J. Linn, C.M. King, R.C. Wendt, Inorg. Chem. (1973) 12, 2770. [5] K.A. Ostoja Starzewski, L. Born, Organometallics (1992) 11, 2701.

Conjugated Complexes and Polymers Derived from Metal–Alkylidyne Building Blocks

Timothy P. Pollagi, Joseph Manna, Timothy C. Stoner,† Steven J. Geib, and Michael D. Hopkins‡
Department of Chemistry and Materials Research Center
University of Pittsburgh
Pittsburgh, Pennsylvania 15260
U. S. A.

ABSTRACT. The syntheses, structures, and properties of the conjugated compounds $W(\equiv CH)L_4(C\equiv CR)$ and $[W(\equiv C\text{–pyr})(OR)_3]_\infty$ are described. These compounds are the first low-dimensional materials to be prepared from metal–alkylidyne building blocks.

Introduction

Despite their obvious similarity to alkenes and alkynes, transition-metal alkylidene and alkylidyne complexes[1] have not been used as building blocks for the synthesis of low-dimensional materials analogous to polyenes and polyynes. We have begun to explore the syntheses, structures, and properties of conjugated complexes and polymers derived from metal–alkylidyne complexes as part of our effort[2] to develop the chemistry of transition-metal analogues of conjugated organic compounds.

Metallabutadiyne Complexes

We have recently prepared and characterized complexes of the type $W(\equiv CH)(dmpe)_2(C\equiv CR)$ (dmpe = 1,2-bis(dimethylphosphino)ethane; R = H, $SiMe_3$, Ph, *p*-$C_6H_4C\equiv CPr^n$),[3] which are internally substituted, metal-containing analogues of the important class of organic butadiynes. X-ray diffraction studies reveal that these pseudooctahedral metallabutadiyne complexes possess nearly linear C≡W–C≡C–R backbones of alternating bond order. Conjugation within this backbone is manifested both structurally and spectroscopically. For example, a π–bonding interaction between the C≡W and C≡CR moieties is suggested by the fact that the HCW–C bond of $W(CH)(dmpe)_2(CCSiMe_3)$ is 0.16 Å shorter than that of $W(CH)(dmpe)_2(n\text{-Bu})$, which is double the difference expected from that between the covalent radii of sp and sp^3-hybridized carbon atoms. π-Electron delocalization is also indicated by the ^1H NMR spectrum of $W(CH)(dmpe)_2(CCH)$, which exhibits 0.8-Hz spin–spin coupling of the terminal hydrogen nuclei over five bonds (*H*–C≡W–C≡C–*H*); by comparison, $^5J_{HH}$ = 2.2 Hz for butadiyne.

The most sensitive probe of π(C≡W–C≡C) conjugation in the metallabutadiynes is electronic-absorption spectroscopy. The lowest-energy absorption band of $W(CH)L_4X$ complexes has been assigned as $[d_{xy} \rightarrow \pi^*(W\equiv C)]$, the terminating orbital of which is of the appropriate symmetry to mix with the π and π* orbitals of the alkynyl ligand. That the

F. R. Kreißl (ed.), Transition Metal Carbyne Complexes, 71–73.

π*(W≡C) and π*(CCR) levels mix considerably is indicated by the fact that the transitions of the metallabutadiynes are significantly red shifted from those of nonconjugated methylidyne complexes, such as $W(CH)(dmpe)_2(n\text{-}Bu)$, and are quite sensitive to the nature of the alkynyl R group (Figure 1).[3] Since interactions between π*(W≡C) and π(CCR) would tend to counteract the shift due to those between π*(W≡C) and π*(CCR), this latter mixing may be greater than that indicated spectroscopically.

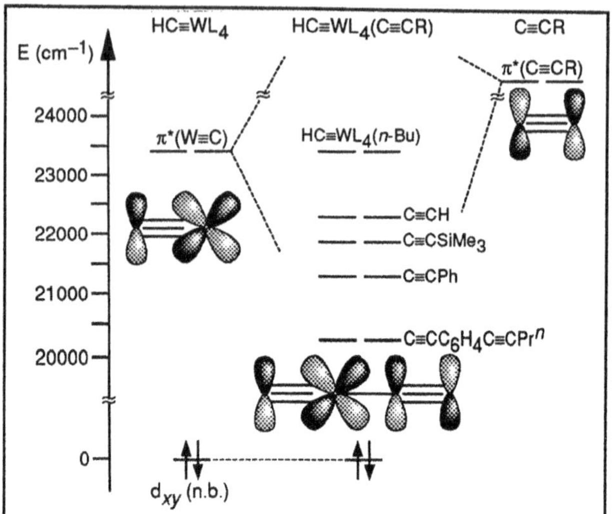

Figure 1. Orbital Interaction Diagram for $W(CH)L_4(CCR)$.

Metal–Alkylidyne Polymers

Our studies of conjugated one-dimensional polymers of the type $[M(\equiv N)(OR)_3]_\infty$[2b] prompted us to investigate the possibility that related polymers might exist in which the nitrido ligand is replaced by the isoelectronic CR ligand (where R is a conjugated N–heterocycle), based on the expectation that this would result in polymers with conjugated backbones of greater covalency. We have prepared and characterized polymers with the general formula $[W(\equiv C\text{–pyr})(OR)_3]_\infty$ (R = CMe_3, CMe_2Et; pyr = 4-pyridinyl, 3-pyridinyl, 4-(3,5-lutidinyl)), which are the first examples of conjugated polymers composed of alternating organic and transition-metal subunits. These are prepared by the metathesis reactions between $W_2(OR)_6$ or $W(CEt)(OCMe_3)_3$ and the appropriate N–heterocyclic alkyne or nitrile (Figure 2), which proceed smoothly at room temperature to give high yields of sparingly soluble materials whose elemental analyses and mass spectra are consistent with the proposed formulation. The poor solubility of these polymers has hindered their characterization by NMR spectroscopy, although we have observed ^{13}C resonances for each in the range δ 250–275 that are attributable to the triply bonded carbon atom.

Under appropriate reaction conditions, crystalline samples of these polymers can be isolated directly from the reaction mixture that are suitable for single-crystal X-ray diffraction studies. These studies reveal that the polymers consist of head-to-tail, W–N bonded assemblies of $W(C\text{–pyr})(OR)_3$ monomers. A striking observation is that both limiting geometries for metal pentacoordination are observed, with the exact nature of the heterocyclic linker appearing to play a dominant role in determining the geometry about the metal center. Specifically, the 3-pyridinyl polymer contains trigonal-bipyramidal tungsten centers with axial W≡C and W–N bonds, while the 4-lutidinyl derivative possesses tungsten atoms coordinated in a square-pyramidal fashion, with the alkylidyne ligand in the axial site (Figure 2). Although the lengths of the W≡C bonds do not appear to be sensitive to the geometry—distances of ca. 1.78 Å are found for both polymers—the W–N bond distance is 2.25 Å for the 4-lutidinyl polymer and 2.53 Å for the 3-pyridinyl derivative, the longer distance for the latter being the result of the strong trans influence of the alkylidyne ligand.

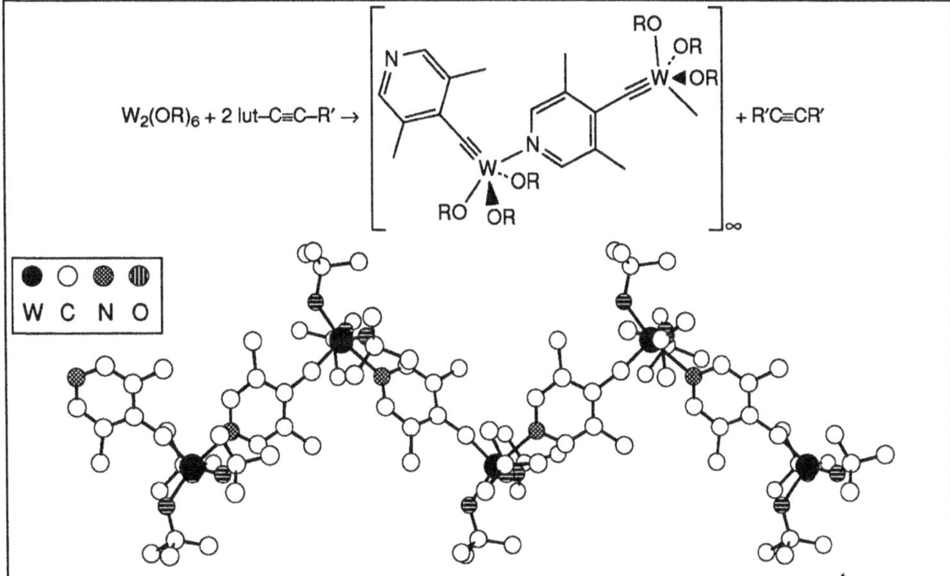

Figure 2. Synthesis and X-ray Crystal Structure of [W(≡C–4-(3,5-lutidinyl))(OBu^t)$_3$]$_\infty$.

A remarkable property of these polymers is that they are luminescent in fluid solution at room temperature, and strongly so in the solid state. Solution-phase emission from LMCT states is rare, and we know of only two other reports involving d^0 transition-metal complexes.[4] Preliminary photophysical studies suggest that the emissive state is a spin triplet ($\tau_{em} \cong 1$ μs). Insight into the bonding of the polymers has been gleaned from the emission spectrum of [W(C–lut)(OBu^t)$_3$]$_\infty$ at 33 K, which exhibits a broad vibronic progression in a ca. 1000-cm^{-1} mode. This appears to indicate that band-gap emission is accompanied by distortions along the ν(W≡C) and/or C–lutidinyl modes, suggesting that the orbitals within the (W≡C–lut)$_\infty$ backbone contribute significantly to these bands. We are continuing our investigations of these and related polymers in order to understand further their molecular and electronic structures.

Acknowledgment. This research was supported by the National Science Foundation.

References

† Andrew W. Mellon Predoctoral Fellow.

‡ NSF Presidential Young Investigator (1987–1992); Dreyfus Foundation New Faculty Awardee (1987–1992); Packard Foundation Fellow (1990–1995).

1. (a) Mayr, A.; Hoffmeister, H. *Adv. Organomet. Chem.* **1991**, *32*, 227. (b) Fischer, H.; Hofmann, P.; Kreissl, F. R.; Schrock, R. R.; Schubert, U.; Weiss, K. *Carbyne Complexes*; VCH Publishers: New York, 1988.
2. (a) Stoner, T. C.; Dallinger, R. F.; Hopkins, M. D. *J. Am. Chem. Soc.* **1990**, *112*, 5651. (b) Pollagi, T. C.; Stoner, T. C.; Dallinger, R. F.; Gilbert, T. M.; Hopkins, M. D. *J. Am. Chem. Soc.* **1991**, *113*, 703. (c) Stoner, T. C.; Geib, S. J.; Hopkins, M. D. *J. Am. Chem. Soc.* **1992**, *114*, 4201.
3. Manna, J.; Geib, S. J.; Hopkins, M. D. *J. Am. Chem. Soc.* **1992**, *114*, in press.
4. (a) Paulson, S.; Sullivan, B. P.; Caspar, J. V. *J. Am. Chem. Soc.* **1992**, *114*, 6905. (b) Pfennig, B. W.; Thompson, M. E.; Bocarsly, A. B. *J. Am. Chem. Soc.* **1989**, *111*, 8947.

Methylidyne Complexes: Structures, Spectra, and Bonding

Joseph Manna, Linda A. Mlinar, Raymond J. Kuk,† Richard F. Dallinger,†
Steven J. Geib, and Michael D. Hopkins‡
Department of Chemistry and Materials Research Center
University of Pittsburgh
Pittsburgh, Pennsylvania 15260
U. S. A.

ABSTRACT. The nature of the M≡C bond has been probed by single-crystal X-ray diffraction and NMR, electronic, and Raman spectroscopic studies of methylidyne complexes of the type *trans*-W(≡CH)L_4X.

Introduction

Methylidyne complexes (M(≡CH)L_n)[1] are the ideal subjects of study from the standpoint of understanding the electronic structures of alkylidyne complexes and the nature of the M≡C bond, since alkylidyne R groups more complex than H have the potential both to interact strongly with the M≡C bond and to mask the effects of the ancillary ligands. Unfortunately, these complexes are rare.[2] Our interest in the bonding of alkylidyne complexes[3] led us to extend Schrock's archetypal class of W(CH)L_4X compounds[1a–c] to include axially substituted derivatives of strong π-donating (X = $OSiMe_3$), σ-donating (*n*-Bu), and π-accepting (C≡$CSiMe_3$) ligands, with the aims of structurally and spectroscopically probing the influence of these ancillary ligands on the electronic structures of alkylidyne complexes and of establishing the intrinsic electronic properties of the M≡C bond. We report herein our preliminary findings on these matters.

Results and Discussion

Complexes of the type W(CH)$(dmpe)_2$X (dmpe = 1,2-bis(dimethylphosphino)ethane; X = Cl, *n*-Bu, $CCSiMe_3$) have provided the first disorder-free X-ray crystal structures for the methylidyne class.[4] These compounds adopt a pseudooctahedral geometry with trans methylidyne and X ligands (Figure 1), as has been found for all other M(CR)L_4X complexes,[2] with short W≡C and long W–X bond distances (Table 1); the latter are 0.1–0.3 Å longer than those found for simple tungsten complexes of these ligands, as a result of the strong trans influence of the alkylidyne ligand. In contrast to nearly all other alkylidyne complexes, however, the methylidyne ligand is found to be significantly bent (∠(W–C–H) ≅ 160°), for those compounds in which we have been able to locate the methylidyne hydrogen atom. We have noted elsewhere that this bend may be intrinsic to these compounds,[4b] although the possibility that it is the result of crystal packing cannot be excluded.

Although the extreme length of the W–X bond would seem to imply that the axial ligand is not likely to strongly perturb the electronic structure of the W≡CH fragment, the structural

F. R. Kreißl (ed.), Transition Metal Carbyne Complexes, 75–77.

and spectroscopic parameters most sensitive to the nature of the W≡C bond suggest otherwise. The most noteworthy structural feature in this regard is the W≡C distance of the *n*-butyl derivative, which is ca. 0.03 Å longer than those of the chloro and alkynyl complexes. This is in keeping with the fact that *n*-butyl is the strongest donor ligand in this series.

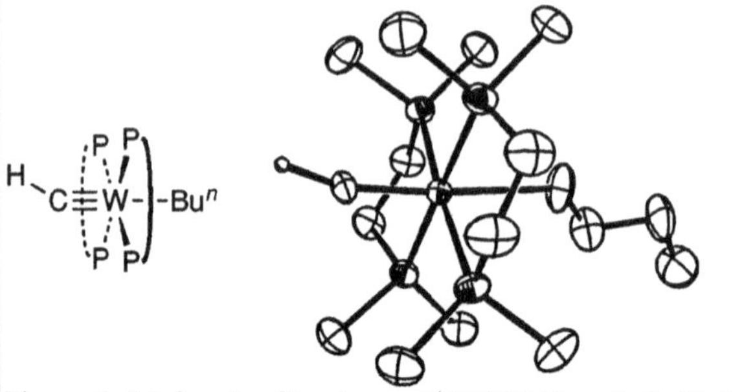

Figure 1. Molecular Structure of W(CH)(dmpe)$_2$(*n*-Bu).

The spectroscopic parameter most straightforward to interpret is the energy of the [$d_{xy} \rightarrow \pi^*$(W≡C)] electronic transition ($\bar{\nu}_{max}$, Table 1), which we have assigned as the lowest-energy band in the electronic-absorption spectrum.[3] The d_{xy} orbital is nonbonding (δ symmetry) with respect to the axial ligand, and can be considered to be of constant energy across this series of complexes, to first order. Since π^*(W≡C) can mix with appropriate axial ligand orbitals, changes in transition energy as a function of X mirror the changes in energy of π^*(W≡C). The trend in transition energy for W(CH)(dmpe)$_2$X (OSiMe$_3$ > Cl > *n*-Bu > C≡CSiMe$_3$) indicates that the π^*(W≡C) level is destabilized by interactions with π-donor ligands and stabilized by π-acceptor ligands, relative to ligands that are σ-donors only, in accord with expectation from the spectrochemical series. That the axial ligand can play a strong role in determining the energy of the LUMO is demonstrated by the 0.5-eV range of these values. The difference in transition energy between W(CH)(PMe$_3$)$_4$Cl and W(CH)(dmpe)$_2$Cl may reflect the splitting of the d_{xz} and d_{yz} components of π^*(W≡C) under C_{2v} symmetry; these orbitals are degenerate under C_{4v} symmetry.

Neither δ W*C*H nor $^1J_{CW}$ appear to correlate in a simple fashion with the energy of the [$d_{xy} \rightarrow \pi^*$(W≡C)] transition. Such correlations may be stronger with the energies of transitions originating from the π(W≡C) levels; we are currently attempting to identify such transitions in the spectra of these complexes. Among the NMR data, it is noteworthy that W(CH)(dmpe)$_2$(*n*-Bu) displays an unusually low W–C coupling constant, consistent with its long W≡C bond.

We have also begun to study W(CH)L$_4$X complexes using Raman spectroscopy (Figure 2),[5] which has revealed that the W≡C stretching frequencies (875–920 cm^{-1}; Table 1) and force constants (ca. 6 mdyne Å^{-1}) of these compounds are substantially lower than those

Table 1. Structural and Spectroscopic Data for W(CH)L$_4$X Complexes

Compound	*d*(W≡C) (Å)	*d*(W–X) (Å)	∠WCH (deg)	δ W*C*H (ppm)	$^1J_{CW}$ (Hz)	ν(W≡C) (cm^{-1})	$\bar{\nu}_{max}$ (cm^{-1})
W(CH)(PMe$_3$)$_4$Cl	1.84[a]	2.419 (2), 2.442 (5)[a]	–	250[b]	200[b]	910	25400
W(CH)(dmpe)$_2$Cl	1.797 (10)	2.606 (3)	158 (8)	246[b]	205[b]	920	24800
W(CH)(dmpe)$_2$(OSiMe$_3$)	–	–	–	239	175	892	25400
W(CH)(dmpe)$_2$(*n*-Bu)	1.828 (5)[c]	2.402 (7)[c]	160 (3)[c]	245[c]	171[c]	875	23500
W(CH)(dmpe)$_2$(CCSiMe$_3$)	1.801 (7)[d]	2.246 (6)[d]	164 (5)[d]	254[d]	184[d]	882	21800[d]

[a] Crystallographically disordered (Churchill, M. R.; Rheingold, A. L.; Wasserman, H. J. *Inorg. Chem.* **1981**, *20*, 3392). [b] Ref 1b. [c] Ref 4b. [d] Ref 4a.

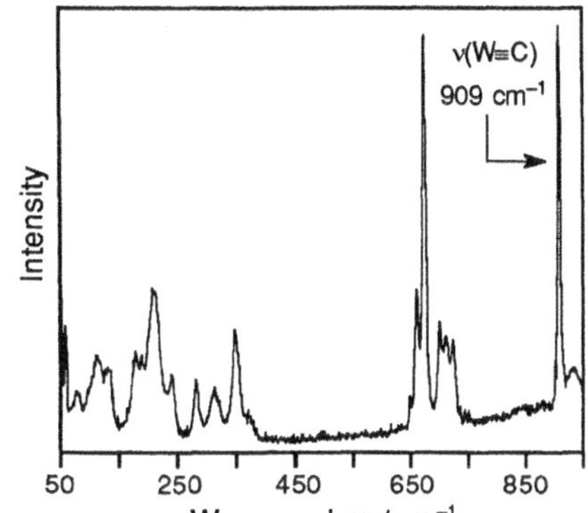

Figure 2. Raman spectrum of $W(CH)(PMe_3)_4Cl$.

reported by Dao, Fischer, and coworkers (ca. 1300 cm^{-1} and 7 mdyne $Å^{-1}$) for $W(CR)(CO)_4X$ (R = Me, Ph; X = Cl, Br, I) complexes.[6] The assignment of ν(W≡C) has been definitively established on the basis of selective isotope labelling of $W(CH)(PMe_3)_4Cl$ (natural abundance = 909 cm^{-1}; $W(CH)(PMe_3\text{-}d_9)_4Cl$ = 910 cm^{-1}; $W(CD)(PMe_3)_4Cl$ = 870 cm^{-1}). We believe these values are a better representation of the W≡C oscillator than are those of $W(CR)L_4X$ (R ≠ H) complexes, for which the diatomic-oscillator approximation is vitiated by mixing between the W≡C and the C–R and internal R coordinates. It is intuitively satisfying that $W(CH)(dmpe)_2(n\text{-Bu})$ possesses the lowest W≡C stretching frequency, since it has the longest W≡C bond. The possibility that mixing between ν(W≡C) and modes of the axial ligand may complicate the correlation between ν(W≡C) and *d*(W≡C) is under investigation.

In view of the fact that the estimated standard deviations of W≡C bond distances determined by X-ray diffraction are relatively large, it is likely that these spectroscopic methods are a more sensitive probe of the nature of the M≡C bond than are structural techniques, particularly for subtle perturbations arising from the ancillary ligands. We are extending our studies in order to develop detailed correlations among these parameters.

Acknowledgment. Support of this research by an NSF Research Opportunity Award to M.D.H. and R.F.D. is gratefully acknowledged.

References

† Department of Chemistry, Wabash College, Crawfordsville, Indiana 47933, U. S. A.

‡ NSF Presidential Young Investigator (1987–1992); Dreyfus Foundation New Faculty Awardee (1987–1992); Packard Foundation Fellow (1990–1995).

1. (a) Sharp, P. R.; Holmes, S. J.; Schrock, R. R.; Churchill, M. R.; Wasserman, H. J. *J. Am. Chem. Soc.* **1981**, *103*, 965. (b) Holmes, S. J.; Clark, D. N.; Turner, H. W.; Schrock, R. R. *J. Am. Chem. Soc.* **1982**, *104*, 6322. (c) Holmes, S. J.; Schrock, R. R.; Churchill, M. R.; Wasserman, H. J. *Organometallics* **1984**, *3*, 476. (d) Chisholm, M. H.; Folting, K.; Hoffman, D. M.; Huffman, J. C. *J. Am. Chem. Soc.* **1984**, *106*, 6794. (e) Jamison, G. M.; Bruce, A. E.; White, P. S.; Templeton, J. L. *J. Am. Chem. Soc.* **1991**, *113*, 5957.
2. (a) Mayr, A.; Hoffmeister, H. *Adv. Organomet. Chem.* **1991**, *32*, 227. (b) Fischer, H.; Hofmann, P.; Kreissl, F. R.; Schrock, R. R.; Schubert, U.; Weiss, K. *Carbyne Complexes*; VCH Publishers: New York, 1988.
3. Manna, J.; Gilbert, T. M.; Dallinger, R. F.; Geib, S. J.; Hopkins, M. D. *J. Am. Chem. Soc.* **1992**, *114*, 5870.
4. (a) Manna, J.; Geib, S. J.; Hopkins, M. D. *J. Am. Chem. Soc.* **1992**, *114*, in press. (b) Manna, J.; Geib, S. J.; Hopkins, M. D., submitted for publication.
5. Manna, J.; Kuk, R. J.; Dallinger, R. F.; Hopkins, M. D., submitted for publication.
6. Dao, N. Q.; Fevrier, H.; Jouan, M.; Fischer, E. O. *Nouv. J. Chim.* **1983**, *7*, 718.

NOVEL CYCLIZATIONS INVOLVING CATIONIC CARBYNE COMPLEXES

H. FISCHER,* C. TROLL, AND J. SCHLEU
Fakultät für Chemie, Universität Konstanz
Postfach 5560, W-7750 Konstanz 1, Germany

ABSTRACT. The highly electrophilic cationic carbyne complex $[Cp(CO)_2Mn{\equiv}C\text{-}Ph]^+$ (**1**) readily reacts with N≡C and electron-rich C≡C bonds. Metallacycles, carbocycles, and heterocycles may be obtained depending on the substrate and the reaction conditions. E.g. the reaction of **1** with diorganylcyanamides, $N{\equiv}C\text{-}NR_2$, affords ansa-carbene complexes containing a chelating η^1,η^5-[(cyclopentadienyl)-(phenyl)methyleneamino](diorganylamino)carbene ligand. When the $N{\equiv}C\text{-}NR_2$ is employed in excess a [2+2+1] cyclization is observed and imidazolium complexes are formed. A metallafulvene finally results from the reaction of **1** with polar electron-rich alkynes such as diethylaminopropyne. In contrast, in the reaction of **1** with symmetrically substituted electron-rich alkynes, e.g. bis(diorganylamino)acetylenes, the carbyne ligand is transferred to the C≡C bond affording cyclopropenyl cations.

1. Introduction

The first synthesis of a complex containing a metal-carbon triple bond was reported in 1973 [1]. Since then, numerous carbyne complexes have been prepared. In recent years, the study of the reactivity of these complexes has attracted considerable interest [2]. E.g. carbyne complexes have extensively been used as building blocks in the synthesis of transition metal clusters [3]. The coupling of carbyne ligands with CO or isocyanide ligands has also been studied in detail [4]. However, the number of reports on the use of carbyne complexes in synthetic organic chemistry is rather limited in contrast to carbene complexes which have found many applications in the synthesis of carbo- and heterocycles [5].

Cationic carbyne complexes, such as $[(CO)_5Cr{\equiv}CNEt_2]^+$ and $[Cp(CO)_2M{\equiv}C\text{-}Aryl]^+$ (M = Mn, Re), were shown to readily add neutral and anionic nucleophiles to the carbyne carbon atom to give the corresponding cationic or neutral carbene complexes. Carbene complexes have also been obtained in the reaction of $[Cp(CO)_2Mn{\equiv}C\text{-}Tol]^+$ with Ph(H)C=NMe and Ph(H)C=N-N=C(H)Ph [6]. The products result from insertion of the imine between the carbyne carbon and a carbon of the Cp ligand. We now report on the reactions of $[Cp(CO)_2Mn{\equiv}C\text{-}Ph]^+$ and substituted derivatives thereof with electron-rich N≡C and C≡C compounds. The aim of our investigations was to study the possibility of transferring the carbyne ligand to these multiple bonds.

2. Results

2.1. Reactions with Cyanamides

The cationic carbyne complex **1** reacts with a slight excess of $N{\equiv}C\text{-}NR_2$ in dichloromethane to give the neutral *ansa*-carbene complexes **2**. In **2** the former carbyne ligand is no longer connected with the metal but attached to the Cp ring whereas the

F. R. Kreißl (ed.), Transition Metal Carbyne Complexes, 79–84.

cyanamide N≡C carbon now constitutes the carbene carbon atom. A possible reaction mechanism is shown in Scheme 1: nucleophilic addition of N≡C-NR_2 to the carbyne carbon gives **A** which then rearranges by 1,3-migration of the $Cp(CO)_2Mn$ fragment (formation of **B**). Intramolecular electrophilic attack of the NCPh carbene substituent at the Cp ring and deprotonation finally gives **2**.

Scheme 1

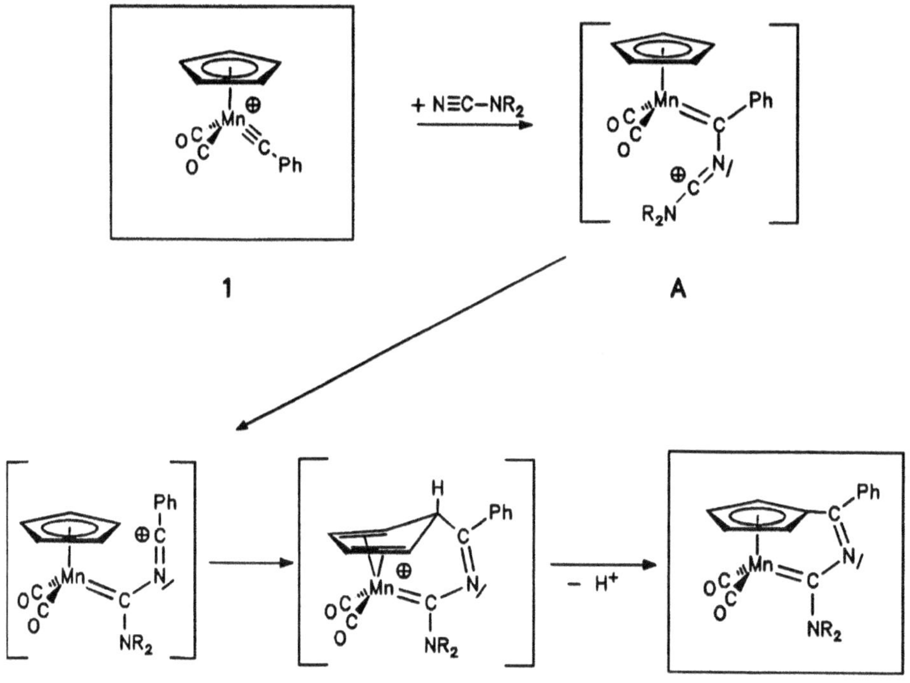

1 **A** **B** **2**

R = Me, Et, i−Bu, CH_2Ph

When the BCl_4^- salt of **1** is used in the reaction with N≡C-NR_2 (R = Me, Et) a long-lived intermediate can be detected at -40°C and characterized IR-spectroscopically. It very likely has the structure of a BCl_4^- adduct of **A** or that of a rotational isomer. This adduct can be trapped with carbanions. When adding LiMe to the solution at -40°C (R = Me), $Cp(CO)_2Mn{=}C(Ph)[N{=}C(Me)NMe_2]$ is isolated from the reaction mixture. $Cp(CO)_2Mn{=}C(NMe_2)[N{=}C(Me)Ph]$, the expected reaction product from addition of Me^- to **B**, cannot be detected. Therefore we believe that the 1,3-migration of $Cp(CO)_2Mn$ (**A** → **B**) is the rate-limiting step in the formation of **2** from **1** and N≡C-NR_2.

The selectivity of the "ring-closure" was tested employing $[(\eta^5\text{-}C_5H_4Me)\text{-}(CO)_2Mn{\equiv}C\text{-}Ph]^+BF_4^-$. Two isomeric carbene complexes (**3** and **4**) were obtained (Scheme 2). Complex **4** resulting from attack of the $NCPh^+$ functionality at the slightly more nucleophilic ß-carbon atom constituted the major product (ratio **3** : **4** = 1 : 5).

Scheme 2

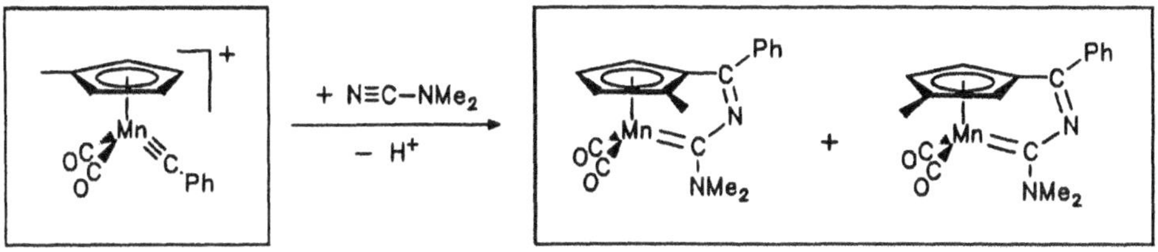

3 **4**

From the reaction of $[(\eta^5\text{-}C_5Me_5)(CO)_2Mn{\equiv}C\text{-}Ph]^+BF_4^-$ with $N{\equiv}C\text{-}NMe_2$ again an ansa-carbene complex was obtained (Scheme 3). At present the exact fate of the Me^+ group displaced from the Cp^* ring is not clear. It is presumably taken over by another $N{\equiv}C\text{-}NMe_2$ molecule (Scheme 3).

Scheme 3

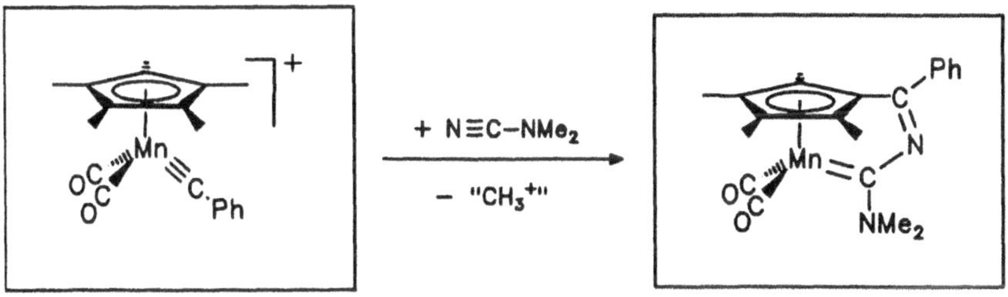

2.2. Reaction with $N{\equiv}C\text{-}NR_2$ in Excess

Apart from LiMe the adduct **A** could also be trapped with cyanamide. The cation **1** reacted with $N{\equiv}C\text{-}NR_2$ in excess to form the novel imidazolium complexes **5** (Scheme 4). The formation of **5** corresponds to a [2+2+1] cyclization. Isomers of **5** have not been detected. The complexes **5** are derived from adduct **A** by nucleophilic addition of $N{\equiv}C\text{-}NR_2$ to the carbon atom of the $NCNR_2$ substituent of **A**. The cyclization possibly proceeds via the sequence (a) formation of a metallacycle, (b) reductive elimination of a cationic imidazolium derivative and (c) formation of **5**. Isomers which are derived from an attack of $N{\equiv}C\text{-}NR_2$ at the CNPh substituent of **B** or at the carbene carbon atom of either **A** or **B** have not been observed.

Scheme 4

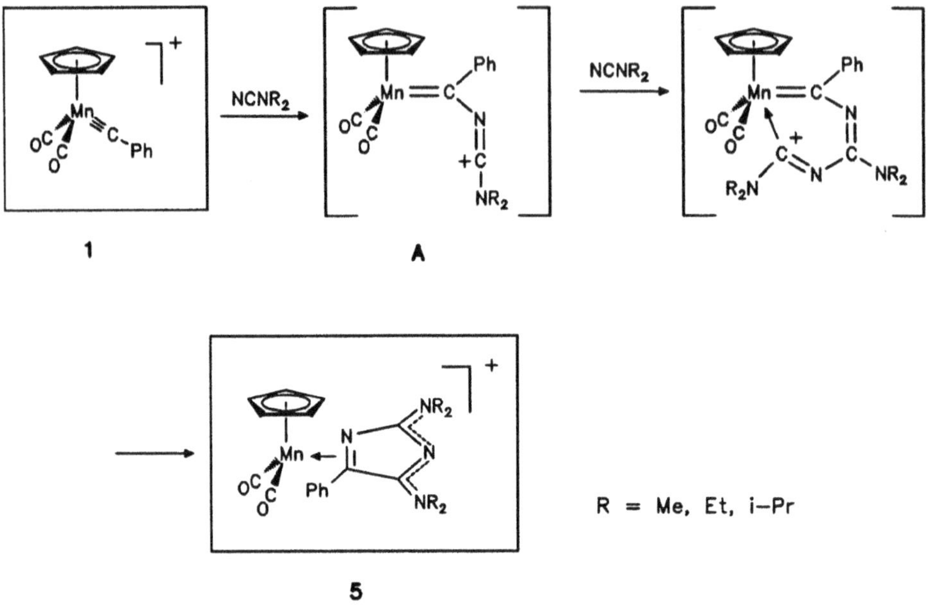

2.3. Reaction with an Ynamine

The course of the reactions of **1** with diorganylcyanamides and with electron-rich alkynes differs considerably. The major product of the reaction of **1** with 1-diethylaminopropyne in excess is the metallafulvene **6** (Scheme 5). The formation of **6** can be explained by the following reaction sequence: (a) [2+2] cycloaddition of the C≡C to the Mn≡C bond, (b) insertion of a coordinated CO ligand into the Mn=C(Ph) bond of the metallacyclobutadiene to give a metallacyclopentadienone followed by (c) cycloaddition of another 1-diethylaminopropyne across the C=O bond of the metallacyclopentadienone and (d) electrocyclic ring-opening to finally give **6**. Support for the proposed steps (a+b) comes from the observation that a metallacyclopentadienone is formed in the reaction of the neutral carbyne complex $Cl(CO)[PMe_3]_3W{\equiv}CPh$ with Ph-C≡C-H [7].

Scheme 5

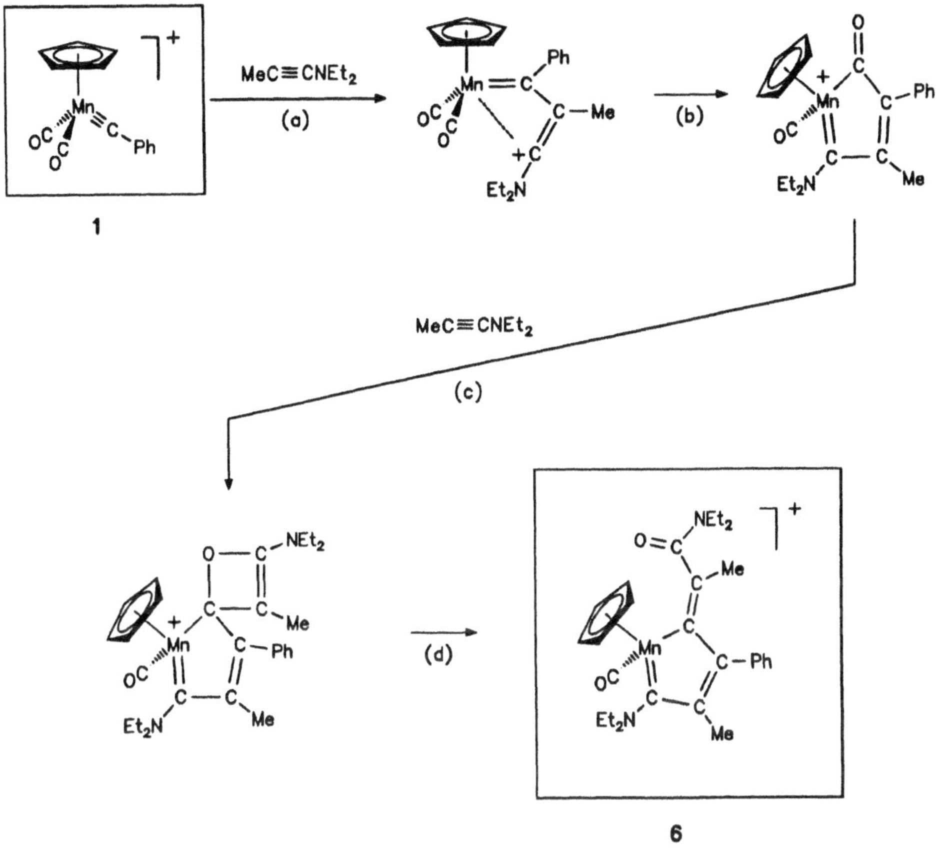

2.4. Reactions with Bis(diorganyl)acetylenes

The transfer of the carbyne ligand to a C≡C bond was finally achieved by using symmetrically substituted electron-rich alkyne substrates. The cation **1** reacts with $R_2NC{\equiv}CNR_2$ (R = Me, Et) to give cyclopropenium compounds (Scheme 6) which can be isolated as the BF_4^- salts in moderate yields. This reaction represents the first transfer of an electrophilic carbyne ligand to a C≡C bond.

Scheme 6

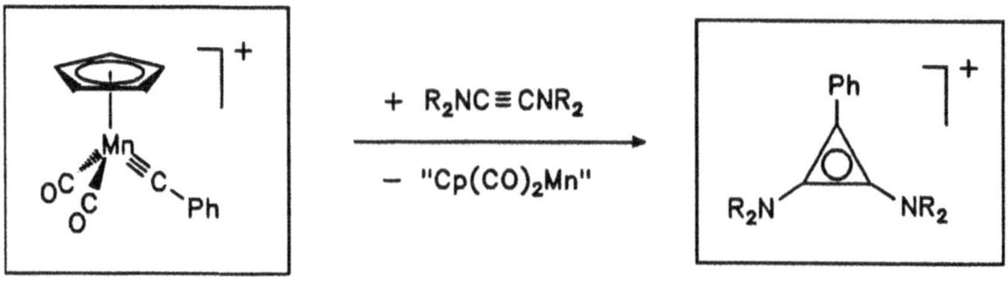

1

NR_2 = NMe_2, NEt_2

3. Conclusion

The reactions of cationic carbyne complexes of manganese with X≡C bonds are strongly substrate- and concentration-dependent. The compounds may be used as C_1 sources in the synthesis of carbo- and heterocycles via cyclization or transfer reactions.

4. References

1 Fischer, E. O.; Kreis, G.; Kreiter, C. G.; Müller, J.; Huttner, G.; Lorenz, H; *Angew. Chem.* **1973**, *85*, 618; *Angew. Chem. Int. Ed. Engl.* **1973**, *12*, 564.

2 See e.g. (a) Fischer, H.; Hofmann, P.; Kreissl, F. R.; Schrock, R. R.; Schubert, U.; Weiss, K.; *Carbyne Complexes*, VCH Verlag, Weinheim, **1988**. (b) Mayr, A.; Hoffmeister, H.; *Adv. Organomet. Chem.* **1991**, *32*, 227.

3 E.g. Stone, F. G. A.; *Pure and Appl. Chem.* **1986**, *58*, 529.

4 (a) Kreissl, F. R.; Frank, A.; Schubert, U.; Lindner, T. L.; Huttner, G.; *Angew. Chem.* **1976**, *88*, 649; *Angew. Chem. Int. Ed. Engl.* **1976**, *15*, 632. (b) Review: Mayr, A.; Bastos, C. H.; *Progr. Inorg. Chem.* **1992**, *40*, 1.

5 E.g. Dötz, K. H.; *Angew. Chem.* **1984**, *96*, 573; *Angew. Chem. Int. Ed. Engl.* **1984**, *23*, 587.

6 Handwerker, B. M.; Garrett, K. E.; Nagle, K. L.; Geoffroy, G. L.; Rheingold, A. L.; *Organometallics* **1990**, *9*, 1562.

7 Mayr, A.; Lee, K. S.; Kahr, B.; *Angew. Chem.* **1988**, *100*, 1798; *Angew. Chem. Int. Ed. Engl.* **1988**, *27*, 1730.

NEW ADDITION AND CYCLOADDITION REACTIONS OF THE CATIONIC CARBYNE COMPLEXES $[Cp(CO)(L)M{\equiv}CR]^+$ (M = Mn, Re; L = CO, PPh_3; R = Me, Tolyl)

GREGORY L. GEOFFROY,* COLLEEN KELLEY, LISA A. MERCANDO, MICHAEL R. TERRY, NOËL LUGAN,† CHAE YI, AND ANNE KAPLAN
Department of Chemistry
The Pennsylvania State University
University Park, PA 16802

ABSTRACT. The electrophilic carbyne complex $[Cp(CO)_2Re{\equiv}CTol]^+$ reacts with azoarenes, epoxides, aziridines, and propylene sulfide to give a variety of new metallacycles and carbene complexes. Vinyl-substituted carbyne complexes of the form $[Cp(CO)LMn{\equiv}C\text{-}CR{=}CR_2]^+$ (L = CO, PPh_3) have been prepared by three different routes, and they react with nucleophiles to give carbene and vinylidene derivatives, depending upon the size of the nucleophile. The ethylidyne complexes $[Cp(CO)_2M{\equiv}CCH_3]^+$ (M = Mn, Re) can be deprotonated to yield the new vinylidene complexes $Cp(CO)_2M{=}C{=}CH_2$, and these complexes react with imines and benzalazine to give new metallacycles and with $Bu^tN{=}C{=}NBu^t$ to induce metathesis of the vinylidene C=C bond to form complexes possessing $Bu^tN{\equiv}C$ isocyanide ligands.

1. Introduction

Cationic carbyne complexes within the $[Cp(CO)_2M{\equiv}CR]^+$ (M = Mn, Re) family were shown by the early studies of E. O. Fischer and co-workers to be highly electrophilic at the carbyne carbon and to add a variety of nucleophiles to yield carbene derivatives [1]. As illustrated in **Scheme I**, we recently reported that new cycloaddition reactions of

Scheme I

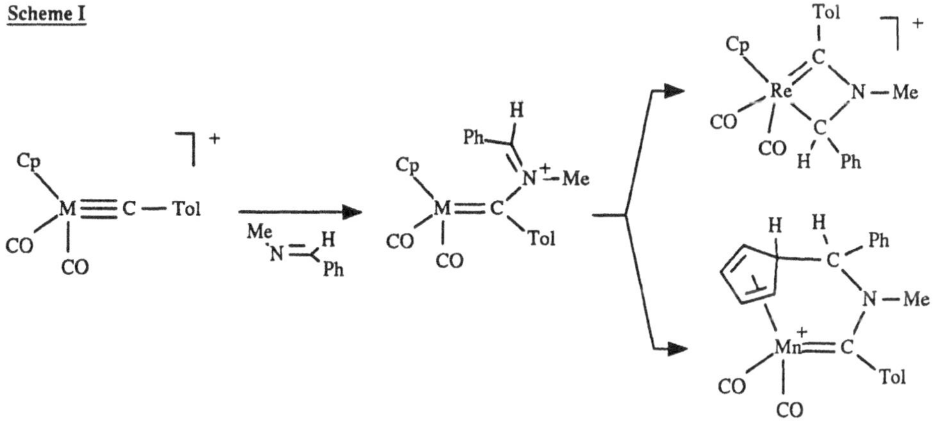

these complexes could be induced via the addition of the nucleophilic and unsaturated

F. R. Kreißl (ed.), Transition Metal Carbyne Complexes, 85–96.

organics MeN=CHPh and Bu^{t}N=O to the carbyne carbon [2]. Herein is described a summary of further addition and cycloaddition reactions of $[Cp(CO)_2Re{\equiv}CTol]^+$ with azoarenes, epoxides, aziridines, and propylene sulfide, the synthesis of new vinyl-substituted carbyne complexes, $[Cp(CO)(L)Mn{\equiv}C\text{-}CR{=}CR_2]^+$, and their nucleophilic addition reactions, and new addition and cycloaddition reactions of the methyl-substituted carbyne complexes $[Cp(CO)_2M{\equiv}CCH_3]^+$ (M=Mn,Re) and the vinylidene complexes $Cp(CO)_2M{=}C{=}CH_2$ which derive from them. The experimental details and full characterization of the products described herein will be described separately in full journal publications [3-5].

2. Results and Discussion

2.1. Addition and Cycloaddition Reactions with $[Cp(CO)_2Re{\equiv}CTol]^+$, 1 [3]

As illustrated in eqs. 1 and 2, azobenzene, azotoluene, and benzo[c]cinnoline have been found to undergo net [2+2] cycloaddition of their N=N double bond across the Re≡C triple bond of the BPh_4^- salt of carbyne complex 1 [3]. These reactions represent the

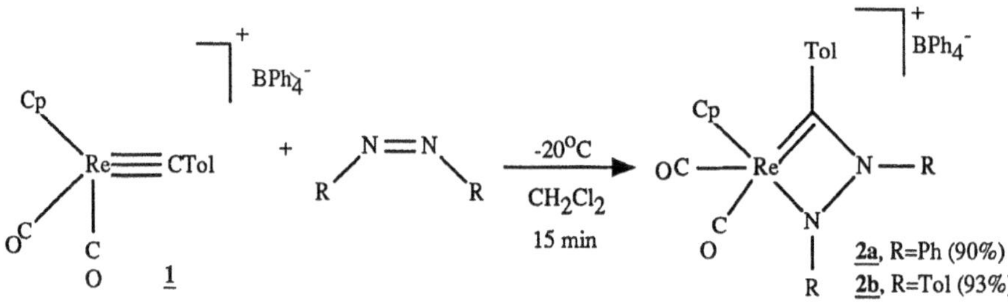

2a, R=Ph (90%)
2b, R=Tol (93%)

(1)

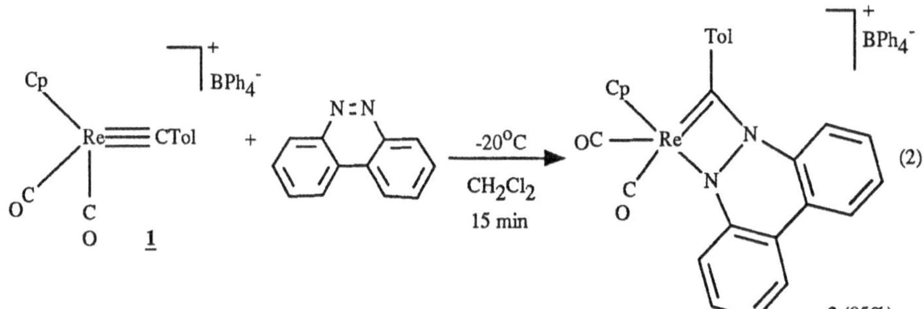

(2)

3 (85%)

first cycloadditions of azoarenes with carbyne complexes, and the new metallacycles 2a,b and 3 were isolated in good yields and have been spectroscopically characterized. These reactions likely proceed via nucleophilic addition of one of the azoarene nitrogen atoms to the electrophilic carbyne carbon followed by subsequent ring closure (see **Scheme I** for a related intermediate). The BCl_4^- salt of carbyne complex 1 was observed to form initially a similar product, but as illustrated in **Scheme II**, further reaction occurred to give the 5-membered metallacyclic complexes 4a,b by chloride abstraction from the BCl_4^- ion and insertion of a CO ligand into the rhenium-carbon bond [3]. A similar reaction was earlier observed for the reaction of Bu^{t}N=O with the BCl_4^- salt of 1 [2].

Scheme II

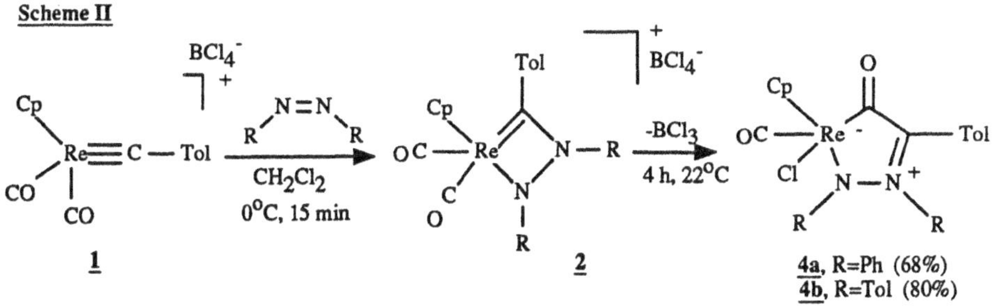

A surprising transformation occurred when an attempt was made to induce the conversion of 2 into 4 by the addition of [PPN]Cl (PPN^+ = $(PPh_3)_2N^+$) to the pre-formed BPh_4^- salt of 2, rather than directly from the BCl_4^- salt as in **Scheme II**. This reaction did not give the same product 4 but instead led to loss of an equivalent of arylisocyanate (detected by IR) and formation of complexes 5a,b, presumably by the sequence of reactions illustrated in **Scheme III**. These products were isolated in the

Scheme III

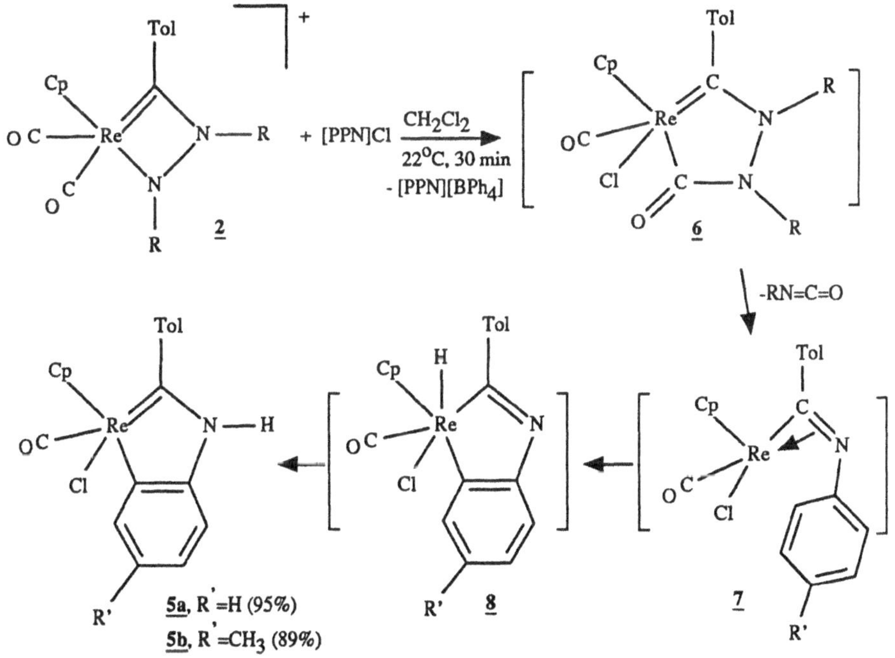

indicated yields, and complex 5a was fully characterized by an X-ray diffraction study. We suggest that chloride coordinates to the rhenium center to induce insertion of CO into the Re-N bond to form intermediate 6 which then loses RN=C=O to form 7. Metallation of the arene substitutent of 7 followed by hydrogen migration to the ring nitrogen atom would give the observed product 5. Note that the essential difference between the reactions illustrated in **Schemes II** and **III** is CO insertion into the Re-C bond of 2 induced by BCl_4^- whereas [PPN]Cl is suggested to induce CO insertion into the Re-N bond of the metallacycle. We do not at present have an explanation for these differing reaction paths, but it is likely that the Lewis acid BCl_3 plays an important role in determining the reaction outcome.

We next considered the possibility of inducing [2+3] cycloadditions of 1 via reaction with epoxides, aziridines, and propylene sulfide. Reactions with these substrates did occur, but the products were not metallacycles but instead the carbene derivatives shown in **Scheme IV**. As illustrated, the first step in each reaction likely involves

Scheme IV

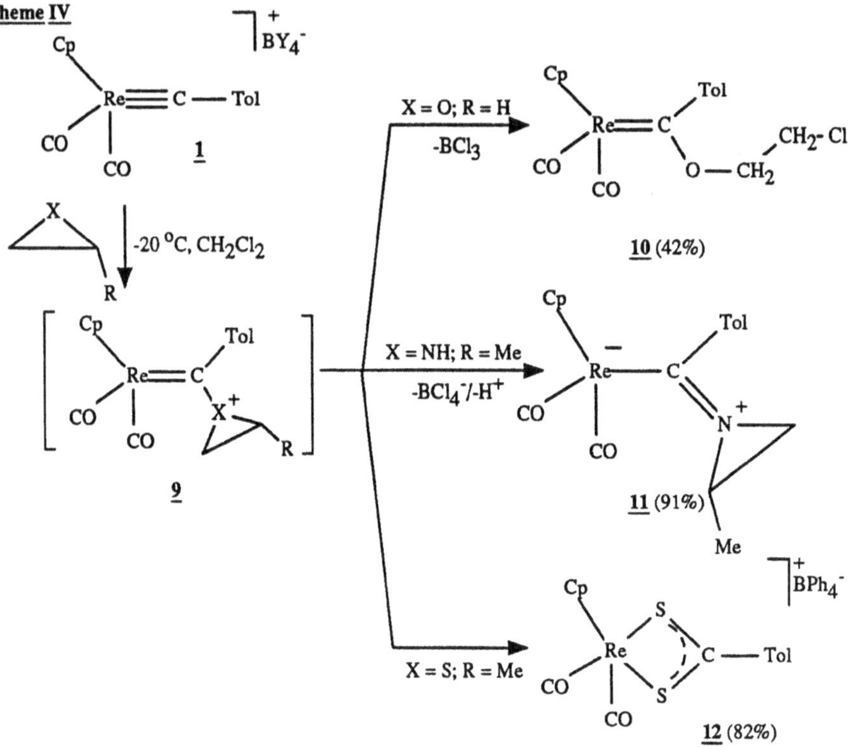

coordination of the organic heterocycle to the carbyne carbon via the nucleophilic heteroatom. In the case of methylaziridine, loss of proton from the nitrogen atom leads directly to carbene complex 11, a reaction which is similar to the addition of other amines to electrophilic carbyne complexes [6]. With propylene sulfide, transfer of the sulfur atom to the carbyne carbon occurs, presumably to give a thioacyl ligand which then reacts with a second equivalent of propylene sulfide to give the dithiocarboxylate complex 12. Other workers have observed similar formation of dithiocarboxylate complexes via reaction of carbyne complexes with sulfur delivery agents [7]. As illustrated in A, with ethylene oxide, addition of chloride to the coordinated epoxide

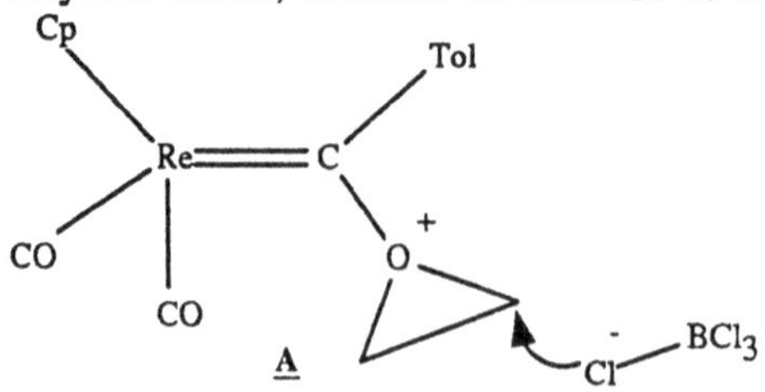

induces ring opening to give carbene complex **10**. Halides are well known to ring open epoxides, a reaction which is also assisted by Lewis acids, a function served by carbyne complex **1** in this transformation. Similar ring-opening reactions leading to carbene derivatives were observed for propylene oxide, isobutylene oxide, and 3,3-dimethyloxetane [3].

2.2. PMe_3-Induced Cyclopentadienyl Ligand Displacement Reactions [3]

As illustrated in **Scheme V**, addition of excess PMe_3 to complexes **2** and **12** results in displacement of the cyclopentadienyl ligand as a phosphonium salt and formation of complexes **13** and **14**. Both of these reactions have precedent in PMe_3-induced Cp

Scheme V

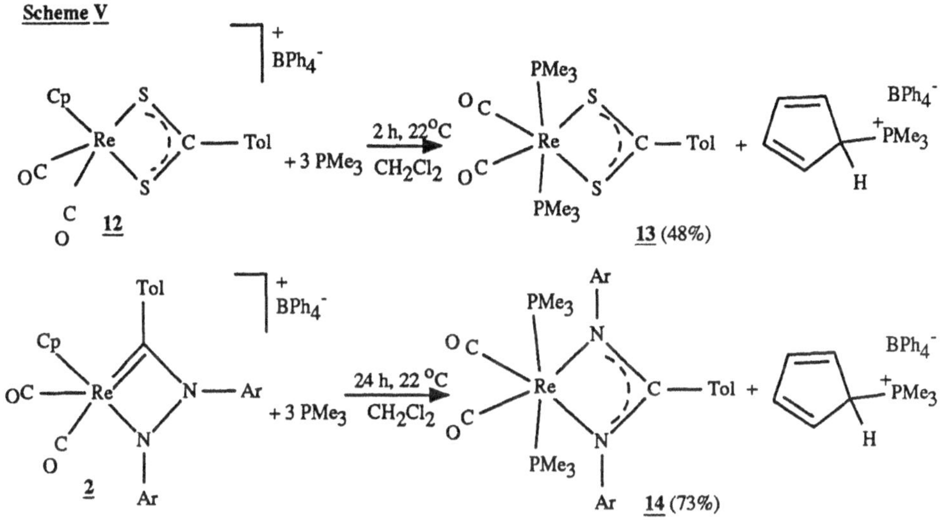

ligand displacement reactions observed by Casey and co-workers for related rhenium complexes [8] and likely proceed via PMe_3 coordination to rhenium to induce slippage of the Cp ligand to an η^3 coordination mode, addition of a second PMe_3 ligand to induce η^1 coordination of the Cp ligand, and finally PMe_3 addition to the η^1-Cp ligand to give displacement of the metal fragment. The metallacycle in complex **2** also undergoes rearrangement during this process to place the tolyl-substituted carbon between the nitrogen atoms. An intermediate has been isolated in this transformation and shown spectroscopically to have three PMe_3 ligands, two CO ligands, and the elements of the original metallacycle present. While we do not yet know the specific structure of this species, logical possibilities are those drawn in **B**-**D** below.

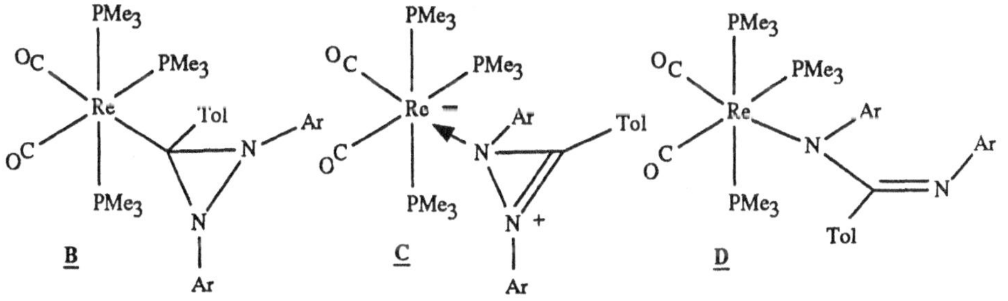

2.3. Synthesis of Vinylcarbyne Complexes and Their Reactions with Nucleophiles [4]

Prior to the studies described below, the only examples of complexes possessing terminal vinyl-substituted carbyne ligands are those prepared by Kolobova, *et al.* via the reversible protonation of the allenylidene complexes shown in eq. 3 [9]. We have

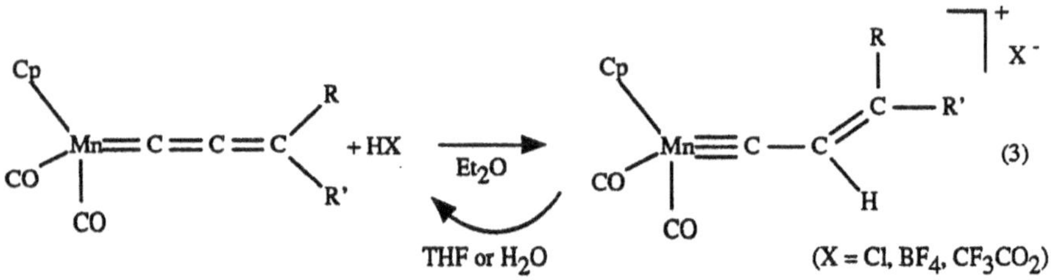

(3)

developed *three* independent routes to complexes within this family, illustrated by the reactions shown in eqs. 4-6. The first (eq. 4) involves treatment of the acetylide

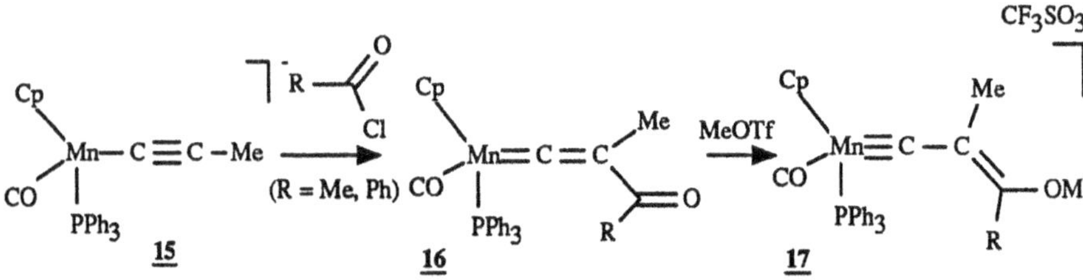

(4)

complex 15 with acid halides to form the vinylidene complex 16 which can be alkylated at the carbonyl oxygen to form the cationic vinylcarbyne complex 17 [10]. The second route (eq. 5) uses the standard Fischer methodology and involves treatment

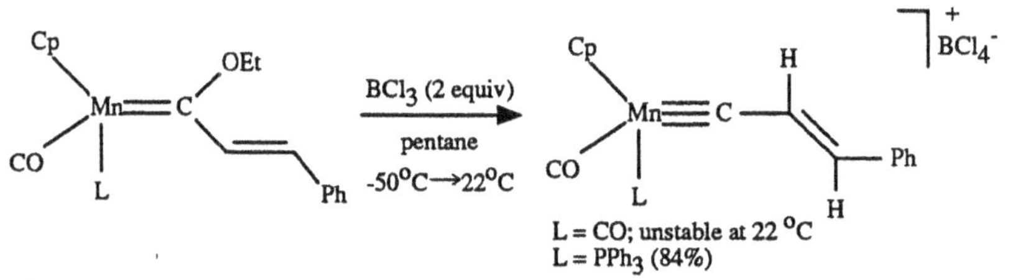

(5)

of a vinylcarbene precursor with BCl_3 to remove the ethoxy substituent [11]. The third route is the most unusual and involves reaction of acetylide complex 15 with ketones, eq. 6 [4]. The mechanism by which this latter reaction occurs is shown in **Scheme VI.**

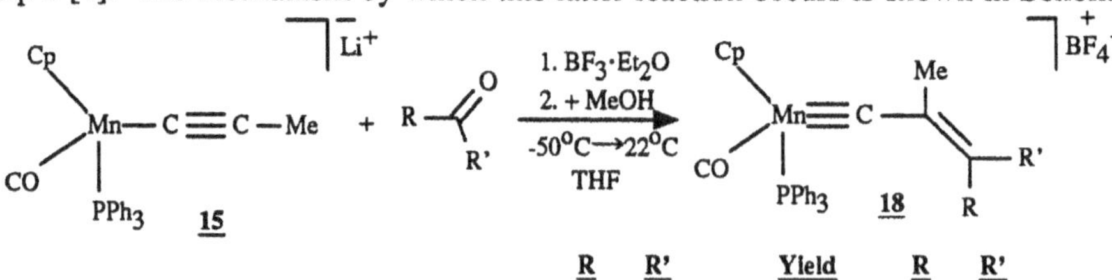

(6)

R	R'	Yield	R	R'	Yield
Me	Tol	90%	Me	Me	85%
Me	t-Bu	91%	Ph	CF_3	60%
Me	c-C_3H_5	93%	Ph	Ph	85%

The presence of BF_3 is necessary for these reactions to occur, and we suggest that it activates the ketone by coordination to the oxygen atom, facilitating attack at the

Scheme VI

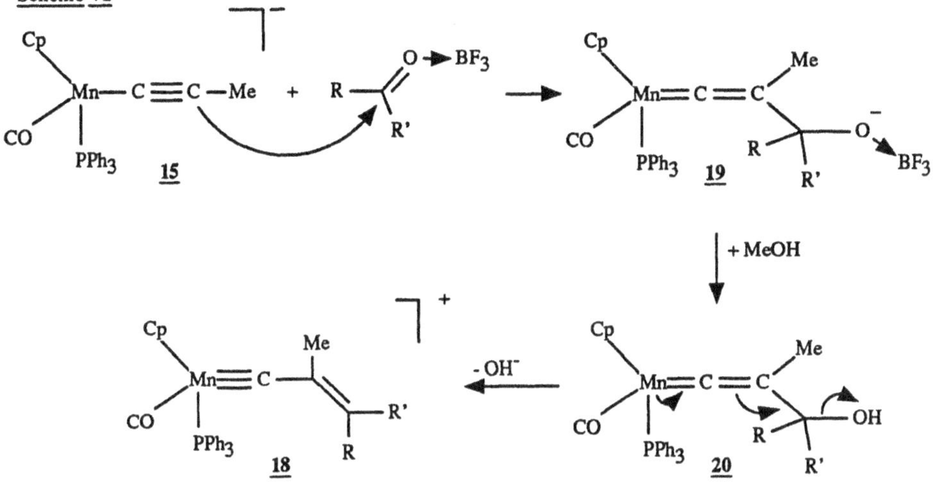

carbonyl carbon by the highly nucleophilic β-carbon of the acetylide complex 15. This leads to vinylidene complex 19 which upon protonation would give vinylidene complex 20, and loss of OH^- from this species would give the observed vinylcarbyne complex 18.

With these vinylcarbyne complexes in hand, and realizing that little was known about the reactivity of vinyl-substituted carbyne ligands, we undertook a brief survey of their chemistry. As illustrated in **Scheme VII**, we have found that nucleophiles will add to both the α- and γ-carbon atoms of the vinylcarbyne ligand to give carbene and vinylidene derivatives, respectively, with the site of the attack controlled by the size of the entering nucleophile [4]. It was also observed that deprotonation of the methyl

Scheme VII

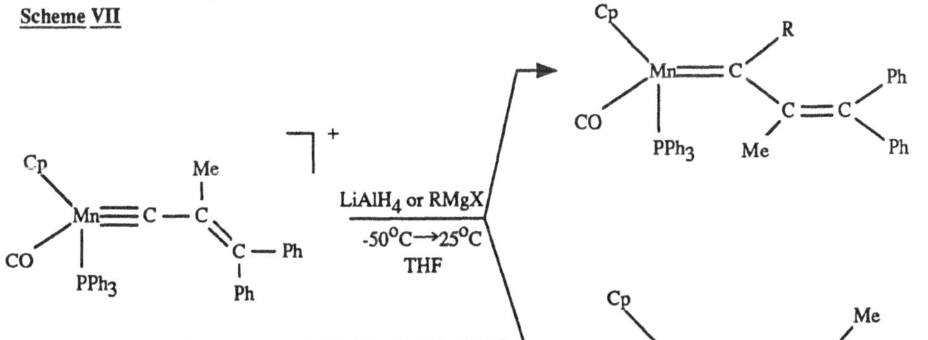

Nucleophile	Vinylidene	Carbene
$LiAlH_4$	0%	100%
MeMgI	26%	74%
EtMgI	65%	35%
Pr^iMgCl	100%	0%

substituent on the β-carbon of the vinylcarbyne ligand readily occurred to give the vinyl-substituted vinylidene complex 21 shown in eq. 7.

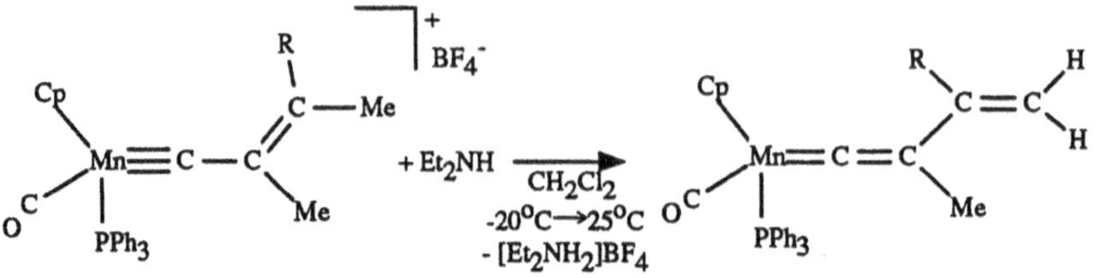

(7)

21a, R = Tol (27%)
21b, R = Bu^t (65%)

2.4. Addition and Cycloaddition Reactions of the Ethylidyne and Vinylidene Complexes $[Cp(CO)_2M{\equiv}CCH_3]^+$ and $Cp(CO)_2M{=}C{=}CH_2$ [5]

As an extension of the studies described above [3] and our previous work [2] with carbyne complexes having a tolyl substituent on the carbyne carbon, we turned to the corresponding ethylidyne complexes 22a,b. It was quickly found that these complexes readily deprotonate to form the neutral vinylidene complexes 23a,b, eq. 8 [5]. For the

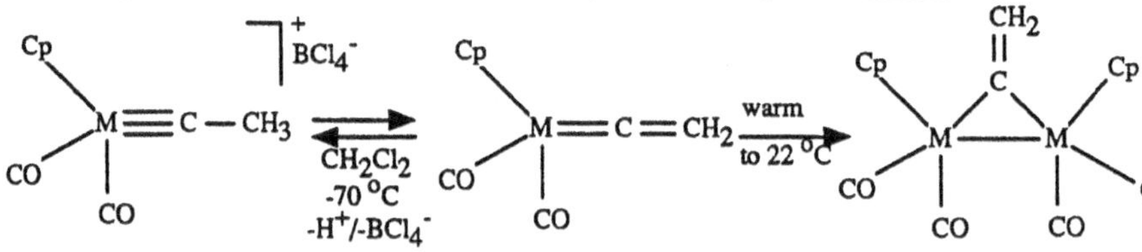

(8)

22a, M=Mn
22b, M=Re

	M	ν(CO)	ν(C=C)
23a	Mn	1993,1928	1624 cm^{-1}
23b	Re	1992,1915	1632 cm^{-1}

^{1}H NMR: δ 5.49 (Cp), 2.34 (CH_2)

24a, M=Mn
24b, M=Re

rhenium complex 22b, IR evidence indicates that deprotonation occurs spontaneously upon dissolution in THF. HCl(g) is given off in this latter reaction as evidenced by the visible fumes which turn litmus paper red above the solution. The manganese complex does not undergo a similar deprotonation upon simple dissolution in THF, but instead forms the THF adduct 25, eq. 9. A similar pyridine adduct of $[Cp(CO)_2Mn{\equiv}CPh]^+$ was

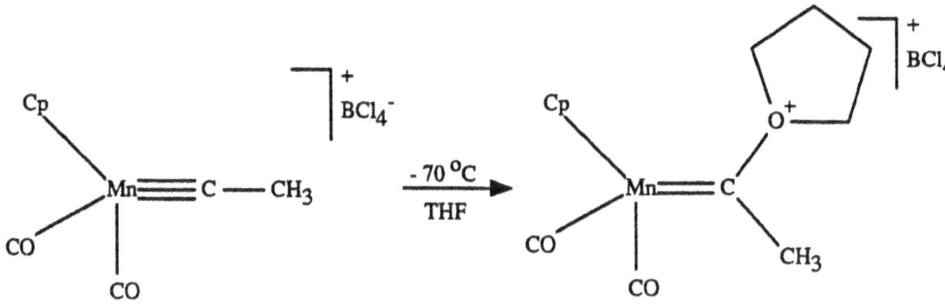

22a 25

described by Meineke in 1975 [12]. However, the manganese ethylidyne complex 22a does form the vinylidene complex 23a when treated with Et_3N. Both 23a and 23b decomposed upon warmup to room temperature to give the binuclear vinylidene complexes **24a,b**, and consequently they could not be isolated but were spectroscopically characterized at low temperature. Neither of the vinylidene complexes **23a,b** has been previously described, but many substituted derivatives of the form $Cp(CO)_2M{=}C{=}CRR'$ (R=aryl, alkyl; R'=H, aryl, alkyl) are known [13].

The equilibrating mixture of ethylidyne complex **22** and vinylidene complex **23** have been found to give a variety of new reactions with imines, azines, and carbodiimides (see **Scheme VIII**) [5]. Although the products of the reactions described below are

Scheme VIII

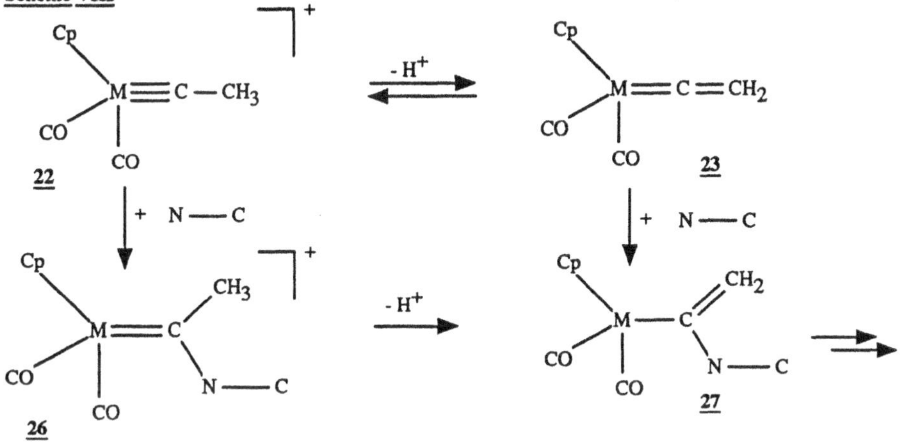

similar for the manganese and rhenium complexes, it appears that the order of deprotonation and nucleophile addition are different in the two cases. Spectroscopic evidence indicates that the manganese complex reacts via addition of the nucleophilic organic to the ethylidyne complex **22a** to give **26** which then deprotonates to give **27**. This latter intermediate can also form via addition of the nucleophilic organic to the vinylidene complex **23**. From this point on, the reactions described below appear to follow a common path.

With imines, both complexes give a net [2+2] cycloaddition reaction to form the cyclic carbene complexes **28a-c**, eqs. 10 and 11 [5]. Note that in reaction 10, the Et_3N base is added after the imine addition, whereas in eq. 11, deprotonation occurs first to form the vinylidene complex **23b**. These complexes were isolated in the indicated yields and have been spectrospically characterized with **28a** fully defined by an X-ray diffraction study. These reactions are proposed to proceed via the paths illustrated in **Scheme VIII** with complex **27** being the key intermediate which gives **28** by ring closure to form the CH_2-CHPh bond. In contrast to these results, the use of an imine with a hydrogen substituent on the nitrogen atom gives for both Mn and Re, via the vinylidene complexes, the imine-substituted carbene complexes **29a,b**, eq. 12.

A particularly interesting reaction is the consecutive [2+3] cycloaddition reactions observed for **22a** and **23b** with benzalazine to give the binuclear complexes **30a,b** shown in **Scheme IX** [5]. These reactions are the first examples of organometallic analogues of the "criss-cross" cycloaddition reaction, eq. 13, which has been known for organic substrates since 1917 [14]. The new complexes **30a,b** were isolated in modest yields and have been spectroscopically characterized, with **30a** fully defined by an X-ray crystallographic study. The mechanism of the organometallic reactions are likely analogous to that established for the organic "criss-cross" cycloadditions [14] and

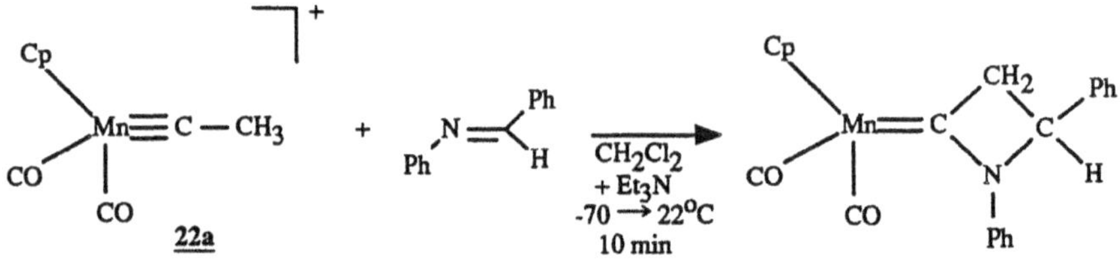

$\underline{22a}$

(10)

$\underline{28a}$ (85%)

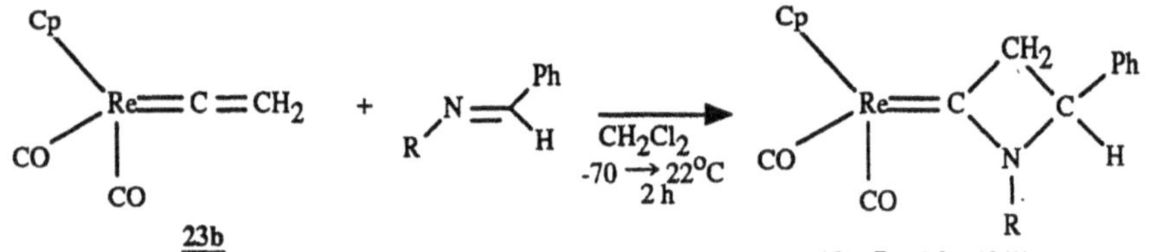

$\underline{23b}$

(11)

$\underline{28b}$, R = Me (48%)
$\underline{28c}$, R = Ph (49%)

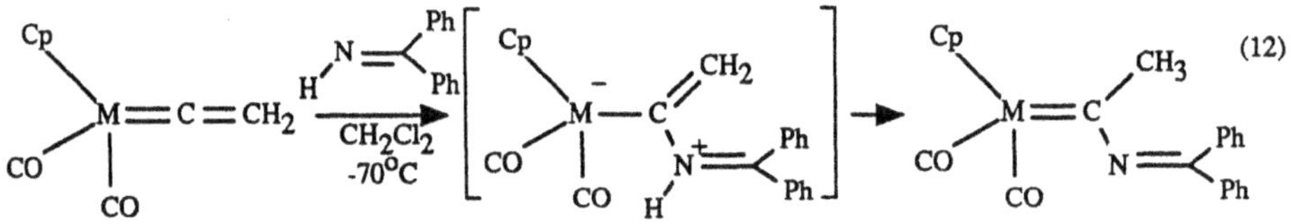

(12)

$\underline{29a}$, M = Mn, 28%
$\underline{29b}$, M = Re, 62%

<u>Scheme IX</u>

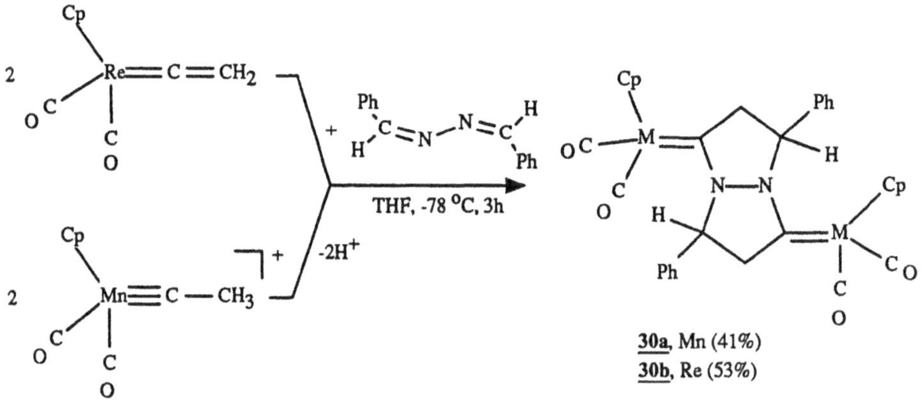

$\underline{30a}$, Mn (41%)
$\underline{30b}$, Re (53%)

is illustrated in **Scheme X** for the vinylidene complex $\underline{23b}$.

Finally, with $Bu^tN{=}C{=}NBu^t$, both complexes $\underline{22a}$ and $\underline{23b}$ have been observed to give net metathesis of the N=C bond of the carbodiimide with the C-C bond of the vinylidene or ethylidyne ligand, **Scheme XI**.

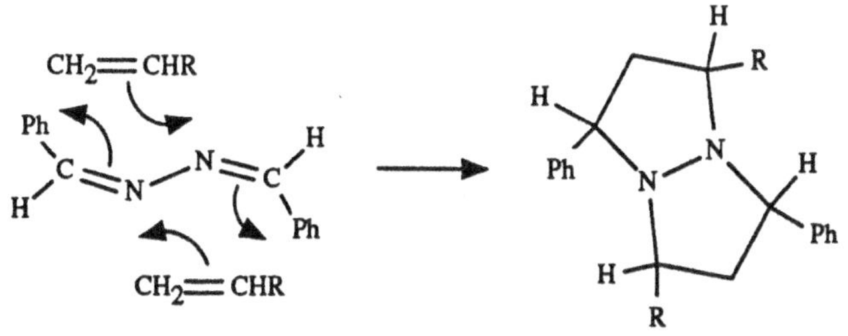

(13)

Scheme X

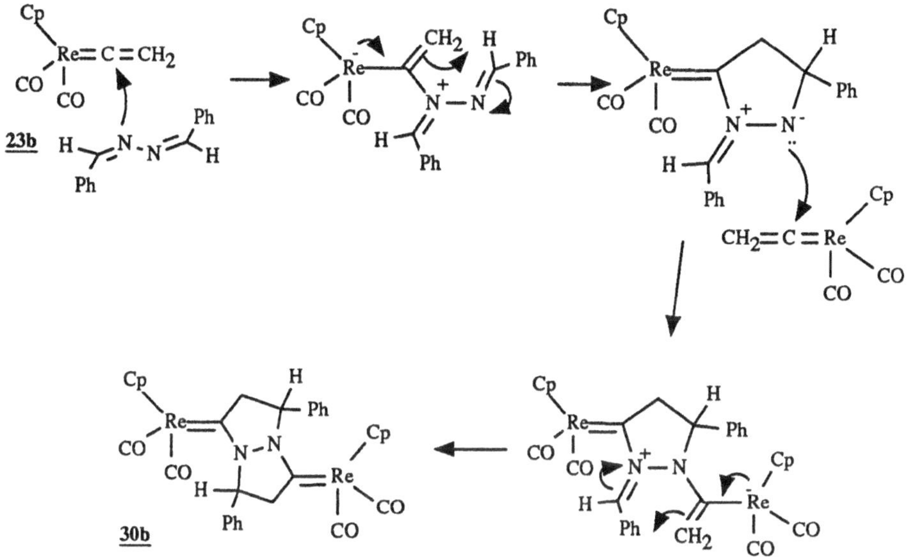

Scheme XI

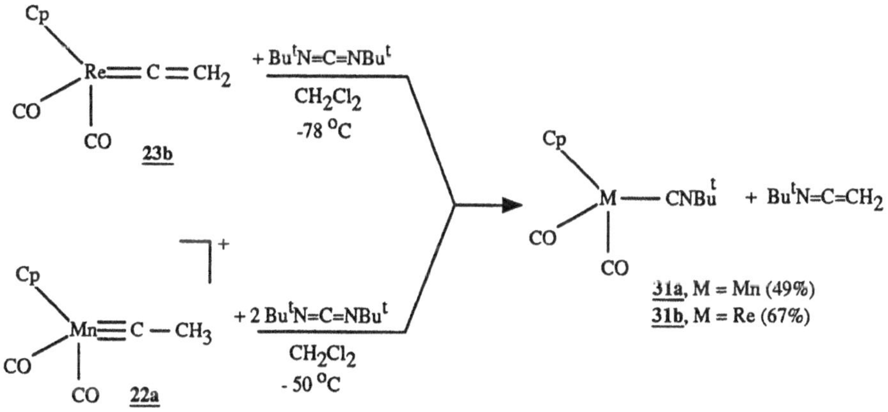

3. Acknowledgements

We thank the US Department of Energy, Office of Basic Energy Sciences for support of this research and NATO for a travel grant which facilitated this collaborative research effort between workers at the Pennsylvania State University and the CNRS, Toulouse labora.

4. References

* Author to whom correspondence should be addressed.
† Permanent address: Laboratoire de Chimie de Coordination du CNRS, Toulouse, France

1. a) Fischer, E. O.; Chen, J.; Scherzer, K. *J. Organomet. Chem.* **1983**, *253*, 231. b) Fischer, E. O.; Wanner, J. K. R. *Chem. Ber.* **1985**, *118*, 2489. c) Fischer, E. O.; Schambeck, W. *J. Organomet. Chem.* **1980**, *201*, 311. d) Fischer, E. O.; Clough, R. L.; Stückler, P. *J. Organomet. Chem.* **1976**, *120*, C6. e) Fischer, E. O.; Frank, A. *Chem. Ber.* **1978**, *111*, 3740.
2. Handwerker, B. M.; Garrett, K. E.; Nagle, K. L.; Geoffroy, G. L.; Rheingold, A. L. *Organometallics* **1990**, *9*, 1562.
3. Mercando, L. A.; Handwerker, B. M.; MacMillan, H. J.; Geoffroy, G. L., Rheingold, A. L. *Organometallics*, submitted for publication.
4. Terry, M. R.; Kelley, C.; Lugan, N.; Geoffroy, G. L.; Haggerty, B. S.; Rheingold, A. L., to be submitted.
5. Kelley, C.; Mercando, L. A.; Terry, M. R.; Lugan, N.; Geoffroy, G. L.; Xu, Z.; Rheingold, A. L. *Angew. Chem., Int. Ed. Engl.*, in press.
6. Fischer, E. O.; Stückler, P.; Beck, H.-J.; Kreißl, F. R. *Chem. Ber.* **1976**, *109*, 3089.
7. a) Kreissl, F. R.; Ulrich, N. *J. Organomet. Chem.* **1989**, *361*, C30. b) Gill, D. S.; Green, M.; Marsden, K.; Moore, I.; Orpen, A. G.; Stone, F. G. A.; Williams, I. D.; Woodward, P.*J. Chem. Soc., Dalton Trans.* **1984**, 1343. **1971**, *104*, 1877.
8. a) Casey, C. P.; O'Conner, J. M.; Haller, K. J. *J. Am. Chem. Soc.* **1985**, *107*, 1241. b) O'Conner, J. M.; Casey, C. P. *Chem. Rev.* **1987**, *87*, 307.
9. Kolobova, N. E.; Ivanov, L. L.; Zhvanko, O. S.; Khitrova, O. M.; Batsanov, A. S.; Struchkov, Y. T. *J. Organomet. Chem.* **1984**, *262*, 39.
10. Kelley, C.; Lugan, N.; Terry, M. R.; Geoffroy, G. L.; Haggerty, B. S.; Rheingold, A. L. *J. Am. Chem. Soc.*, **1992**, in press.
11. Yi, C.; Kaplan, A.; Geoffroy, G. L., unpublished observations.
12. a) Meineke, E. W., Dissertation Technische Universität München, 1975. b) Fischer, H.; Hofmann, P.; Kreissl, F. R.; Schrock, R. R.; Schubert, U.; Weiss, K. *Carbyne Complexes* VCH Publishers, New York, **1988**.
13. a) Bruce, M. I. *Chem. Rev.* **1991**, *91*, 197. b) Bruce; M. I.; Swincer, A. G. *Adv. Organomet. Chem.* **1983**, *22*, 59.
14. a) Bailey, J. R.; Moore, N. H. *J. Am. Chem. Soc.* **1917**, *39*, 279. b) Bailey, J. R.; McPherson, A. T. *ibid.* **1917**, *40*, 1322. c) Padwa, A. *1,3-Dipolar Cycloaddition Chemistry* Wiley, New York, **1984**. d) Wagner-Jauregg, T. *Synthesis* **1976**, 349.

ALKYNYLCARBENE COMPLEXES OF TRANSITION METALS AS SUITABLE SUBSTRATES FOR STEREOSELECTIVE CYCLOADDITIONS.

L. JORDI, A. LLEBARIA, S. RICART, J.M. VIÑAS, J.M.MORETO*.
Departament de Química Orgànica Biològica. C.I.D. (C.S.I.C.)
Jordi Girona 18-26 E-08034 Barcelona (Spain).

ABSTRACT. Strong electronic effects of the carbene-metal unit on the adjacent functionalities in alkynylcarbene metal (Cr,W) complexes may be regarded as the main cause for easy and stereocontrolled cycloadditions with a variety of substrates.

Alkynylalkoxycarbene complexes of Cr and W present intrinsic peculiarities when compared to conventional Fischer type carbene complexes, since their distal electrophilic center (the β-acetylene carbon atom) competes advantageously for external nucleophiles that add to the triple bond instead of substituting the alkoxy group at the carbene center[1]. In this sense, they have been regarded as alkynes having an internal Lewis acid activation.Their anomalous behaviour may be explained taking into account the charge interaction between the triple bond, the carbene atom and the metal. Thus, the remarkable low field chemical shift for the β carbon atom in the ^{13}C NMR may be considered a physical indication of a partial contribution of the vinyl cation character on their structural nature in close analogy to the activation of alkynes by Lewis acids (Scheme 1).

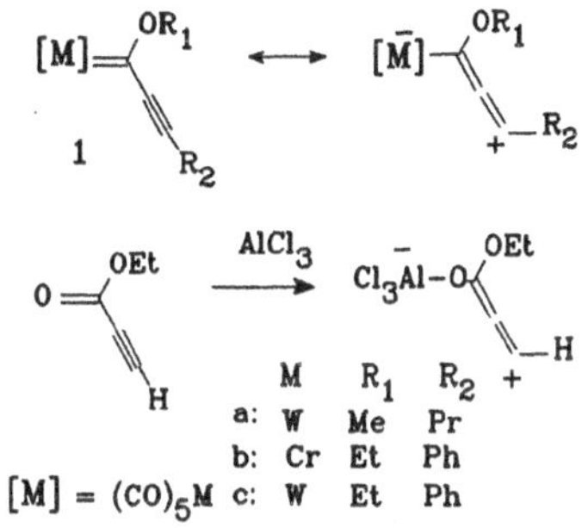

Scheme 1. Alkynylalkoxycarbene metal complexes as vinyl cation analogs.

These systems are known to behave as vinyl cations displaying, as a consequence, a potentially interesting chemistry towards a variety of olefinic substrates[2]. Provided that this similarity applies, it will confer a promising outlook to alkynylcarbene metal complexes in cycloaddition chemistry concerning versatility and, what is even of the utmost importance, a strict control of the stereochemistry of the products may be expected[3].

[2+2] Cycloadditions.

Several electron rich olefins (C- and Si- enol ethers) were reacted with different alkynylalkoxycarbene metal (Cr,W) complexes. These reactions were found to proceed at room temperature with significant differences in rate. The results (see Table featuring a few representative cases) led to the following conclusions:

a) The more electron rich is the olefin the faster is the reaction, being the [2+2] cycloadduct the exclusive reaction product.

b) When the reaction rate slows down other type of products accompany the [2+2] cycloadduct. These products seem to arise from conjugate linear addition of the enol ether (further cyclization on the carbene center may follow; Scheme 2).

F. R. Kreißl (ed.), Transition Metal Carbyne Complexes, 97–99.

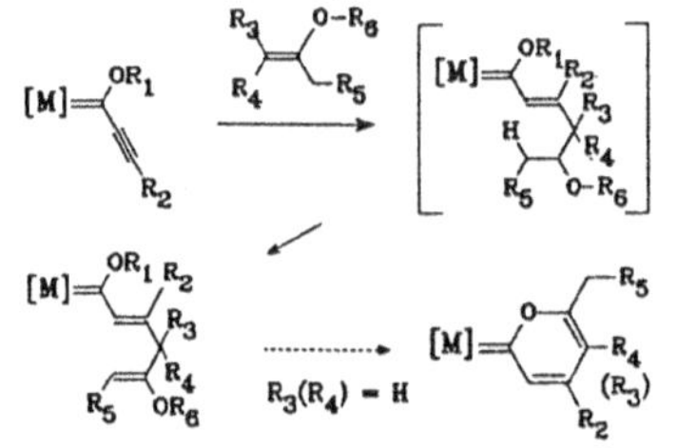

Scheme 2. Side reactions accompanying [2+2] cycloadditions.

c) The relative stereochemistry of the olefin substituents is retained .
d) When the adducts come from a considerable polarized enol ether, the original cycloadducts tend to suffer a more or less extensive conrotatory ring opening[4] and further reactions give the final products (see Scheme 3). This conclusion was achieved after we realized that, although α-pyranyliden metal complexes were formed from different enol ethers, those from diethoxyacrylate did not keep the original atom connectivity. A careful insight into the reaction pathway allowed to trap reaction intermediates pointing out to a [2+2] cycloaddition as the initial step.

Table

Reaction of electron-rich olefins with alkynylalkoxycarbene metal (Cr,W) complexes.

Olefin	Complex	Time	Products (Yield %)
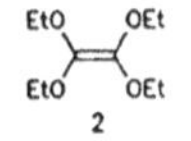 2	1a	15 m	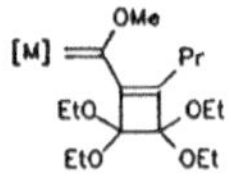 7 (65)
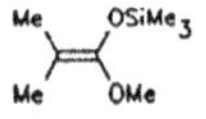 3	1b	2 h	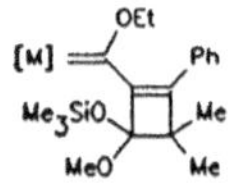 8 (90)
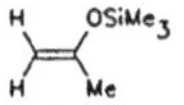 4	1b	48 h	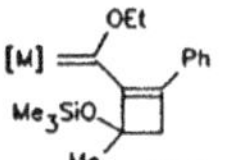 (44) 9; 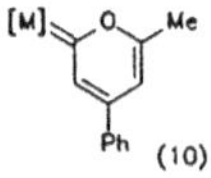 (10) 10
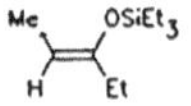 5	1b	5 d	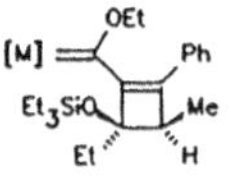 11 (60); 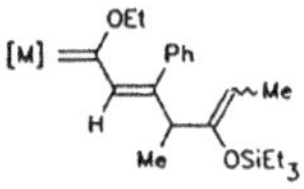 12 (10)
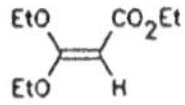 6	1c	3 h	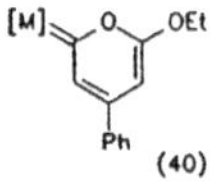 (40) 13; 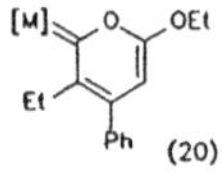 (20) 14

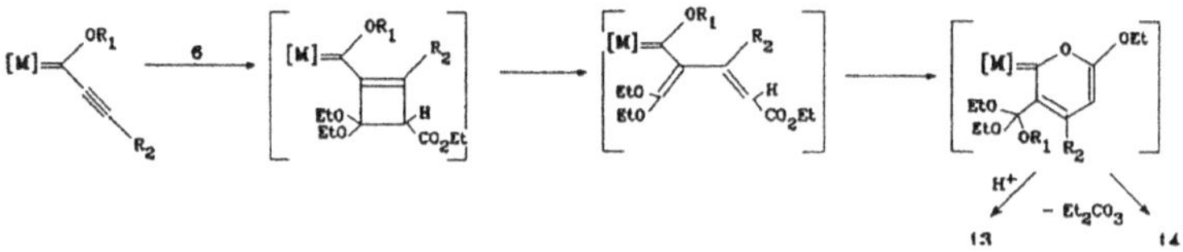

Scheme 3. Proposed mechanism for formation of pyranylidene complexes.

[2+2+1] Cycloadditions

We tried to take advantage of the mentioned activation in other type of reactions[5]. Since the Co mediated carbonylative cycloaddition of alkynes and olefins (Pauson-Khand reaction) is under extensive study, we advanced that the mentioned triple bond activation might allow reaction completion under milder temperatures than those usually required. While no reaction was found after treating alkynylallyloxycarbene metal complexes with $Co_2(CO)_8$, a very mild reaction with allylaminocarbene analogs rendered the expected cyclopentenone adducts with final release of the $Co_2(CO)_6$ moiety (Scheme 4). Again, the reaction proved to be stereospecific affording the stereoadducts expected from application of the proposed reaction mechanism to this case. However, the presence of bulky sustituents on the triple bond caused the interruption of the reaction at the stage of the alkynyl-Co complex. All efforts to force the reaction to go to completion were unsuccesful. Additional evidences from substituting the carbene-heteroatom moiety ment that, in the present case, either electronic and/or steric properties inherent to the heteroatom branch are more decisive than the triple bond activation itself, since cobalt complexation may be considered to bring the triple bond to an activated stage. A special interaction of the metal center with the heteroatom through the carbene center has been recognized since the early days of Fischer-carbene complexes and it rather appears as the responsible factor for these [2+2+1] cycloadditions.

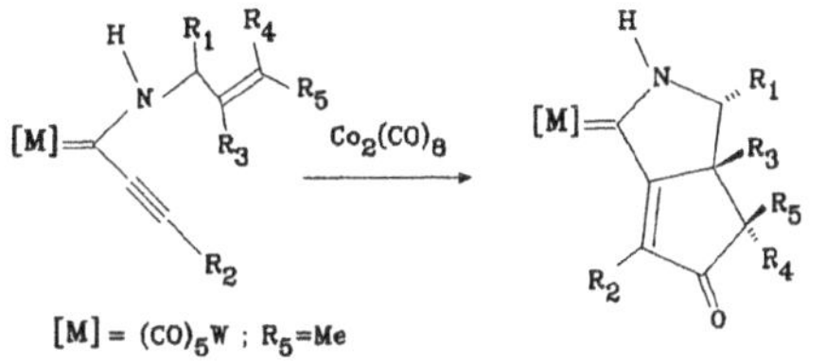

[M] = $(CO)_5W$; R_5=Me

15 R_2=Ph ; R_1=R_3=R_4=H 17 (72 %)

[M]=$(CO)_5W$; $-R_1$--- R_4- =$-(CH_2)_3-$

16 R_2=Ph ; R_3=R_5=H 18 (35 %)

Scheme 4. "Pauson-Khand" reaction in alkynylallylaminocarbene metal complexes

Conclusions

At the present, our studies with alkynylcarbene complexes of Cr and W suggest that their cycloadditions with electron-rich olefins follow, mainly, a concerted [2+2] pathway by activation of the triple bond[6], whereas, the intramolecular Co-induced carbonylative cycloaddition of the corresponding allylamino complexes seems to be facilitated by the alternative influence of the metal on the heteroatom side. A strict control of the stereochemistry in these reactions has been observed making these complexes valuable auxiliaries in organic synthesis. Efforts to broaden the scope of application are under way.

References

[1] E.O. Fischer, H.J. Kalder, *J. Organomet. Chem. 131* (1977) 57. F. Camps, A. Llebaria, J.M. Moretó, S. Ricart, J.M. Viñas, *J. Organomet. Chem. 401* (1991) C-17.

[2] B.B. Snider, D.M. Roush, D. Gonzalez, D. Spindell, *J. Org. Chem. 45* (1980) 2773. B.B. Snider, D.J. Rodini, R.S.E. Conn, S. Sealfon, *J. Am. Chem. Soc. 101* (1979) 5283.

[3] R.B. Woodward,R. Hoffmann, *Angew. Chem. Int. Ed. Engl. 8* (1969) 781

[4] K.L. Faron, W.D. Wulff, *J. Am. Chem. Soc. 110* (1988) 8727. *ibid. 112* (1990) 6419.

[5] F. Camps, J.M. Moretó, S. Ricart, J.M. Viñas, *Angew. Chem. Int. Ed. Engl. 30* (1991) 1470.

[6] See also: R. Pipoh, R. van Eldik, S.L.B. Wang, W.D. Wulff, *Organometallics 11* (1992) 490.

REACTION OF ALKYNOLS WITH ALKYNYLALKOXYCARBENE METAL (Cr,W) COMPLEXES.

J.M. VIÑAS, J.M. MORETO, S. RICART*.
Departament de Química Orgànica Biològica. CID. (CSIC).
C/J. Girona 18-26. 08034-Barcelona.

ABSTRACT. The reaction of alkynols with the title complexes follows different pathways depending on the type of acetylene (mono or disubstituted) and the mutual distance between functionalities giving, as a consequence, products structurally different.

As it is known, the reaction of alkynylalkoxycarbene metal complexes with simple alcohols leads to conjugate addition onto the triple bond giving alkoxyvinyl(alkoxy)carbene complexes[1].However, when the alcohol has another functionality attached to the main chain at an appropriate distance, both functionalities may interact with either the triple bond and the metal carbene producing changes in the expected reaction course.

We report here on the behaviour of alkynols with 2 or 3 methylene spacers between alcohol and alkyne functionalities , and the differences observed for terminal or internal alkynes in their reaction with the mentioned carbene complexes.

Results and discussion.

TERMINAL ALKYNES.

The reaction of propargyl alcohol **2** (n=1) with complexes of type **1** gave, after stirring for 15 min., at room temperature in the presence of a catalytic amount of DBU, the expected product from conjugate addition **3** in a fairly good yield (75%).(see Table)

No reaction was observed, under the same conditions, with its homologue 3-butin-1-ol with either **1a** or **1b**. However, when the temperature was raised to 45°, the starting product disappeared in three hours and an orange solution was obtained. Examination of the resulting mixture by TLC (hexane:dichloromethane, 9:1) revealed the presence of a yellow complex as major product together with a trace of starting material. After removal of the solvent, the residue was chromatographed (flash chromatography), using the same eluent, on silica. A yellow band developped which fraction was collected and, after evaporation of the solvent, afforded lemon yellow crystals of **4a** (or **4b**) identified as oxacycloalkylidene complexes[2][3][4]. With chromium complexes (entry 3), further elution of the crude product with hexane:dichloromethane 2:1

F. R. Kreißl (ed.), Transition Metal Carbyne Complexes, 101–103.

allowed the obtention of a white solid **6**, whose spectroscopic data (IR, ^1H and ^{13}C NMR, MS) evidenced an angular tricyclic structure, one of the rings corresponding to a cyclopentenone .

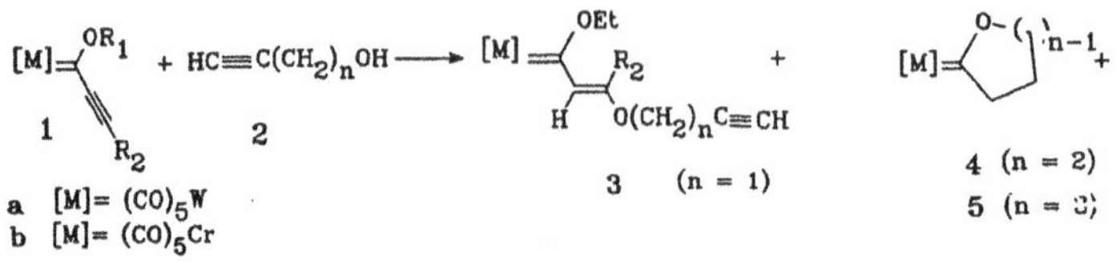

Table. Reaction of different alkynols with alkynylalkoxycarbene metal complexes.

Entry	1	2 n =	Time	Temp.	Main Products Yield (%)	Other Products
1	**a**	1	15m.	r.t.	**3a** (75)	-
2	**a**	2	3h.	45°C	**4a** (50)	Oligomer mixt.
3	**b**	2	3h.	45°C	**4b** (63)	**6** (35)
4	**a**	3	3h.	45°C	**5a** (52)	Oligomer mixt.
5	**b**	3	3h.	45°C	**5b** (56)	Organic adduct

Further confirmation for this structure was brought about by a single crystal diffractometric determination. Formally, this product corresponded to a head to tail coupling of two units of the carbene ligand with one of alkynol (see Figure 1). From the same reaction with the tungsten analog (**1a**) a mixture of minor unidentified organic products was accompanying **4a** (entry 2).

It is noteworthy, in the present case, that, unlike with simple alcohols, there is no reaction between the alcohol function and the triple bond of the carbene metal complex. Similarly, 4-pentyn-1-ol (entry 5) gave **5b** from its reaction with **1b** as the main product and a major organic adduct related to **6** as side product (not yet fully characterized).

INTERNAL ALKYNES.

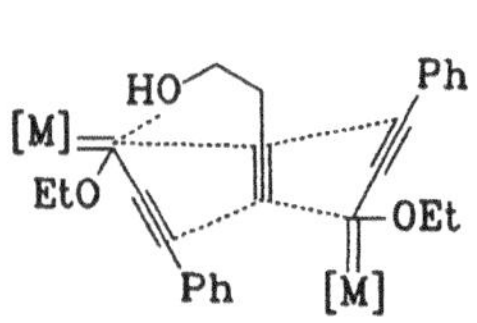

Figure 1. Atom connectivity in **6**.

As above, conjugated addition of 2-butyn-1-ol onto the triple bond of **1** was accomplished in 12 hours in the presence of a catalytic amount of DBU at room temperature (40%). However, its homologe 3-pentyn-1-ol behaved in a very different way. Under the same experimental conditions, a new complex was obtained as red crystals after flash chromatography on silica (hexane:dichloromethane 4:1). Its infrared spectrum evidenced the presence of the $M(CO)_5$ unit. In addition, the ^1H NMR spectrum showed the existence of an ethoxy group bonded to the metal

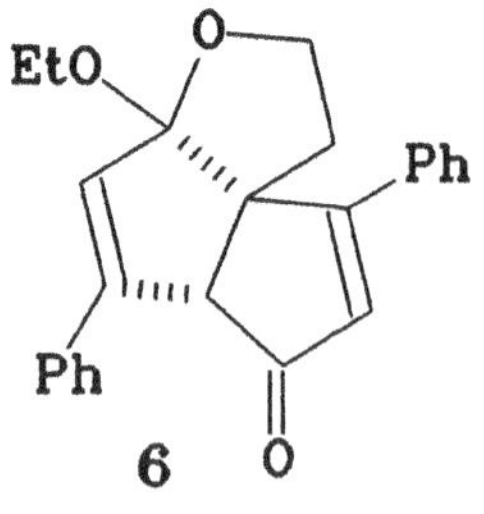

6

carbene unit. From all the other collected data its structure was established as **7a** (75% yield). Thus, contrary to our expectations the product obtained was formally that resulting from the [2+2] cycloaddition reaction between **1a** and the cyclic enol ether resulting from the internal addition of the alcohol to the triple bond[5]. This unexpected result prompted us to investigate the behaviour of 3-pentyn-1-ol in the presence of DBU. ^{1}H NMR experiments carried out at different ratios alkynol/base did not show any 2-methyl-4,5-dihydrofurane formation. On the other hand, monitoring the reaction by ^{1}H NMR spectroscopy showed the absence of any conjugated attack of the alcohol to the triple bond, since no vinylic proton could be observed.

Conclusions.

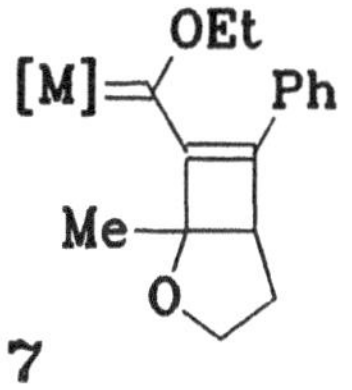

7

Although alcohols readily add to the triple bond of alkynylalkoxycarbene metal complexes affording alkoxy vinyl carbene complexes this is not a general reaction. Alcohols with monosubstituted alkynes spaced 2 or 3 carbon atoms undergo ligand exchange with complexes **1** giving the corresponding oxacycloalkylidene complexes and organic adducts in which the alkynol is inserted between molecules of the original carbene fragment.

Alcohols with an internal triple bond undergo a reaction that apparently seems to be a [2+2] cycloaddition between the corresponding enol ether of the alkynol and the alkynylalkoxycarbene metal complexes. Further studies in order to clarify the mechanism for such a different behaviour are currently underway in our laboratories.

References.

[1]. Camps,F., Llebaria, A., Moretó, J.M., Ricart, S., Viñas, J.M., Ros, J., Yañez, R., *J. Organomet. Chem.* **1991**, *401*, C7.

[2]. Casey, C.P., Anderson, R.L. *J. Am. Chem. Soc.* **1971**, *93*, 3554.

[3]. Dötz, K.H., Sturm, W., Alt, H.G., *Organometallics* **1987**, *6*, 1424.

[4]. Le Bozec, H., Ouzzine, K., Dixneuf, P.H., *Organometallics* **1991**, *10*, 2768.

[5]. Faron, K.L., Wulff, W.D., *J. Am. Chem. Soc.* **1990**, *112*, 6419.

CHEMISTRY AND ELECTROCHEMISTRY OF ALKYNE- AND ISOCYANIDE-DERIVED CARBYNE COMPLEXES OF RHENIUM, MOLYBDENUM OR TUNGSTEN

ARMANDO J.L. POMBEIRO
Centro de Química Estrutural, Complexo I,
Instituto Superior Técnico, Av. Rovisco Pais,
1096 Lisbon codex, Portugal

ABSTRACT. Alkynes, when activated by an electron-rich d^6 Re, Mo or W phosphinic centre, undergo hydrogen shift reactions (to give, e.g., vinylidene species) or oxidatively add to the metal (forming alkynyl-hydrido or alkynyl complexes). These alkyne-derived products undergo β-protonation to afford a variety of carbyne-fluoro or -chloro complexes, whereas aminocarbynes are obtained upon β-electrophilic attack (e.g., by a protic or a Lewis acid) at isocyanides when ligating such metal sites. Mechanistic studies, by stopped-flow spectrophotometry, are also indicated.
The structural and electronic features of the carbyne species are discussed on the basis of X-ray, spectroscopic and extended Hückel MO studies, as well as of their electrochemical behaviour which also involves the electrocleavage of N-H or C-H bonds.

1. INTRODUCTION

A number of synthetic strategies has been followed [1] for complexes with multiple metal-carbon bonds, commonly involving a nucleophilic attack, an α-abstraction or a scission process, an elimination or a rearrangement reaction at a suitable C-bonded species.

The route described herein consists mainly of a β-electrophilic attack at an unsaturated alkyne-derived or isocyanide ligand when activated by an electron-rich group 6 or 7 heavy metal phosphinic centre. Metal ligating carbynes of the types $\equiv CCH_2R$, $\equiv CNH_2$, $\equiv CNHR$, $\equiv CN(AlEt_3)R$ or $\equiv CN(\underline{M})R$ ($\underline{M}$ = transition metal Lewis acid) have thus been obtained and are the object of this discussion.

2. ALKYNE-DERIVED CARBYNES

2.1. Rhenium Phosphinic Complexes

In contrast to the well developed coordination chemistry of carbyne

F. R. Kreißl (ed.), Transition Metal Carbyne Complexes, 105–121.

complexes of the groups 5 (Nb and Ta) and 6 (Mo,W) transition metals [1], only a relatively limited number of clearly established carbyne complexes has been quoted for rhenium [2-4], via routes which involve, namely, nucleophilic attack at a carbonyl ligand or α-abstraction from a C-bonded ligand such as a neopentylidene [3].

We have shown that carbyne complexes of rhenium can be conveniently prepared by β-protonation at vinylidene ligands [5], a route which has been described for other systems [6]. The vinylidene species have been obtained by 1,2-hydrogen shift at 1-alkynes (HC≡CR) when activated by an electron-rich d^6 Re centre, typically trans-{ReCl(dppe)$_2$} (dppe=$Ph_2PCH_2CH_2PPh_2$) provided by a parent dinitrogen complex [5].

Thus, treatment of trans-[ReCl(N_2)(dppe)$_2$] with an excess of the appropriate 1-alkyne affords the corresponding vinylidene complex trans-[ReCl(=C=CHR)(dppe)$_2$] (R = Ph, C_6H_4Me-4, Et,Bu^t, CO_2Me,CO_2Et,etc.) (eq. 1) [5].

Moreover, reactin of the vinylidene complex with [Et_2OH][BF_4] gives the corresponding complexes trans-[ReX(≡CCH_2R)(dppe)$_2$][BF_4] (1, X=Cl or F) (eqs.2 and 3, respectively) [7].

The formation of the fluoro complex (eq. 3) involves not only the expected β-protonation at the vinylidene ligand, but also the displacement of chloride by fluoride from BF_4^-. Such a replacement is noteworthy in view of the rather limited number of known carbyne-fluoro complexes [6a] (see also below), in contrast with the well documented carbyne stabilization by a trans-chloro ligand. The chloride displacement at the metal centre can be promoted by Tl[BF_4] and, in the presence of this salt, the carbyne-fluoro complexes trans-[ReF(≡CCH_2R)(dppe)$_2$][BF_4] (1, X=F) can be more conveniently prepared, in a single pot experiment, directly from the reaction of RC≡CH with trans-[ReCl(N_2)(dppe)$_2$], as indicated by eq. (4), in which the ammonium salt, NH_4BF_4, behaves as the protonating agent [7b,c].

The carbyne complexes (1) can be readily deprotonated by base, [Bu_4N]OH, to give the corresponding vinylidene compounds, and this constitutes a convenient route for the synthesis of the vinylidene-fluoro species trans-[ReF(=C=CHR)(dppe)$_2$] (eq.5) which, upon protonation, can regenerate the carbyne complexes (eq. 6) [7c].

In all these compounds, the carbyne or the vinylidene ligand is trans to the strongest net electron-donor co-ligand, F or Cl, in agreement with their expected π-electron acceptor character. Moreover, the HOMOs appear to be less stabilized in energy for the fluoro complexes compared with those for the homologous chloro compounds, as suggested by some electrochemical studies (see section 4).

The molecular structure of trans-[ReF(≡CCH_2Bu^t)(dppe)$_2$][BF_4] has been authenticated by an X-ray diffraction analysis and is depicted in figure 1 [7a].

The rhenium-carbon distance, 1.772(7) Å, is, as expected, considerably shorter than those in the related vinylidene and aminocarbyne complexes trans-[ReCl(=C=CHPh)(dppe)$_2$], 2.046(8) Å [5], and trans-[ReCl(≐C≐NHMe)(dppe)$_2$][BF_4], 1.80(3) Å [8] (see below), respectively, in which there is a π-electron delocalisation along the

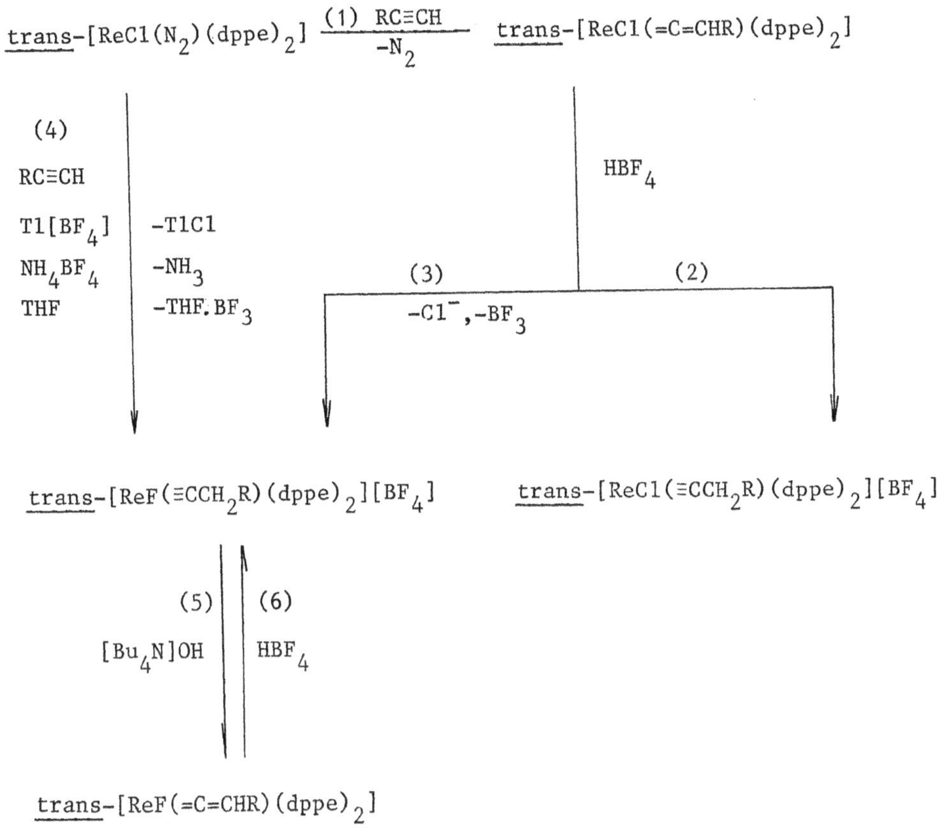

ligand framework. The distance is, nevertheless, somewhat above the range predicted, 1.75-1.72 Å, from the sum of the triple-bonded covalent radii of Re and sp-C. It is also slightly longer than those reported for Re≡C in [Sn(TPP){Re(CO)$_3$C}$_2$] (1.75 Å, TPP = 5,10,15,20-tetraphenylporphinato) [4] and [Re(≡CBu^t)(=CHBu^t)(C$_5$H$_5$N)$_2$I$_2$] [1.742(9)Å] [3]; in the latter complex, a shorter bond would be expected since the formally Re(VII) metal should have a smaller radius than the formally Re(V) metal of our carbyne compound.

The formation of the carbyne complexes (1) (eqs. 2,3 or 6) involves an apparently regiospecific attack of the acid at the β-carbon atom of a vinylidene ligand. However, the mechanism of this reaction has been investigated by stopped-flow spectrophotometry which has demonstrated that such protonation can occur by three different pathways, two of which involving an initial protic attack at the metal, as indicated by scheme 1, for the reaction of trans-[ReCl(=C=CHPh)(dppe)$_2$] with $NHEt_3^+$, in THF [9]. The hydride intermediate would then either convert directly into the carbyne or undergo further protonation to an hydride-carbyne species which by reaction with a base (NEt$_3$) would lead to the final carbyne product [9]. Complexes of these types (hydrides and hydride-carbynes)

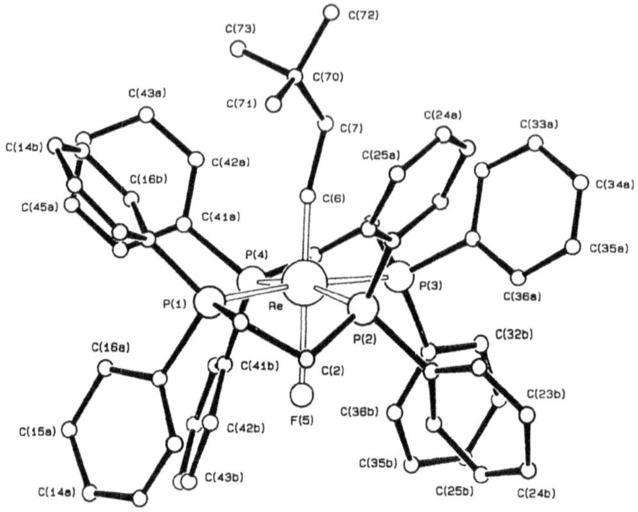

Figure 1. Molecular structure of trans-[ReF(≡CCH$_2$Bu^t)(dppe)$_2$]$^+$ [7a].

are also involved in the protonation reactions of isocyanides at related Mo or W centres, as discussed below.

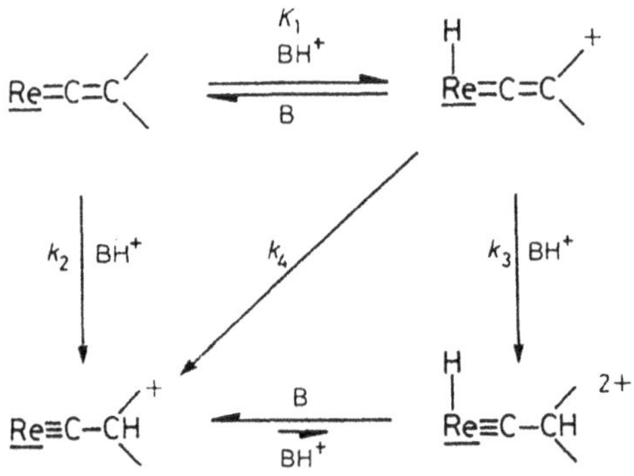

Scheme 1 - Mechanistic pathways for the formation of a carbyne complex (1) by protonation of a vinylidene compound [Re stands for trans-{ReCl(dppe)$_2$} and B for NEt$_3$][9].

The mechanism of the formation of the vinylidene complexes via 1,2-hydrogen migrations at the 1-alkynes (eq. 1) has not been investigated but extended Hückel calculations performed by others [10] for d^6 ML$_5$ metal centres suggest that it involves a di- to a mono-hapto slippage of the alkyne ligand followed by 1,2-proton migration. However, in our systems, no alkyne complex has been isolated, conceivably as a result of a destabilizing repulsive four electron $d\pi$(metal)-$\pi_{\perp}$(alkyne) interaction involving the filled pseudo-t_{2g} set of the d^6 Re atom in the electron rich {$ReCl(dppe)_2$} site [5]. Similar observations have been reported for the phenylpropyne($PhC{\equiv}CCH_3$)-to-allene ($PhCH{=}C{=}CH_2$) conversion (by a 1,3-hydrogen migration) at this site [11], as well as for the 1-alkyne-to-vinylidene conversion at some group 6 metal sites [12, 6a].

2.2. Molybdenum or Tungsten Phosphinic Complexes

In contrast with the {$ReCl(dppe)_2$} site, in which the dominant reactions of 1-alkynes or $PhC{\equiv}CCH_3$ involve hydrogen migration along the alkyne carbon skeleton to afford vinylidene [5] or allene [11] species (see above), at the more electron-rich {$M(dppe)_2$} (M = Mo or W) sites C-H oxidative addition reactions of $RC{\equiv}CH$ occur to form dihydrido-dialkynyl, dialkynyl or trihydrido-alkynyl (for a bulky alkynyl) complexes [$MH_2(C{\equiv}CR)_2(dppe)_2$] [13], [$M(C{\equiv}CR)_2(dppe)_2$] [13] or [$MoH_3(C{\equiv}CBu^t)(dppe)_2$] [14], respectively. These complexes have been prepared from the reactions of 1-alkynes with <u>trans</u>-[$M(N_2)_2(dppe)_2$] and the easier oxidation of the Mo(0) or W(0) metal centres, compared with the Re(I) site, as well as the higher lability of the former towards N_2 loss and the absence of the π-donor halide ligand (which would help stabilisation of metal-C multiple bonds) are factors which conceivably favour the C-H oxidative addition at the group 6 metal sites.

The molecular structures of the following alkynyl complexes have been determined by X-rays: <u>trans</u>-[$WH_2(C{\equiv}CCO_2Me)_2(dppe)_2$] [13], <u>trans</u>-[$Mo(C{\equiv}CPh)_2(dppe)_2$] [13] and [$MoH_3(C{\equiv}CBu^t)(dppe)_2$] (figure 2a) [14].

They undergo β-protonation, by [Et_2OH][BF_4], to give carbyne- or carbene-fluoro complexes, <u>i.e.</u>, <u>trans</u>-[$WF({\equiv}CCH_2CO_2Me)(dppe)_2$] (2) from [$WH_2(C{\equiv}CCO_2Me)_2(dppe)_2$] (reaction 7) [13], <u>trans</u>-[$WF({=}CHCH_2Ph)(dppe)_2$][BF_4] from [$WH_2(C{\equiv}CPh)_2(dppe)_2$] — reaction (8), possibly <u>via</u> protonation of the postulated carbyne intermediate [$WF({\equiv}CCH_2Ph)(dppe)_2$] [13] — and <u>trans</u>-[$MoF({\equiv}CCH_2Bu^t)(dppe)_2$][$BF_4$] (3) from [$MoH_3(C{\equiv}CBu^t)(dppe)_2$] (reaction 9) [15].

$$[WH_2(C{\equiv}CR)_2(dppe)_2] \xrightarrow[(R=CO_2Me)]{HBF_4,\ (7)} \textit{trans}\text{-}[WF({\equiv}CCH_2R)(dppe)_2]$$

$$[WH_2(C{\equiv}CR)_2(dppe)_2] \xrightarrow[(R=Ph)]{HBF_4,\ (8)} \textit{trans}\text{-}[WF({=}CHCH_2R)(dppe)_2][BF_4]$$

$$[MoH_3(C{\equiv}CBu^t)(dppe)_2] \xrightarrow{HBF_4} \textit{trans}\text{-}[MoF({\equiv}CCH_2Bu^t)(dppe)_2][BF_4] \quad (9)$$

The molecular structure of (3) has been authenticated by an X-ray analysis (figure 2b) [15], and this appears to provide the first example of a stable paramagnetic carbyne complex to be structurally characterized.

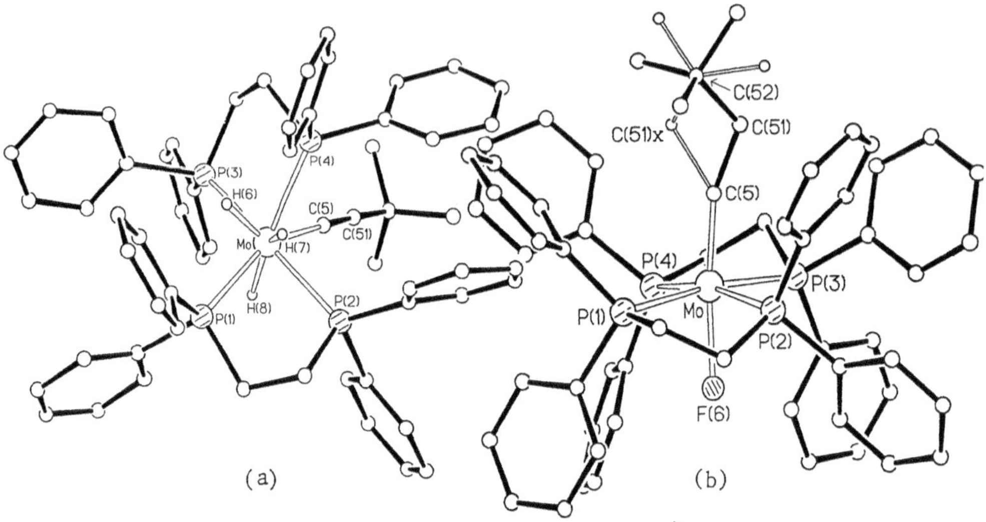

Figure 2. Molecular structures of (a) $[MoH_3(C{\equiv}CBu^t)(dppe)_2]$ [14] and (b) the cation of the derived carbyne complex trans-$[MoF({\equiv}CCH_2Bu^t)(dppe)_2][BF_4]$ [15].

The carbyne is trans to the fluoride ligand and is disordered with two possible orientations. The Mo≡C bond length, 1.821(7) Å, of this Mo(V) d^1 complex is very close to that of trans-$[MoBr({\equiv}CSiMe_3)(dppe)_2]$, 1.819(12) Å [16], with a Mo(IV) centre, suggesting that the fluoride ligand presents a significant trans influence which compensates for the expected shortening of the Mo≡C bond on passing from the latter to the former complex as a result of the increase in the metal oxidation state [15]. Moreover, the expected shortening of the metal-carbon distance, from 2.175(10) to 1.821(7) Å, with a concomitant elongation of the adjacent carbon-carbon bond length, from 1.209(14) to 1.542(13)Å, results from the protonation of the alkynyl ligand in $[MoH_3(C{\equiv}CBu^t)(dppe)_2]$ to give the carbyne species (3).

3. ISOCYANIDE-DERIVED AMINOCARBYNES

3.1. Rhenium Phosphinic Complexes

In the isocyanide complexes trans-$[ReCl(CNR)(dppe)_2]$ (R = alkyl or aryl), prepared by replacement of N_2 by CNR at trans-$[ReCl(N_2)(dppe)_2]$, the ligating isocyanides exhibit strong IR ν(CN) bands at very low wavenumbers, well below those shown by the free species, e.g., ca. 1800 cm^{-1} for trans-$[ReCl(CNMe)(dppe)_2]$ [17,18]. Moreover, an X-ray analysis of this complex has just been performed [19], showing a bent geometry for the methylisocyanide ligand, with a bending angle at the N atom of $139.9(2.5)^{\circ}$ [19].

The bending at the isocyanide is electronic in origin (see below), but a linear geometry has been quoted [20] for the bulkier $CNBu^t$ ligand at the same metal site. In addition, the Re-C bond length, for trans-$[ReCl(CNMe)(dppe)_2]$, 2.015(17) Å [19] is quite short, even slightly shorter than that exhibited by the vinylidene complex trans-$[ReCl(=C=CHPh)(dppe)_2]$, 2.046(8) Å [5].

These features are accounted for by considering an extensive π-electron release from the electron-rich d^6 metal to a C≡N π^* orbital of the isocyanide ligand, with resulting weakening of this bond, a concomitant strengthening of the metal-carbon bond and a localization of electronic charge at the N atom. In agreement with this interpretation, the isocyanide ligand undergoes ready protonation which occurs at the N atom (for either an alkyl or an aryl isocyanide) to give the aminocarbyne species trans-$[ReCl(CNHR)(dppe)_2]^+$ (4) (eq. 10) [18].

$$\text{trans-}[ReCl(CNR)(dppe)_2] + H^+ \longrightarrow \text{trans-}[ReCl(CNHR)(dppe)_2]^+ \qquad (10)$$

This type of activation of an isocyanide ligand and, therefore, the corresponding route for aminocarbyne species (eq.11) were previously discovered for trans-$[M(CNR)_2(dppe)_2]$ (M=Mo or W) [21], but the derived aminocarbyne complexes commonly present a much lower stability than (4) (see section 3.2).

$$M{=}C{=}\ddot{N}{-}R + H^+ \longrightarrow \left[M \overset{\dots}{\equiv} C \overset{\dots}{=} N{<}^{H}_{R}\right]^+ \qquad (11)$$

Moreover, organocyanides (NCR), at the $\{ReCl(dppe)_2\}$ site, are also susceptible to a similar β-protonation reaction to give methylene-amido type species (N=CHR), i.e., $[ReCl(N{=}CHR)(dppe)_2]^+$ derived from the reaction of $[ReCl(NCR)(dppe)_2]$ with acid [22].

The molecular structure of (4, R=Me) has been determined by X-ray analysis (figure 3) [18] which indicates that the aminocarbyne ligand should be represented, in a V.B. formulation, as a hybrid of the forms

$$\overset{+}{M}{\equiv}C{-}\ddot{N}{<}^{H}_{R} \longleftrightarrow M{=}C{=}\overset{+}{N}{<}^{H}_{R}$$

with a significant weighting of the carbene form as also suggested by the detection of an IR ν(C=N) band (e.g., at 1575 cm^{-1} for the CNMe ligand).

As a result of the protonation of the CNMe ligand at

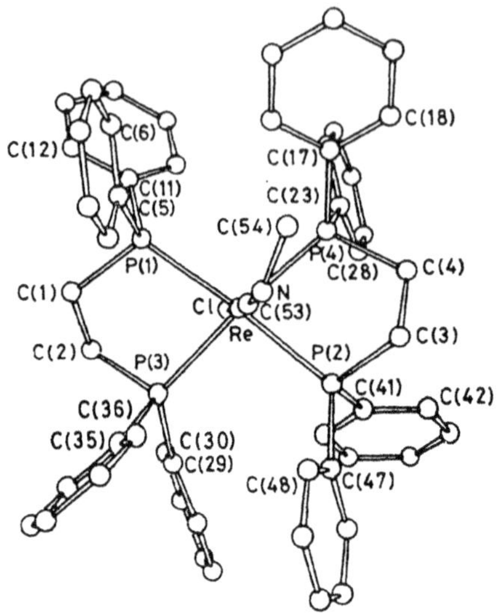

Figure 3. Molecular structure of the methylaminocarbyne species trans-$[ReCl(CNHMe)(dppe)_2]^+$ [18].

trans-$[ReCl(CNMe)(dppe)_2]$, the expected shortening of the metal-carbon bond occurs, from 2.015(17) (at the isocyanide ligand) [19] to 1.798(30) Å (at the carbyne ligand) [18], with a concomitant elongation of the unsaturated C-N bond, from 1.136(27) [19] to 1.347(32) Å [18]. Moreover, the Re-Cl distance decreases, from 2.654(5) [19] to 2.484(6) Å [18] and the Re-P average distance increases, from 2.399(7) [19] to 2.457(7) Å [18], upon protonation at the isocyanide, in accord with the stronger π-accepting ability of the aminocarbyne compared with the isocyanide ligand, which is also indicated by electrochemical studies (see section 4).

The abovementioned behaviour has been rationalized, at least partly, by simplified qualitative π-MO schemes [23] and also by means of extended Hückel calculations [24] which indicate that, although the HOMO of the isocyanide complex is mainly localized at the chloride ligand and at the metal atom, the nitrogen atom carries a considerable negative net atomic charge (in contrast with the metal and the ligating isocyanide carbon atom). Therefore, the β-protonation appears to be charge-controlled rather than frontier-orbital controlled [24].

Moreover, protonation of the isocyanide ligand results in an increase of both the σ- and the π-overlap populations for the Re-C bond, with the expected decrease for the C-N bond [24], in accord with the strengthening of the former and the weakening of the latter bond which occur with the isocyanide to aminocarbyne conversion.

The simplest aminocarbyne group, CNH_2, has been generated at the $\{ReCl(dppe)_2\}$ centre by reactions (12-14) [25] which involve the N-Si bond cleavage at the $CNSiMe_3$ ligand (derived from the cyano isomer, $NCSiMe_3$) and protonation at the N atom, to give trans-$[ReCl(CNH_2)(dppe)_2][BF_4]$ (4, R=H), the first example of a

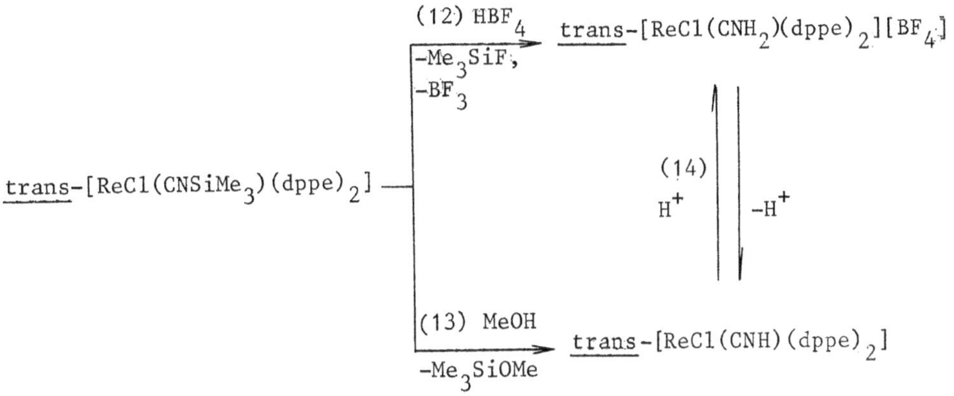

complex with the CNH_2 ligand. In its synthesis, the protonation is assisted by the formation of a Si–F or a Si–O bond, upon reaction of the trimethylsilyl isocyanide ligand with HBF_4 or MeOH, respectively.

The X-ray structure of (4, R=H) has been determined (figure 4) [25], and the Re–C and C–N bond lengths of the CNH_2 ligand, 1.802(4) and 1.309(5) Å, respectively, are comparable to those quoted (see above) for CNHMe in (4, R=Me). Moreover, the location of the H atoms of the CNH_2 group shows that it is roughly planar indicating a delocalization of the N lone pair electrons as accounted for by the above mentioned V.B. carbene structure.

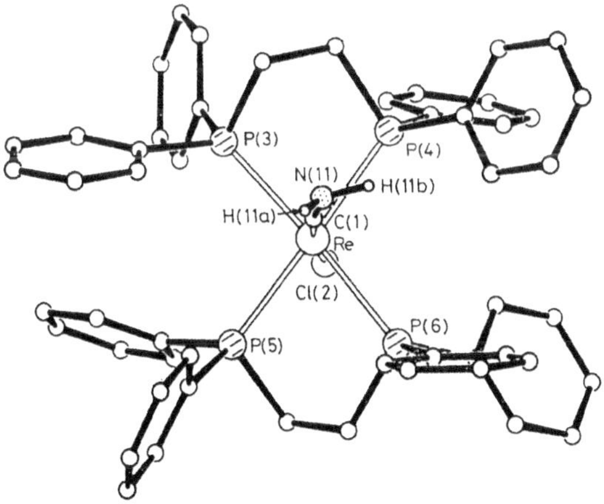

Figure 4. Molecular structure of trans-$[ReCl(CNH_2)(dppe)_2]^+$ [25].

In the reactions discussed above, the isocyanide undergoes a single protonation. However, further attack by acid can occur when the binding metal site presents labile ligands, namely at mer-[$ReCl(N_2)(CNR)\{P(OMe)_3\}_3$] (R=Me, Bu^t or C_6H_4Me-4) in which the isocyanide undergoes protic cleavage to amine (in the acidic methylammonium form) [26], a subject which will be discussed below in more detail for Mo or W complexes.

The activation of the isocyanide ligand by the {$ReCl(dppe)_2$} centre towards electrophiles, other than the proton, has also been studied. Unstable adducts are derived from the addition of an organo--aluminum Lewis acid, e.g., Al_2Et_6, to the CNMe ligand, and the dinuclear adducts [$ReCl\{CN(M)Me\}(dppe)_2$] [5; M = $CoCl_2(THF)$, $ReOCl_3(PPh_3)$, $WCl_4(PPh_3)$ or $WCl_4(PEtPh_2)$] are formed via electrophilic β-addition of the appropriate transition metal Lewis acid, i.e., $CoCl_2(THF)_{1.5}$, [$ReOCl_3(PPh_3)_2$], [$WCl_4(PPh_3)_2$] or [$WCl_4(PEtPh_2)_2$], respectively (eq. 15) [24].

$$\text{trans-}[ReCl(CNMe)(dppe)_2] \xrightarrow{M} [ReCl\{CN(M)Me\}(dppe)_2] \qquad (15)$$

3.2. Molybdenum or Tungsten Phosphinic Complexes

When ligating the electron-rich d^6 metal sites {$M(dppe)_2$} (M = Mo or W), and in a similar way to that discussed above for the related Re(I) centre, isocyanides exhibit very low ν(CN) wavenumbers in their IR spectra, a bent geometry and a short metal-carbon bond length as observed for trans-[$Mo(CNMe)_2(dppe)_2$] [$\angle C_\alpha NC$ = 156(7)° and Mo-C = 2.101(7) Å, respectively] [21,27]. In accord with these features, the isocyanides are then activated towards a variety of electrophiles such as H^+ [21,27-30], Me^+, Et^+ [31] or a Lewis acid (Al_2Et_6) [28a], to give aminocarbyne complexes of the types trans-[$M(CNER)(CNR)(dppe)_2$]$^+$ (6; E^+=H^+, Me^+ or Et^+; R = Me or Bu^t) (eq. 16) or the unstable adduct [$W\{CN(AlEt_3)Me\}_2(dppe)_2$].

$$\text{trans-}[M(CNR)_2(dppe)_2] + E^+ \longrightarrow \text{trans-}[M(CNER)(CNR)(dppe)_2]^+ \quad (16)$$

These reactions provided the first examples of electrophilic attack at an isocyanide ligand and constituted a novel route for carbyne complexes.

The complexes derived from a single protonation, trans-[$M(CNHMe)(CNMe)(dppe)_2$]$^+$ (6, E=H, R=Me) are susceptible to undergo further protic attack which occurs either at the still unreacted isocyanide ligand or at the metal, to give the dicarbene-type species trans-[$M(CNHMe)_2(dppe)_2$]$^{2+}$ (7) (eq.17) [28a] or the hydrido-aminocarbynes [$MH(CNHMe)(CNMe)(dppe)_2$]$^{2+}$ (8) [29,30] which are formed via intermediate hydride-diisocyanide complexes derived from β-proton migration from the aminocarbyne ligand to the metal (eqs. 18 and 19) [30]. in fact, these aminocarbyne complexes present a much lower stability than those obtained for Re and discussed above, (4) conceivably owing to the

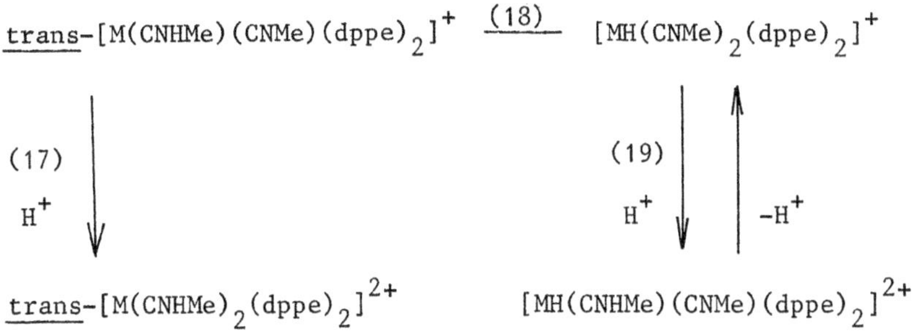

absence in the former of the π-donor Cl ligand with a stabilizing effect on the trans metal-carbyne bond (in contrast, the trans isocyanide ligand is a competitor for the available metal d electrons). In agreement with this interpretation, the dialkylaminocarbyne complexes trans-[M(CNER)(CNR)(dppe)$_2$]$^+$ (6; E=Me or Et; R=Me or Bu^t) are also unstable in solution, undergoing a trans to cis isomerization [31], thus converting into species in which the ligand trans to the carbyne is a phosphine rather than the stronger π-acceptor isocyanide ligand.

Moreover, on the basis of the abovementioned nucleophilic character of the BF_4^- anion, which readily fluorinates our metal centres, as indicated by our studies, and the recognition, by others [28b], of the involvement of aminocarbyne intermediates in the proton-induced reductive coupling of isocyanides, we are investigating the possibility of rearrangement (or reformulation) of compounds (7), derived from the protonation, by HBF_4, of the parent diisocyanide species, to the bis(methylamino)acetylene-fluoro complexes trans-[MF(η^2-MeHNC≡CNHMe)(dppe)$_2$][BF_4] or related compounds [28c].

If the starting isocyanide complexes present labile co-ligands, as previously mentioned for the Re systems, a more extensive protonation can occur, via aminocarbyne species, until cleavage of the unsaturated bond, e.g., at trans-[Mo(CNMe)$_2$L$_4$] (L=PMe$_2$Ph) or mer-[W(CNMe)$_3$L$_3$], to give methylamine, ammonia and methane, in a maximum yield corresponding to the consumption of the six valence electrons at the metal [32]. Such an extensive protonation is promoted by the replacement, along the course of the reaction, of the labile phosphines by stronger electron donor ligands (the anion of the acid).

The bonding, structure and reactivity of trans-[Mo(CNR)$_2$(dppe)$_2$] and the derived aminocarbyne complexes has also been investigated by EHMO-SCCC calculations [33].

In agreement with earlier simplified qualitative π-MO schemes [23], the HOMO of the isocyanide complex presents a metal-C bonding character, in accord with the short bond length, and a C-N antibonding character which accounts for the low IR ν(CN) [33]. Although the HOMO is mainly metal d-centred, the net atomic charge appears to be more negative at the N atom, for a bent isocyanide, thus indicating that the observed protonation at this atom appears to be charge-controlled rather than frontier-orbital controlled [33], as indicated above for the rhenium isocyanide complexes.

As a result of the protonation, significant changes are observed in both the σ- and the π-overlap populations of the axial bonds, which increase for the M-C bond but decrease for the C-N bond [33]. Therefore, they account for the strengthening of the former and the weakening of the latter bond as expected to occur for the isocyanide-to-aminocarbyne conversion.

Moreover, in accordance with the expected decrease of the π-electron acceptance of the isocyanide ligand which did not undergo protonation, at trans-[Mo(CNHMe)(CNMe)(dppe)$_2$]$^+$, a strengthening of the corresponding unsaturated C-N bond has occurred (the π-overlap population has increased) [33], also as indicated by IR data [a shift of ν(CN) to a higher value, i.e., from ca. 1860 in the parent diisocyanide complex to ca. 2160 cm^{-1} in that aminocarbyne product].

4. ELECTROCHEMICAL BEHAVIOUR

In contrast with the extensive chemistry developed for complexes with multiple metal-carbon bonds, only very few studies of their redox properties have been reported [34].

The electrochemical behaviour of the carbyne complexes of rhenium trans-[ReX(≡CCH$_2$R)(dppe)$_2$]$^+$ (1, X = F or Cl) has been investigated by cyclic voltammetry and controlled potential electrolysis, in aprotic medium (e.g., 0.2 M [Bu$_4$N][BF$_4$]/THF or NCMe), at a Pt or a Hg electrode (the latter only for the cathodic processes) [34-36].

A reversible single-electron oxidation is observed at $E_{1/2}^{ox}$ ca. 1.5 to 1.6 V vs. SCE [34-36], the fluoro complexes appearing to exhibit oxidation potentials at slightly lower values than the corresponding chloro compounds [36], thus suggesting that the HOMO is less stabilized in energy for the former complexes. Moreover, the reversible character of the one-electron oxidation of (1) indicates the formation of the paramagnetic carbyne complexes trans-[ReX(≡CCH$_2$R)(dppe)$_2$]$^{2+}$ which appear to be stable at least for the time scale of the cyclic voltammetry.

The aminocarbynes trans-[ReCl(CNHR)(dppe)$_2$]$^+$ (4) are oxidized at lower potentials (e.g., at $E_{p/2}^{ox}$ = 0.90 V for the CNH$_2$ complex) [34], indicating that the CNHR ligand presents a weaker net electron π-acceptor minus σ-donor character than the carbyne CCH$_2$R. Such an electronic character can be measured by the electrochemical P_L ligand parameter which has been estimated for these and related ligands of this study, thus allowing them to be ordered as follows (with the estimated P_L values in parentheses):

Carbynes (ca. 0.27) [34, 35] > aminocarbynes (ca. 0.1) [34, 35] > > CO (0) [37] > isocyanides (-0.1 to -0.2) [17, 38, 39] > vinylidenes (ca. -0.27) [34, 35] >> alkynyl (-1.22 V) [13, 40].

Carbynes and, to a lesser extent, aminocarbynes behave as stronger net electron acceptors than carbonyl or isocyanides which, in turn, are better acceptors than vinylidenes. Alkynyls are the weakest net acceptors, i.e., they behave as the strongest net electron donors in this series.

Within the ligands with known P_L, and with the exception of NO$^+$[37], carbynes appear to be the strongest net electron acceptors.

The anodic behaviour of the aminocarbyne complexes (4) involves a much more complex process than the simple electron-transfer step. In fact, (4, R=Me) undergoes, by cyclic voltammetry, a two-electron oxidation with electroinduced proton loss to give the corresponding isocyanide complex, in the dioxidized form, trans-$[ReCl(CNMe)(dppe)_2]^{2+}$, which can be reversibly reduced, in single-electron steps, to the neutral parent species trans-$[ReCl(CNMe)(dppe)_2]$ (eqs. 20) [41]. This complex can also be generated upon cathodic reduction of the aminocarbyne compound [41].

$$
\begin{array}{ccc}
[ReCl(CNHMe)(dppe)_2]^+ & \xrightarrow{-2e/-H^+} & [ReCl(CNMe)(dppe)_2]^{2+} \\
{\scriptstyle +e(-H)\ (or\ base)}\ \downarrow\uparrow\ {\scriptstyle H^+} & & {\scriptstyle -e}\ \uparrow\downarrow\ {\scriptstyle +e} \\
[ReCl(CNMe)(dppe)_2] & \underset{+e}{\overset{-e}{\rightleftharpoons}} & [ReCl(CNMe)(dppe)_2]^+
\end{array}
\qquad (20)
$$

A similar general anodic behaviour is followed by the abovementioned adducts $[ReCl\{CN(\underline{M})R\}(dppe)_2]$ (5) in which, upon electrooxidation, liberation of the electrophile (a transition metal Lewis acid, M) also occurs [24]. Moreover, trans-$[ReCl(CNH_2)(dppe)_2]^+$ (4, R=H) exhibits a related anodic process which, however, involves a single electron transfer to give trans-$[ReCl(CNH)(dppe)_2]^+$ (upon H^+ loss) [34, 35, 42].

The anodically induced heterolytic N-H bond cleavage at the aminocarbyne ligand agrees with the expected increase of the acidic caracter of this ligand as a result of the electron removal from the complex, and can also be accounted for by some theoretical studies which indicate [24] the appearance of a positive charge on the N atom upon oxidation of the complex. Related electrode processes have been observed for the nitrile-derived methyleneamide complexes trans-$[ReCl(N{=}CHR)(dppe)_2]^+$ [43] and for some aminocarbene compounds of Pd or Pt [44].

The electroactivation towards cleavage of the N-H bond of the aminocarbyne ligands can also occur by cathodic reduction to give the corresponding metal ligating isocyanide with possible formation of H_2. In fact, (4, R=H or Me) undergo a single-electron irreversible reduction to form the corresponding neutral isocyanide complexes (H_2 evolution has been detected [34] for the former compound).

A related cathodic behaviour has been detected for the carbyne complexes (1) which, upon reduction, generate the corresponding vinylidene compounds, thus following a process which involves the cathodic C-H bond cleavage at the carbyne ligand CCH_2R to give C=CHR [34, 36].

However, the paramagnetic species trans-$[MoF(\equiv CCH_2Bu^t)(dppe)_2]^+$ (3)

exhibits, by cyclic voltammetry, two single-electron reversible reductions (at $E^{red}_{1/2}$ = -1.17 and -1.52 V) to the corresponding Mo(IV) and Mo(III) carbyne complexes which are stable at least on the usual cyclic voltammetric time scale [15].

5. FINAL COMMENTS

The activation of adequate small unsaturated-C molecules, typically isocyanides or alkyne-derived vinylidenes and alkynyls, towards β-electrophilic attack by electron-rich group 6 (Mo or W) or 7 (Re) transition metal centres constitutes a convenient and established route for the synthesis of carbyne complexes, which has been discussed in this paper.

In particular, a considerable number of carbyne-fluoro complexes have been prepared (apart from related chloro species) and fluoride is shown to present a good stabilizing effect on the *trans*-carbyne ligand, in spite of the previously limited number of known examples of such a type of complexes.

Stable paramagnetic carbynes have also been obtained, usually by electrochemical methods which proved to be successfully applicable to the investigation of the electronic properties of the carbyne ligands (which are shown to behave as rather strong net electron acceptors) and to their activation, in particular towards N-H or C-H bond cleavage. Although electrochemical studies of carbyne complexes have been reported only very rarely, this study illustrates some of their potentialities in this field.

The possibility of application of stopped flow spectrophotometry to the study of the mechanisms of the formation of the carbyne complexes, involving not a too fast protic attack, has also been demonstrated and indicates that the apparent regiospecific β-protonation at a vinylidene can in fact occur *via* different and less straightforward pathways, in particular involving H^+ addition to the metal.

In addition, extended Hückel calculations have also been of significance for the understanding of the electronic factors governing the formation of the carbyne ligands, indicating, *inter alia*, that the β-protonation at an isocyanide to give an aminocarbyne ligand is charge controlled, whereas the α-nucleophilic attack at the latter species is frontier-orbital controlled.

The studies discussed herein can also be of some biological meaning since metal ligating carbynes of the types described above can be postulated as intermediates in the enzymatic reduction of alkynes and isocyanides by nitrogenases.

ACKNOWLEDGEMENTS

Thanks are due to the co-authors indicated in the citations, particularly Dr. R.L. Richards (Nitrogen Fixation Laboratory, Univ. Sussex), for some laboratory facilities and stimulating discussions, Dr. D.L. Hughes (Nitrogen Fixation Laboratory, Univ. Sussex) and Dr. P.B. Hitchcock

(Univ. Sussex) for X-ray diffraction anlyses of carbyne complexes, Dr. M.T. Duarte (Instituto Superior Técnico) for the X-ray diffraction analysis of one of the isocyanide complexes, Dr.R. Henderson (Nitrogen Fixation Laboratory, Univ. Sussex) for stopped-flow kinetic studies, Dr. E.G. Bakalbassis and Prof. C.A. Tsipis (Aristotle Univ., Thessaloniki) for the extended Hückel MO calculations, as well as, from our laboratory, Dr. M.F.N.N. Carvalho (some isocyanide and aminocarbyne complexes, and stopped-flow studies), Lic. M.A.N.D.A. Lemos (electrochemical studies), Lic. S.P.R. Almeida and Lic. M.F.C. Guedes da Silva (some vinylidene and carbyne complexes, and electrochemical studies).

The author also gratefully acknowledges Prof. J.J.R. Fraústo da Silva (Instituto Superior Técnico) for laboratory facilities and general support. The work has been mainly supported by INIC and JNICT (Portugal).

REFERENCES

[1] See, e.g.: (a) Kim H.P., and Angelici, R.J. (1987), Adv. Organometal. Chem. 27, 51; (b) Wilkinson G., Stone F.G.A., and Abel E.W. (Eds.) (1989) Comprehensive Organometallic Chemistry 3, Chaps. 25-28, Pergamon Press Ltd; (c) Schubert U. (Ed.) (1989) Advances in Metal Carbene Chemistry, NATO ASI Series, Kluwer Academic Publ., Dordrecht.

[2] See, e.g.: [1] (b) vol. 4, Chap. 30

[3] Edwards D.S., Biondi C.V., Ziller J.W., Churchill M.R., and Schrock R.R. (1983), Organometallics 2, 1505.

[4] Noda I., Kato S., Mizuta M., Yasuoka N., and Kasai N. (1979), Angew. Chem. Int. Ed. Engl. 18, 83

[5] Pombeiro A.J.L., Almeida S.S.P.R., Silva M.F.C.G., Jeffery J.C., and Richards R.L. (1989), J. Chem. Soc., Dalton Trans., 2381.

[6] (a) Birdwhistell K.R., Tonker T.L., and Templeton J.L. (1985), J. Am. Chem. Soc. 107, 4474; (b) Beevor R.G., Green M., Orpen A.G., and Williams I.D. (1983), J. Chem. Soc., Chem. Commun., 673; (c) Mayr A., Scharfer K.C., and Huang E.Y. (1984), J. Amer. Chem. Soc. 106, 1517

[7] (a) Pombeiro A.J.L., Hills A., Hughes D.L., and Richards R.L. (1988), J. Organometal. Chem. 352, C5; (b) Almeida S.S.P.R., and Pombeiro A.J.L. (1992), II Italian-Portuguese-Spanish Meeting in Inorganic Chemistry, Algarve, Portugal, OM 3; (c) Almeida S.S.P.R., and Pombeiro A.J.L., unpublished results.

[8] Pombeiro A.J.L., Carvalho M.F.N.N., Hitchcock P.B., and Richards R.L. (1981), J. Chem. Soc., Dalton Trans., 1629.

[9] Carvalho M.F.N.N., Henderson R.A., Pombeiro A.J.L., and Richards R.L. (1989), J. Chem. Soc., Chem. Commun., 1796.

[10] Silvestre J., and Hoffmann R. (1985), Helv. Chim. Acta 68, 1461.

[11] Hughes D.L., Pombeiro A.J.L., Pickett C.J., and Richards R.L. (1984), J. Chem. Soc., Chem. Commun., 992.

[12] Birdwhistell K.R., Burgmayer S.J.N., and Templeton J.L. (1983), J. Am. Chem. Soc. 105, 7789.

[13] Hills A., Hughes D.L., Kashef N., Lemos M.A.N.D.A., Pombeiro A.J.L. and Richards R.L. (1992), J. Chem. Soc., Dalton Trans, 1775.

[14] Pombeiro A.J.L., Hills A., Hughes D.L., and Richards R.L. (1990), J. Organometal. Chem. 398, C15.
[15] Lemos M.A.N.D.A., Pombeiro A.J.L., Hughes D.L., and Richards R.L. (1992), J. Organometal Chem. 434, C6.
[16] Ahmed K.J., Chisholm M.H., and Huffman J.C. (1985), Organometallics 4, 1168.
[17] Pombeiro A.J.L., Pickett C.J., and Richards R.L. (1982), J. Organometal. Chem. 224, 285.
[18] Pombeiro A.J.L., Carvalho M.F.N.N., Hitchcock P.B., and Richards R.L. (1981), J. Chem. Soc., Dalton Trans., 1629; Carvalho M.F.N.N., and Pombeiro A.J.L., unpublished results.
[19] Carvalho M.F.N.N., Duarte M.T., Galvão A.M., and Pombeiro A.J.L., unpublished results.
[20] Carrondo M.A.A.F.C.T., Domingos A.M.T.S., and Jeffrey G.A. (1985), J. Organometal. Chem. 289, 377.
[21] Chatt J., Pombeiro A.J.L., Richards R.L., Royston G., Muir K., and Walker R. (1975), J. Chem. Soc., Chem. Commun., 708.
[22] Pombeiro A.J.L., Hughes D.L., and Richards R.L. (1988), J. Chem. Soc., Chem. Commun., 1052.
[23] Pombeiro A.J.L. (1979), Rev. Port. Quím. 21, 90
[24] Carvalho M.F.N.N., Pombeiro A.J.L., Bakalbassis E.G., and Tsipis C.A. (1989), J. Organometal. Chem. 371, C26.
[25] Pombeiro A.J.L., Hughes D.L., Pickett C.J., and Richards R.L. (1986), J. Chem. Soc., Chem. Commun., 246.
[26] Carvalho M.F.N.N., Pombeiro A.J.L., Schubert U., Orama O., Pickett C.J., and Richards R.L. (1985), J. Chem. Soc., Dalton Trans., 2079.
[27] Chatt J., Elson C.M., Pombeiro A.J.L., Richards R.L., and Royston G.H.D. (1978), J. Chem. Soc., Dalton Trans., 165.
[28] (a) Chatt J., Pombeiro A.J.L., and Richards R.L. (1980), J. Chem. Soc., Dalton Trans., 492. (b) Vrtis R.N., and Lippard S.J. (1990), Isr. J. Chem. 30, 331; Filippou A.C. (1990), Polyhedron 9, 727. (c) Pombeiro A.J.L., and Richards R.L., unpublished.
[29] Chatt J., Pombeiro A.J.L., and Richards R.L. (1979), J. Chem. Soc., Dalton Trans., 1585.
[30] Pombeiro A.J.L., and Richards R.L. (1980), Transition Met. Chem. 5, 55; (1979), Rev. Port. Quím. 21, 132.
[31] Chatt J., Pombeiro A.J.L., and Richards R.L. (1980), J. Organometal. Chem. 184, 357.
[32] Pombeiro A.J.L., and Richards R.L. (1980), Transition Met. Chem. 5, 281.
[33] Bakalbassis E.G., Tsipis C.A., and Pombeiro A.J.L. (1981), J. Organometal. Chem. 408, 181.
[34] Lemos M.A.N.D.A., and Pombeiro A.J.L. (1988), J. Organometal. Chem., 356, C79.
[35] Almeida S.S.P.R., Lemos M.A.N.D.A., and Pombeiro A.J.L. (1989), Portugaliae Electrochimica Acta 7, 91.
[36] Almeida S.S.P.R., and Pombeiro A.J.L. (1992), VI Meeting of the Portuguese Electrochemical Society, Vila-Real.
[37] Chatt, J., Kan C.T., Leigh G.J., Pickett C.J. and Stanley D.R. (1980), J. Chem. Soc., Dalton Trans., 2032.
[38] Pombeiro A.J.L. (1985), Inorg. Chim. Acta 103, 95
[39] Carvalho M.F.N.N., and Pombeiro A.J.L. (1989), J. Chem. Soc., Dalton Trans., 1209.

[40] Hills A., Hughes D.L., Kashef N., Richards R.L., Lemos M.A.N.D.A., and Pombeiro A.J.L. (1988), J. Organometal. Chem. 350, C4.
[41] Pombeiro A.J.L., Lemos M.A.N.D.A., and Carvalho M.F.N.N. (1990), IX Iberoamerican Congress on Electrochemistry, La Laguna, Tenerife, 4-12, p. 453; unpublished work.
[42] Pombeiro A.J.L. (1989), XI Meeting of the Electrochemistry Division of the Spanish Chemical Royal Society, Valladolid, CP-2.
[43] Silva M.F.C.G., and Pombeiro A.J.L. (1991), Portugaliae Electrochimica Acta 9, 189.
[44] Bertani R., Mozzon M., Michelin R.A., Benetollo F., Bombieri G., Castilho T.J., and Pombeiro A.J.L. (1991), Inorg. Chim. Acta 189, 175; Castilho T.J., Guedes da Silva M.F.C., Pombeiro A.J.L., Bertani R., Mozzon M., and Michelin R.A. (1992) in 'Molecular Electrochemistry of Inorganic, Bioinorganic and Organometallic Complexes', Pombeiro A.J.L. (ed.), and McCleverty J.A. (co-ed.), NATO ASI Series, Kluwer Academic Publ., Dordrecht.

Note added in proof

After this paper was presented at the NATO ARW, further NMR spectroscopic data for the complexes previously formulated [21, 28a] as trans-[$M(CNHMe)_2(dppe)_2][BF_4]_2$ (7, M = Mo or W) (see also eq. 17) and an X-ray structural analysis (for the Mo compound) became available (Yu Wang, A. J. L. Pombeiro, J. J. R. Fraústo da Silva, R. L. Richards, M. A. Pellinghelli and A. Tiripicchio, unpublished), indicating that such species should be reformulated as the bis(amino)acetylene compounds trans-[$MF(\eta^2$-MeHNC≡CNHMe)$(dppe)_2][BF_4]$. They are derived from reaction of HBF_4 with the corresponding aminocarbyne complexes trans-[$M(CNHMe)(CNMe)(dppe)_2][BF_4]$, the coupling step being promoted by nucleophilic attack at the metal by BF_4^-. Moreover, base-induced CC cleavage of the acetylene ligand (e. q., by an organo-lithium or NaOMe [30]) leads to the diisocyanide complexes trans-[$M(CNMe)_2(dppe)_2$]. These reactions are summarized below and provided, although unknowingly, the first reductive coupling of isocyanide ligands - a reaction which, later on, was unambiguously recognized by others (Lam C. T., Corfield P. W. R. and Lippard S. J., 1977, J. Am. Chem. Soc. 99, 617; [28b]) - and the reverse of this process.

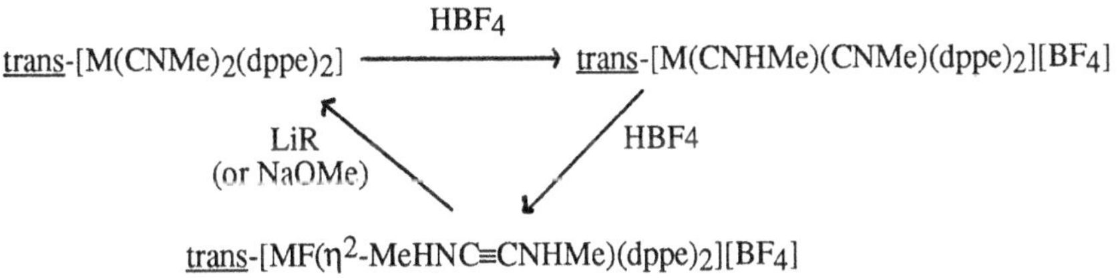

PHOTOOXIDATION OF MOLYBDENUM AND TUNGSTEN CARBYNES

Lisa McElwee-White, Kevin B. Kingsbury, and John D. Carter
Department of Chemistry
Stanford University
Stanford, California 94305
USA

ABSTRACT. The low valent carbyne complexes $CpL_2M{\equiv}CR$ [M = Mo, W; L = $P(OMe)_3$, CO; R = alkyl, aryl] undergo photochemical electron transfer to halogenated hydrocarbons. The resulting 17-electron metal carbyne cation radicals exhibit reactivity that differs significantly from that of the neutral precursors. By manipulating the reaction conditions, we can switch the system from "inorganic" reactivity (the characteristic reactions of metal radicals) to "organic" reactivity (the production of organic products by rearrangement and loss of the carbyne ligand). In the "inorganic" mode, ligand exchange and halogen abstraction yield new complexes with the carbyne ligand intact. In one example of the "organic" mode, photooxidation of the cyclopropyl carbyne $Cp[P(OMe)_3](CO)W{\equiv}C(c\text{-}C_3H_5)$ produces cyclopentenone. Another "organic" reaction is cyclization of the butenyl carbyne $Cp[P(OMe)_3](CO)Mo{\equiv}C(CH_2)_2CH{=}CH_2$ to form cyclohexenone.

Introduction

Photoinduced electron transfer involving organometallic complexes has been of recent interest [1] due to the tremendous increase in reactivity upon oxidation of 18-electron species to the corresponding 17-electron cation radicals [2]. Photooxidation of the low valent carbyne complexes $CpL_2M{\equiv}CR$ (**1**) [M = Mo, W; L = $P(OMe)_3$, CO; R = alkyl, aryl] in the presence of the strongly donating ligand PMe_3 results in exchange of both ancillary ligands L for PMe_3 and abstraction of a chlorine atom from the solvent to yield the modified carbyne complexes **2** in a net 2-electron oxidation (Eq. 1) [3].

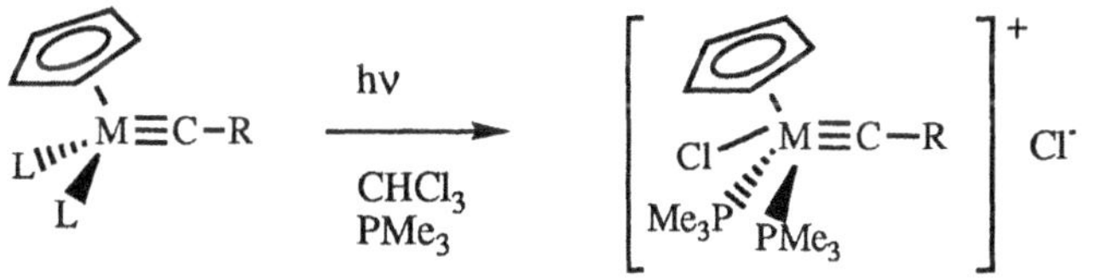

(1)

1 **2**

F. R. Kreißl (ed.), Transition Metal Carbyne Complexes, 123–125.

However, for photooxidation of alkyl-substituted carbynes in the absence of PMe_3, the reactivity shifts to the carbyne ligand itself and free organic products are formed.

Results and Discussion

Although the **1** → **2** conversion occurs for $Cp[P(OMe)_3](CO)W{\equiv}C(c\text{-}C_3H_5)$ (**3**) upon photooxidation in $CHCl_3/PMe_3$, in the absence of the phosphine, rearrangement and carbonylation of the carbyne ligand ultimately lead to cyclopentenone (Eq. 2) [4].

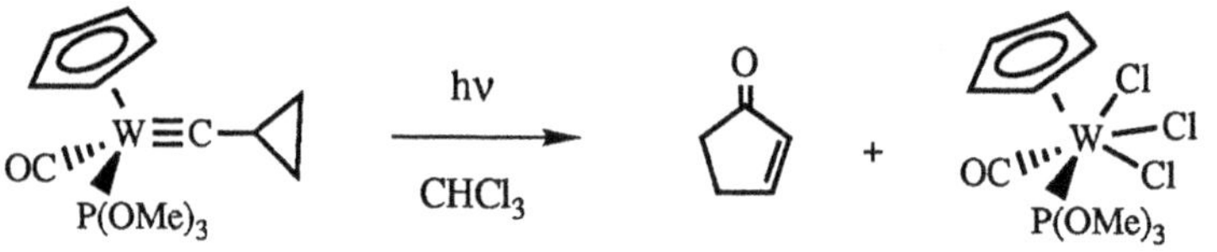

3

(2)

For unsymmetrically substituted cyclopropyl carbyne complexes such as **4**, the formation of cyclopentenones is highly regioselective (Eq. 3). In the case of C2,C3-

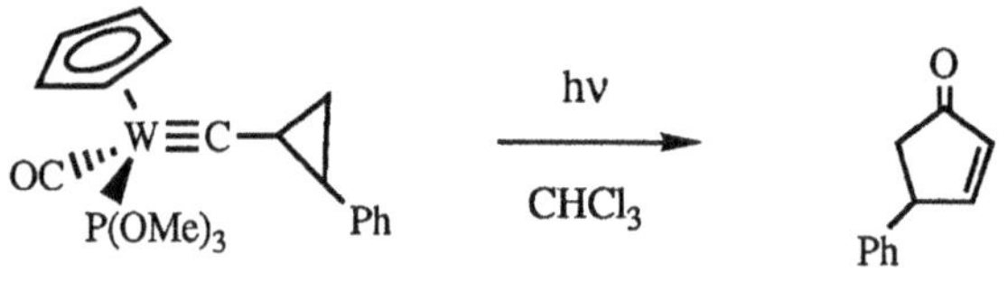

4 only

(3)

disubstituted complexes, stereoselectivity in product formation is provided by photochemical equilibration of the isomeric starting materials (Eq. 4). One of the *trans* diastereomers is rapidly converted to *trans*-4,5-dimethylcyclopentenone, resulting in formation of *trans*-product from either *cis*- or *trans*-**5**.

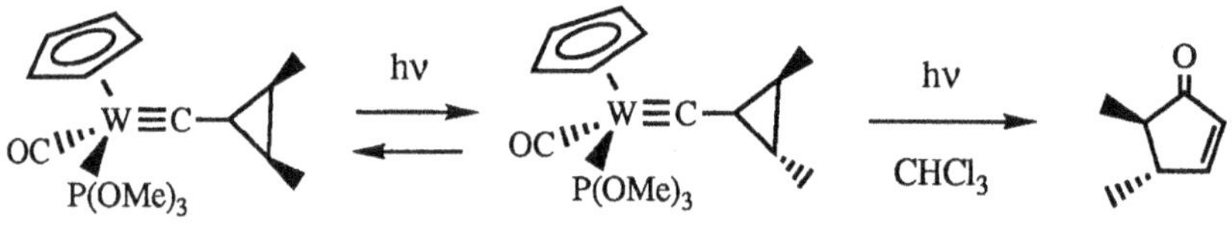

cis-5 **trans-5**

(4)

Formation of cyclopentenones appears to be general for C2- and/or C3-substituted cyclopropyl carbynes. However, substitution at C1 introduces additional complications into the system. As an example, the highly electron-deficient acyl-substituted complexes **6** exhibit a new mode of reactivity [5]. In competition with the formation of cyclopentenones, nucleophilic ring opening by chloride (produced upon electron transfer to $CHCl_3$) results in formation of the oxymetallacycles **7** (Eq. 5).

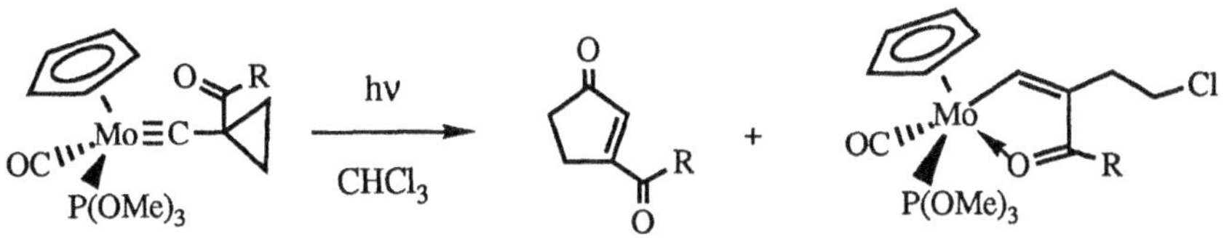

(5)

6 7

In addition to the ring opening/carbonylation pathways illustrated in Eqs. 2-5, photooxidation of metal carbynes can result in other types of ligand rearrangement. An example is the cyclization of butenyl carbyne complex **8** depicted in Eq. 6. Deuterium

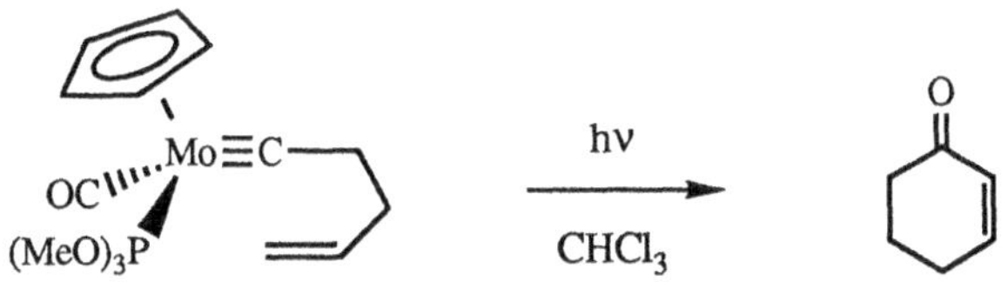

(6)

8

labeling experiments have revealed this to be a complex rearrangement involving intramolecular hydride migration as well as cyclization and carbonylation of the carbyne ligand.

Acknowledgment

We thank the National Science Foundation and the Petroleum Research Fund for financial support of this work.

References

1. C. Gianotti, S. Gaspard, and P. Kramer in *Photoinduced Electron Transfer, Part D. Photoinduced Electron Transfer Reactions: Inorganic Substrates and Applications*, M.A. Fox and M. Chanon, Eds., Elsevier, Amsterdam, **1988**, 200-240.

2. (a) W.C. Trogler, Ed., *Organometallic Radical Processes*, Elsevier, Amsterdam, **1990**. (b) D. Astruc, *Acc. Chem. Res.* **1991**, *24*, 36-42. (c) M. Baird, *Chem. Rev.* **1988**, *88*, 1217-1227.

3. (a) C.J. Leep, K.B. Kingsbury, and L. McElwee-White, *J. Am. Chem. Soc.* **1988**, *110*, 7535-7536. (b) J.D. Carter, K.B. Kingsbury, A. Wilde, T.K. Schoch, C.J. Leep, E.K. Pham, and L. McElwee-White, *J. Am. Chem. Soc.* **1991**, *113*, 2947-2954.

4. K.B. Kingsbury, J.D.Carter, L. McElwee-White, *J. Chem. Soc., Chem. Commun.* **1990**, 624-625.

5. J.D. Carter, T.K. Schoch, and L. McElwee-White, *Organometallics*, in press.

THE DETERMINATION OF THE CHEMICAL BONDS IN SOME METAL-CARBYNE COMPLEXES BY STRUCTURAL STUDIES.

NGUYEN QUY DAO
Laboratoire PCM, U.R.A.441 du CNRS, ECOLE CENTRALE PARIS, Grande Voie des Vignes, F-92295 CHATENAY-MALABRY Cedex, FRANCE.

ABSTRACT. Metal carbyne complexes ($XMCR_3$ where X = Cl, Br ; M = W, Cr ; R = CH_3, CD_3, C_6H_5, C_6D_5) were studied by both experimental techniques (IR, polarised Raman spectroscopy, X-Ray and neutron diffraction) and theoretical methods (*ab initio* and semi-empirical calculations). The vibrational spectra show that a conjugation effect (or hyperconjugation effect in the case of the methylcarbynes) exists in these compounds. Moreover, the temperature-dependant spectra of the partially deuterated derivative (R = CHD_2) show that two conformers due to a librational effect of the methyl group exist above 100 K. The electronic deformation density using combined X-ray and neutron diffraction data and a multipolar atom model also demonstrates this conjugation effect. The comparison of the experimental and ab-initio electronic deformation density maps shows a very good agreement and proves that both experimental and theoretical results are reliable for such complexes.

Introduction.

Some of the carbyne complexes are particularly suitable for a precise determination of the chemical bonds. Their molecular and crystal structures possess high symmetry [1,2]. The molecules themselves are placed on high symmetrical sites in the crystal. In addition, big single crystals of some complexes can be obtained. These conditions are very favorable for electron deformation density studies and for other physical techniques such as polarised Raman spectroscopy in order to understand the nature of the chemical bonds existing in this class of compounds. It is also possible to compare these experimental results with *ab initio* calculations. This paper relates some results of these studies, essentially focused on the conjugation effect of carbynes.

1. Demonstration of the M ≡ C carbyne vibration by polarised Raman spectroscopy.

Using the polarisation effect of $Br(CO)_4WCR$ [R = CH_3 (I), CD_3 (II)] as observed in solution and in single crystals [3] the W≡C stretching frequency was found at 1315 cm^{-1} for (II) and the force constant of the W≡C bond, based on a complete normal coordinate calculation, was found to be equal to 7.0 mdyn.$Å^{-1}$. This stretching vibration is strongly coupled with the $\nu(C\text{-}CH_3)$ stretching mode and to some extent with the symmetrical deformation mode $\delta_s(CH_3)$. These results suggest that there is a hyperconjugation effect between the methyl and the M≡C triple bond.

F. R. Kreißl (ed.), Transition Metal Carbyne Complexes, 127–129.

2. Demonstration of the hyperconjugation effect in the methylcarbyne using pure partial deuterated derivative.

Raman spectra of Br $(CO)_4CrCCHD_2$ (III) were examined at 77 K in the CH stretching region [4]. It was found that two types of CH bonds exist in the methyl group : one strong CH_A bond and two weaker CH_B bonds, situated respectively at 2910 cm^{-1} and 2887cm^{-1} (intensity ratio R = 1:2). There is a non-equivalence of the CH bonds in the methyl group and the difference in strength is very weak : 25 cm^{-1} or 0.05 mdyn.$Å^{-1}$. The hyperconjugation effect exists but is weak in this compound.

3. Existence of two conformers of (III) as a fonction of the temperature.

By varying the temperature from 77 K to 100 K, an extra peak appears on the spectra at 2903 cm^{-1}, in addition to the two previous Raman peaks [5]. At 100 K, three Raman lines are present: 2887, 2903, 2910 cm^{-1} (R = 2:1:1). At 150 K, the intensity of the 2910 cm^{-1} line decreases while the one at 2903 cm^{-1} increases (R = 2:1 for these two peaks). The intensity of the peak at 2887 cm^{-1} remains approximately equal to the sum of the two preceding peaks. At 230 K, the two peaks at 2903 and 2910 cm^{-1} can only be separated with difficulty. The intensity of the peak at 2887 cm^{-1} is equal to the sum of the broadening peaks. The peaks at 2887 and 2910 cm^{-1} have already been interpreted. The presence of the peak at 2903 cm^{--1} is more troublesome. The interpretation of these results is that there are two conformers of (III), the first being the eclipsed configuration E and the second being a non-eclipsed configuration NE which has been rotated by 15° from the E configuration about the C-CH_3 axis.

4. The conjugation effect as seen by electron deformation density in phenyl carbyne (IV).

(IV) crystallises in the non-centrosymmetrical space group I 4_1md. In order to determine reliable phases for the observed structure factors, Hansen and Coppens multipole atomic models are used for the least squares refinement of the X-ray diffraction intensities. It is then possible to visualise the chemical bonds which are due to the valence electron clouds on the electron deformation density maps [6]. It is interesting to note that a comparison of experimental and *ab initio* calculation results concerning the electronic deformation density maps for both the phenyl and methyl carbyne complex shows a very good agreement between the two types of maps.

An examination of the static deformation densities along the Cr≡C-C axis in two planes respectively containing the phenyl ring and perpendicular to it (Fig. 1) shows that the electronic density is not distributed isotropically around the symmetry axis. First, the accumulation between Cr and the carbynic carbon atoms appears slightly more extended in the plane containing the phenyl group. Secondly, an opposite tendency is observed for the accumulation located along the C-C bond, which appears broader in the plane perpendicular to the phenyl group. This loss of cylindrical symmetry which is reminiscent of the organic cumulenes [7], can be attributed to the conjugation between the phenyl ring and the triple bond.

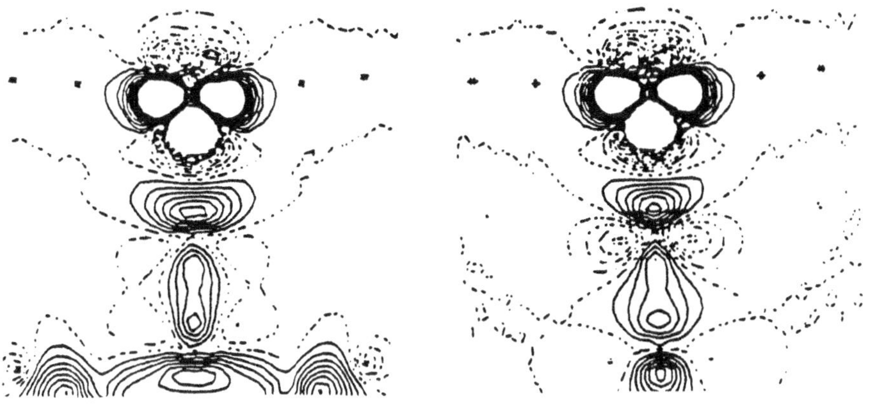

(a) (b)

Figure 1. Conjugation effect in the phenylcarbyne complex.
a) Static deformation map in the plane containing the phenyl ring
b) Static deformation map in the plane perpendicular to the phenyl ring (contour 0.1e.Å^{-3}).

References.

1. Neugebauer, D. Fischer, E.O., Nguyen Quy Dao and Schubert, U. (1978) "Strukturuntersuchungen an trans-Halogeno-Tetracarbonyl-methylcarbin - Wolfram - Komplexen", J. Organomet. Chem., 153, C41-C44.
2. Nguyen Quy Dao, Neugebauer, D., Février, H., Fischer E.O., Becker, P.J. and Pannetier, J. (1982) "Neutron diffraction study on single crystal of *trans*-chloro (Tetracarbonyl) Phenylcarbyne Chromium", Nouv. J. Chim., 6, 359 - 364.
3. Nguyen Quy Dao, Fischer, E.O., Wagner, W.R. and Neugebauer, D. (1979) "Schwingungsspektroskopische Untersuchungen an *trans* - Halogeno - tetracarbonyl - methylcarbin - Wolfram - Komplexen", Chem. Ber., 112, 2552 - 2564.
4. Foulet-Fonseca, G.P., Jouan, M., Nguyen Quy Dao, Fischer, H., Schmid, J. and Fischer, E.O. (1990) "Etude vibrationnelle du composé partiellement deutérié pur $Br(CO)_4CrCCHD_2$ et détermination du champ de force des complexes carbyniques de la famille $Br(CO)_4CrCR(R=CH_3$, CH_2D, CHD_2 et $CD_3)$", Spectrochim. Acta, 46A, 339-354.
5. Nguyen Quy Dao, Foulet-Fonseca, G.P., Jouan, M., Fischer, E.O., Fischer, H. and Schmid, J. (1988). "Mise en évidence de deux conformères du complexe $Br(CO)_4CrCCHD_2$ par spectrométrie de diffusion Raman", Comptes Rendus Acad. Sci. Paris, 307, Série II, 341-346.
6. Spasojevic' de Biré, A., (1989) "Etude de la densité electronique de déformation des complexes méthyl et phénylcarbyniques de chrome", Thèse de Doctorat de l'Université de Paris VI.
7. Irngartinger, H. (1982) "Electron density distribution in the bonds of cumulenes and small ring compounds" P. Coppens and M.B. Hall (Eds), Electron distribution and the Chemical Bond, Plenum Press, New-York, p.36.

Planar Tetracoordinate Carbon -Experimental Determination of the Charge Density of $Cp_2Zr(\mu\text{-}\eta^1,\eta^2\text{-}Me_3SiCCPh)$ $(\mu\text{-}Cl)AlMe_2$ (I) and $Cp_2Zr(\mu\text{-}\eta^1,\eta^2\text{-}MeCCPh)$ $(\mu\text{-}CCPh)AlMe_2$ (II)

C. Krüger, S. Werner

Max-Planck-Institut für Kohlenforschung
Kaiser-Wilhelm-Platz 1
D-4330 Mülheim a. d. Ruhr

INTRODUCTION

The title compounds presented are chosen from a series of alkine complexes

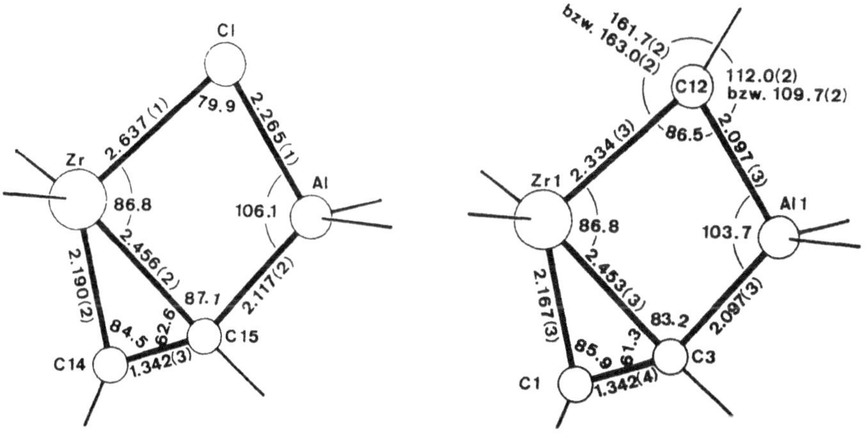

Fig. 1. Bonding parameters in the central molecular planes of **I** and **II**

of zirconocene fragments stabilized by aluminium organyls [1]. One alkine carbon atom bridges the two metal centers, its substitution geometry is best described by forming four σ-bonds in one plane (Fig. 1). The conformation of D_{4h}-symmetry of methane was calculated by Hoffmann. Six valence electrons form four σ-bonds, the two remaining electrons are located in a π-orbital perpendicular to the mean

F. R. Kreißl (ed.), Transition Metal Carbyne Complexes, 131–135.

molecular plane [2]. This bonding situation is therefore favoured by σ-donating and/or π-accepting substitutents [3].

EXPERIMENTAL

The charge density distribution was deduced from high resolution X-ray data introducing a multipole expansion [4] of the atomic scattering factors [5]. Atomic coordinates and ADP's of the non-hydrogen atoms were fixed at the values resulting from standard atom refinement of the high-order data. Hydrogen atom positions were corrected to account for the aspherical shift (d(C-H) 1.08Å).

Independent treatment of the multipole parameters of the two independent molecules in the crystal structure of **II** did not result in any significant difference in the population of corresponding multipole functions. Further experimental details are summarized in Table 1.

The electron deformation density (EDD) of **I** and **II** was calculated by numerical integration in direct space on the basis of the multipole model.

RESULTS AND DISCUSSION

In the vincinity of the planar tetracoordinate carbon atom three maxima of EDD appear (Fig. 2. results for compound **II** are comparable). The valence electron density associated with the maximum located inside the triangle described by Zr, the carbon atom itself and Al indicates a 3-center-2-electron-type of bonding between these atoms.

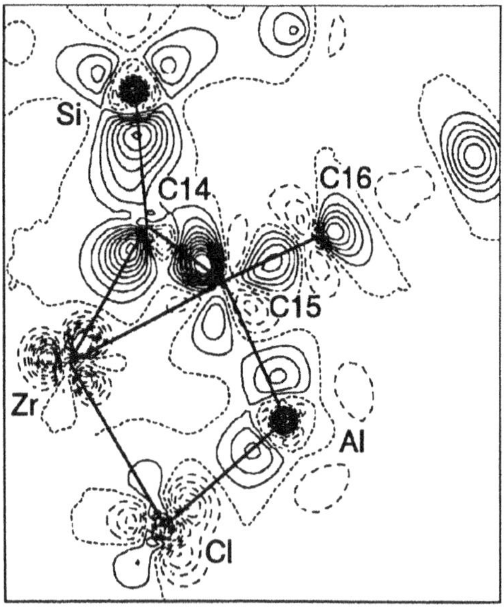

Fig. 2. Static EDD in the central molecular plane of **I**

The extremely short bonding distances Zr-C(14) (**I**) and Zr-C(1) (**II**) at 2.190(2) and 2.167(3)Å (cf. Fig. 1) reflect an additional π-stabilization. In fact, the shape of the positive contour lines in a plane perpendicular to the central molecular plane (Fig. 3) reveals a substantial accumulation of valence electron density in the π-bonding region between Zr and C(1). The charge distribution - as calculated from the monopole populations - is depicted in Fig. 4. This distribution is in complete agreement with results of a recent *ab-initio* study of a model compound of **I** [6]. The metal atoms bear positive charge, whereas the planar tetracoordinate carbon atom is negatively charged.

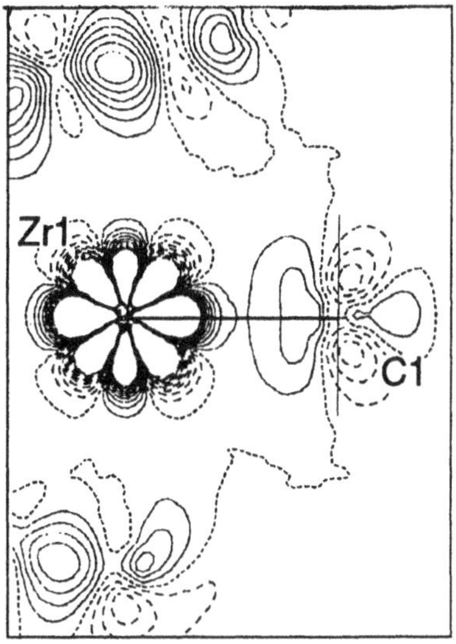

Fig. 3.
Static EDD in the π-bonding region between Zr and C(1)

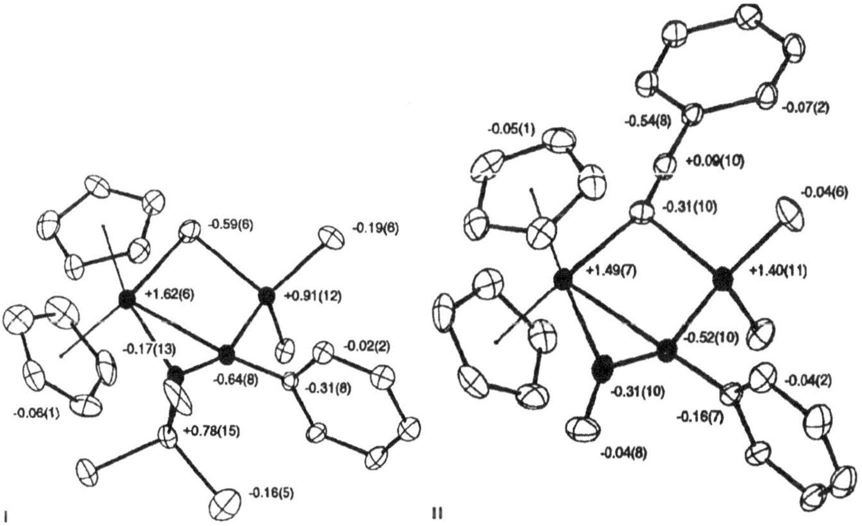

Fig. 4. Partial charges as calculated from the monopole populations

Table 1

	I	II
sumformula	$C_{23}H_{30}AlClSiZr$	$C_{29}H_{29}AlZr$
a b c [Å]	18.138(1) 9.724(2) 27.221(4)	9.693(2) 16.249(4) 16.935(2)
α β γ [°]	90 102.98(1) 90	67.03(1) 84.26(2) 80.91(2)
space group Z F(000)	C2/c 8 2016	P$\bar{1}$ 4 1024
T [K]	100	100
μ_{Mo} [mm^{-1}] absorption correction (min/max)	0.67 1.194/1.197	0.49 1.105/1.263
measured intensities	θ = 1-45.3°, Ψ = 0, 15°: ±h ±k l; θ = 1-16°, Ψ = -30, -10, 10, 30°: ±h ±k± l	θ = 1-37.2°, Ψ = 0, 15°: ±h± k l; θ = 1-20°, Ψ = -45,-30,-15,...45°: ±h ±k l
total number of reflections unique reflections internal consistency	57353 18298 0.040	79282 24933 0.031
STANDARD ATOM REFINEMENT (0 < sin(θ)/λ < 0.85) refined parameters observed reflections R, R_w, e.o.f. (0.7 < sin(θ)/λ < 0.85) refined parameters observed reflections R, R_w, e.o.f. MULTIPOLAR REFINEMENT (0 < sin(θ)/λ < 0.85) refined parameters observed reflections R, Rw, e.o.f.	 334 8720 0.043, 0.038, 2.38 244 3553 0.044, 0.044, 1.09 103 8719 0.036, 0.039, 2.37	 733 19396 0.042, 0.054, 1.68 559 7605 0.047, 0.052, 1.32 118 (223) * 19396 (19396) * 0.039, 0.049, 1.50 (0.039, 0.049, 1.50) *

*Corresponding populations of the multipole functions of the two independent molecules constrained (not constrained)

REFERENCES

[1] G. Erker, R. Zwettler, C. Krüger, R. Noe, S. Werner, *J. Am. Chem. Soc.* **112** (1990) 9620

G. Erker, M. Albrecht, C. Krüger, S. Werner, *Organometallics* **10** (1991) 3791

[2] R. Hoffmann, R. W. Alder, C. F. Wilcox, Jr., *J. Am. Chem. Soc.* **92** (1970) 4992

[3] J. B. Collins, J. D. Dill, E. D. Jemmis, Y. Apeloig, P. v. R. Schleyer, R. Seeger, J. A. Pople, *J. Am. Chem. Soc.* **98** (1976) 5419

[4] N. K. Hansen, P. Coppens, *Acta Cryst.* **A34** (1978) 909

[5] International Tables for X-ray Crystallography IV (1974), Eds. J. A. Ibers, W. C. Hamilton, The Kynoch Press, Birmingham, England, pp 102-147

[6] I. Hyla-Kryspin, R. Gleiter, *unpublished results*

CARBYNE TO CARBENE LIGAND CONVERSION IN DINUCLEAR COMPLEXES

L. BUSETTO, V. ZANOTTI, S. BORDONI, L. CARLUCCI, A. PALAZZI.
Dip. Chimica Fisica ed Inorganica , Universita'
Viale Risorgimento 4, 40136 Bologna, Italy

ABSTRACT. Easily accessible cationic compounds with heteroatoms substituted alkylidyne ligands bridging metal-metal bond provide an opportunity to study the reactions of the μ-CX (X= SR,NR_2) groups with C, N, P, S, nucleophiles. Reactivity patterns are strongly influenced by the nature of both the nucleophile and the alkylidyne ligand. Discussion will focus on the reaction in which the alkylidyne fragment is transformed into a Fischer type carbene ligand bridging the metal-metal bond. Spectroscopic (IR, NMR) and structural investigations together with reactivity studies on these complexes have greatly contributed to better understanding the factors which favour bridging vs. terminal coordination of heteroatom substituted carbene ligands.

Introduction

The objective of this work was to study the chemistry of dinuclear complexes bearing bridging carbene ligands containing π-donor heteroatoms (**A**) which, by contrast with related terminal carbene complexes (**B**) [1] exist in limited numbers when X = OR, SR, but are absent with X = NR_2 or PR_2

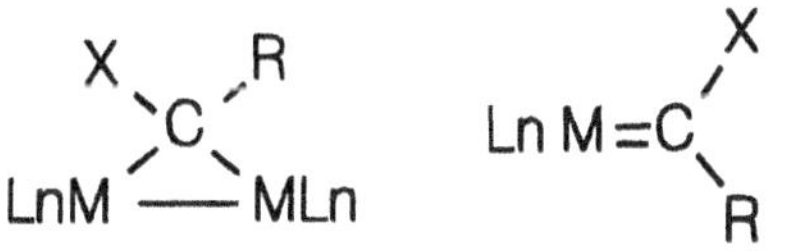

A **B** X = OR, SR, NR_2, PR_2.

In order to achieve our goal the approach to the synthesis of type (**A**) complexes was the simplest one, that is nucleophilic addition at the μ-C atom in cationic thio (**C**) or aminocarbyne (**D**) complexes. Among these classes, the known diiron and diruthenium $[Fe_2(\mu\text{-CSR})(\mu\text{-CO})(CO)_2Cp_2]^+$(**1**) [2], $[M_2(\mu\text{-CNR}_2)(\mu\text{-CO})(CO)_2Cp_2]^+$(**2**) [3,4] have been chosen as starting derivatives.

F. R. Kreißl (ed.), Transition Metal Carbyne Complexes, 137–147.

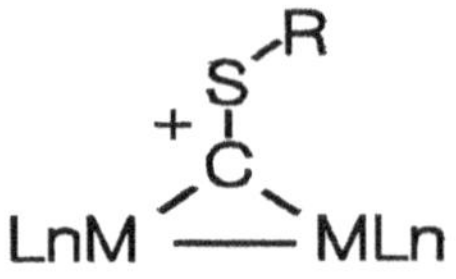

C

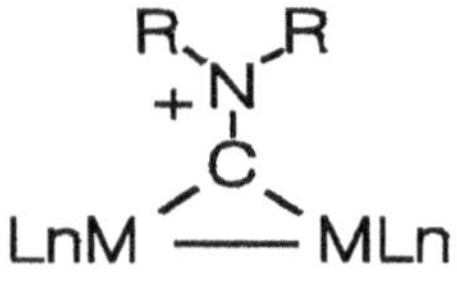

D

Herein we report on their reactions with a variety of nucleophiles and the transformation of the μ-carbyne groups into the Fischer type carbene ligands. Our attention will be focused on the methods that lead to aminocarbenes and to the electronic and steric factors which stabilize this currently unknown class of complexes.

1. Reactions of $[Fe_2(\mu\text{-}CSR)(\mu\text{-}CO)(CO)_2Cp_2]^+$(1) with nucleophiles.

Like other dinuclear complexes containing bridging alkylidyne ligands [5], the chemistry of the Angelici's thiocarbyne $[Fe_2(\mu\text{-}CSR)(\mu\text{-}CO)(CO)_2Cp_2]^+$(**1**) is dominated by the electrophilic character of the μ-carbyne carbon atom. In fact **1** reacts with a variety of nucleophiles such as RS^- (R = Me, PhBz), $PhSe^-$, BH_4^- [6] and CN^- [7] to give the corresponding carbene derivatives **2**, **3** and **4** (Scheme 1).

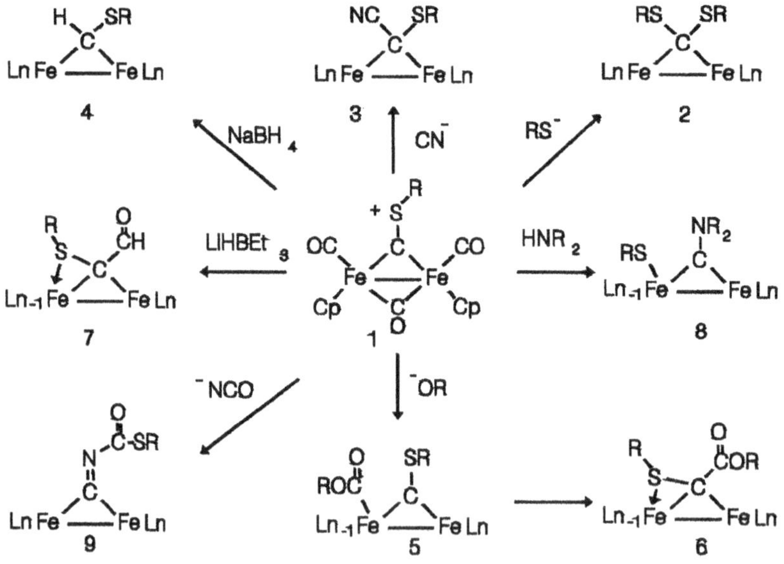

Scheme 1

The attempts to extend the μ-C addition to RO^- and HNR_2 nucleophiles failed to form the expected μ-carbene complexes [6]. As the matter of fact the reactions with a variety of RO^- at -20 °C in THF result in the formation of the neutral thiocarbyne complex **5** containing a Fe-bonded alkoxycarbonyl ligand. This latter derivative decomposes on standing in CH_2Cl_2 solution to form the μ-carbene complex **6** *via* migration of the COOR group to the bridging carbon atom. It is of interest to note that other dinuclear complexes containing a μ-C(R')COOR group have been similarly generated by reacting the cationic alkylidenes $[MPt(\mu\text{-}CPh)(CO)_4(PMe_3)]^+$ (M = Cr, W) with alkoxydes (MeO^-, EtO^-) [8].

A similar example of C-C bond formation *via* "migratory coupling" between thiocarbyne and carbonyl ligands has been observed by reacting **1** with $LiHBEt_3$ at -60 °C in THF [9]. The complex **7** resulting from this reaction, which does not stop at the metal formyl intermediate $[Fe_2(\mu\text{-CSR})(\mu\text{-CO})(CO)(CHO)Cp_2]$, has been characterized, like complex **6** by an X-ray study (Fig.1). The main features of the reaction of **1** with X^- (X = H, RO) consist in the C-X and C-C bonds formation which are key steps in the Fischer-Tropsch synthesis. Therefore the formation of type **5**, **6** and **7** derivatives may represents a model, analogous to that claimed by Keim et al. [10], for explaining the formation of oxygenated species in the catalytic process.

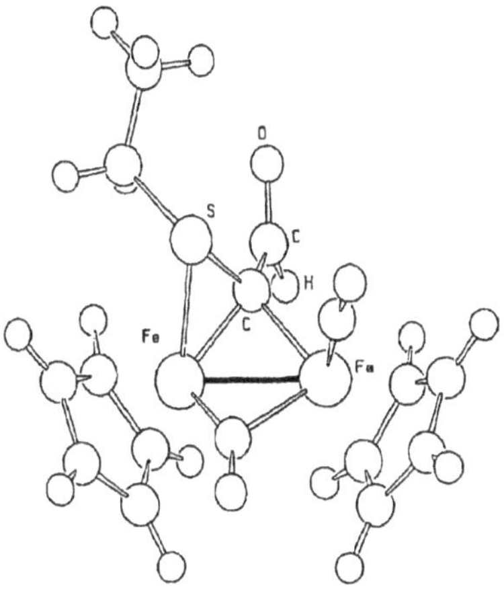

Fig.1 X-ray structure of complex **7**

A further example which indicates that the μ-C addition of a second π-donor heteroatoms, apart from sulfur as in **2**, results in a destabilization of the carbene complex is given by the reaction of **1** with $NHMe_2$. The formation of the spectroscopically characterized complex **8** has been tentatively interpreted as due to the conversion of the aminocarbene intermediate $[Fe_2\{\mu\text{-}C(NMe_2)(SMe)\}(\mu\text{-CO})(CO)_2Cp_2]$ [6] through migration of the SMe group from the bridging carbon to a terminal position. Similar rearrangements have been also observed on type **2** derivatives.

The tendency of N-bearing nucleophiles to add at the μ-C carbon and to cause C-S bond rupture is confirmed by the reaction of **1** with cyanate ion [11]. The NCO^- inserts into the C-S bond of **1** to give the isocyanide bridged complex **9**. The mechanism of this unprecedented reaction could imply a 1-2 cycloaddition of NCO^- at the C-S bond trough attack of the cyanate ion at the electrophilic μ-thiocarbyne carbon. The thus formed spirocyclic intermediate:

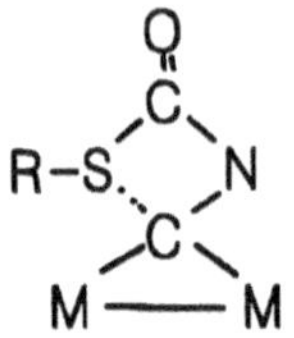

undergoes ring opening *via* C-S bond breaking to yield the functionalized isocyanide **9.**

The NCO^- insertion reaction has been successfully applied to convert the CSMe ligand of both the *cis* and *trans*-$[Fe_2(\mu\text{-CSMe})(\mu\text{-CS})(CO)_2Cp_2]^+$ (**10**) isomers into the corresponding μ-acylisocyanide group. All the complexes prepared have been spectroscopically characterized and their most informative ^{13}C NMR data are given in the following Scheme 2:

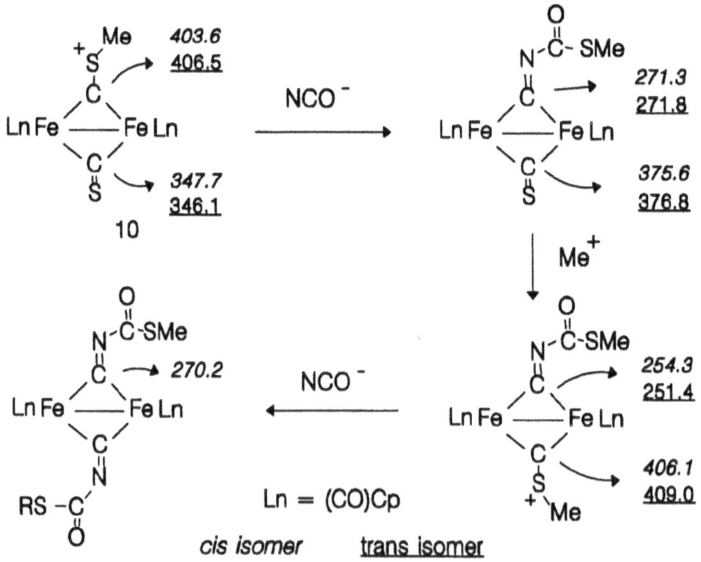

Scheme 2.

2. From Dinuclear μ-Aminocarbyne to Aminocarbene Complexes and Vice versa.

The nitrogen atoms of the isocyanides ligands CNR in $M_2(CO)_3(CNR)Cp_2$ (M = Fe, Ru) has been proved to be very susceptible to alkylation reactions [12]. The cationic products contain the CN(R')R group invariably located in the bridging position. The carbyne character of these ligands has been evidenced from their μ-C nucleophilic addition reactions of H^- and CN^- which may form either bridging or terminal aminocarbene derivatives. The results of our studies, collected in Scheme 3, can be summarized as follows:

i) the $C(X)NR_2$ ligands are bridging when the X substituent at the μ-C is the CN group.

ii) when X = H, the same ligands are terminally coordinated, but they become bridging if the electron-withdrawing C(O)SR group bounds the nitrogen atom [11].

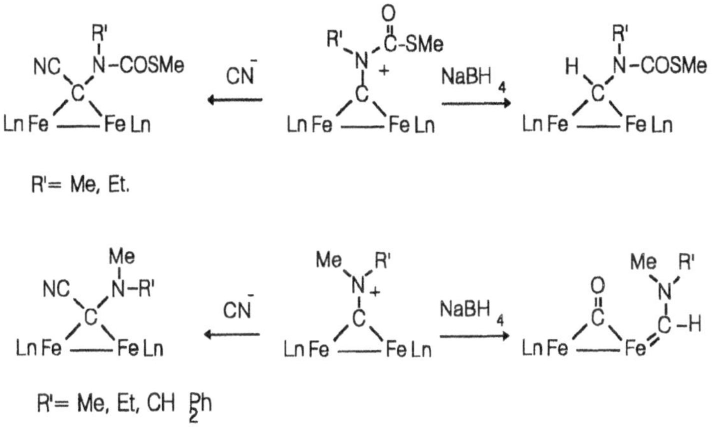

Scheme 3

Recently Adams in reviewing [13] the metal cluster complexes containing heteroatom-substituted carbene ligands, attributed the absence of bridging aminocarbene derivatives (**A**) to the destabilizing effect of the strong π-donation from the lone pair of the nitrogen atom to the empty p-orbital on the carbene carbon atom.

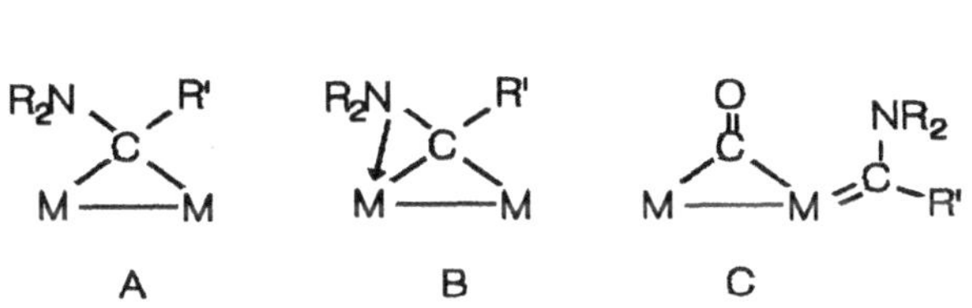

This conclusion was supported by the existence of both type (**B**) [14] and (**C**) [15] coordination modes of the aminocarbene ligands in cluster chemistry. The results depicted on scheme 3 are in agreement with this view. Therefore the effect of the CN^- may be that of reinforcing backbonding from the metal to the empty p-orbital on the carbene carbon, resulting in a partial subtraction of this orbital from the π-interaction with the nitrogen atom. In the absence of this effect (H^- case) the destabilization due to reduced M_2C π-bonding is sufficiently large to cause conversion from bridging to terminally coordinated ligands (**C**). The crystal structures of the two μ-aminocarbene complexes Fe_2\{μ-C(CN)N(Me)C(O)SMe\}(μ-CO)$(CO)_2Cp_2$ (**11**) [11] and Fe_2\{μ-C(CN)NHPh\}(μ-CO)$(CO)_2Cp_2$ (**12**) [16] (see next section) reinforce our view . In both cases the appropriate C-N bond lengths indicate that the stronger interaction are those involving the π-orbitals of the C(O)SR or that of the phenyl ring substituents rather than the μ-C-N distances (Fig 2).

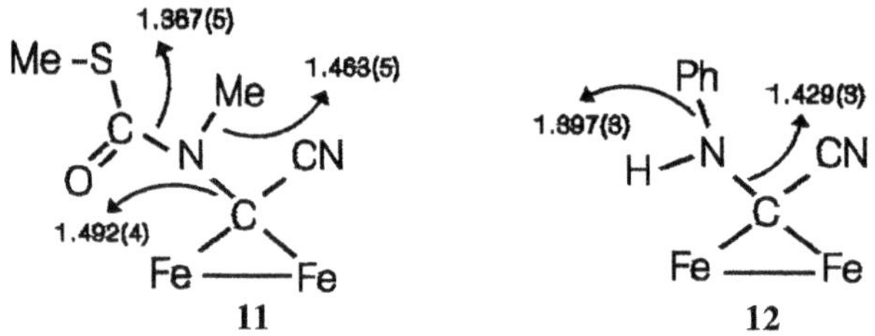

11 **12**

Figure 2.

The aminocarbyne to carbene conversion *via* H^- or CN^- μ-C addition, has been successfully extended to diruthenium complexes (Scheme 4). Interestingly, the complex (**13**) has been shown to afford bridging carbene derivatives in reacting both with CN^- or H^- [17].

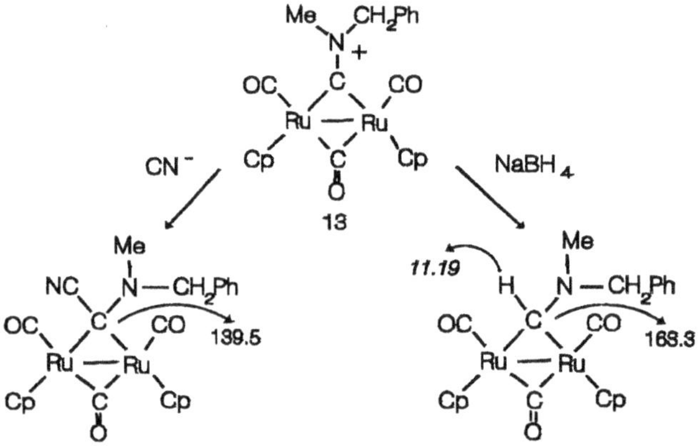

Scheme 4

The differences between the analogous iron and ruthenium complexes can be attributed to the higher basicity of the Ru as compared to the Fe atom [18]. Thus the stronger π-interaction (metal to carbene) prevents the expected bridging to terminal migration in the μ-C(H)NCH_2Ph case. The high stability of the aminocarbene of ruthenium is evidenced by the stepwise synthesis of the (bis) μ-aminocarbene (Scheme 5).

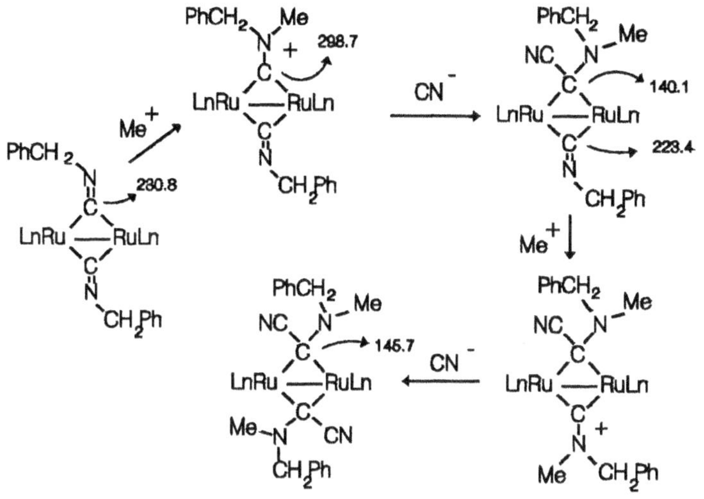

Scheme 5

In the previous section we have briefly reported on the tendency of the nitrogen bearing nucleophiles, such as NHR_2 or NCO^-, to cause the C-S bond breaking in the thiocarbyne complex **1**. The formation of **8** and **9** derivatives has been supposed to occur via nitrogen containing bridging carbene intermediates which can be considered as the analogous of the stable μ-aminocarbene complexes. The tendency to form stronger π interactions between the μ-carbon and the nitrogen atom, is confirmed by the reaction of the diiron aminocarbenes with alkylating or Lewis acid reagent as well as under photochemical condition.

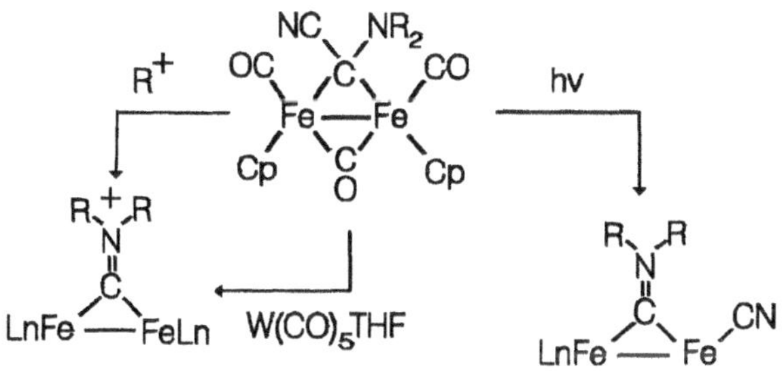

14

These reactions promote the C-C bond breaking and the formation of neutral or cationic aminocarbyne precursors of the aminocarbene complexes. Interestingly the reaction with

$W(CO)_5(THF)$ forms the cationic complex **14** in which the counter anion is the CN bridging derivative $[(CO)_5WCNW(CO)_5]^-$.

3. The " pseudoalkylidyne" μ-carbon atom of the sulfonium complex $[Fe_2\{\mu\text{-}C(CN)SMe_2\}(\mu\text{-}CO)(CO)_2Cp_2]SO_3CF_3$ (15).

The carbyne to carbene conversion routes presented in the previous sections are obviously applicable to the synthesis of thio and aminocarbene derivatives. Despite this limitation, it has been well documented that the electronwithdrawing CN group is able to stabilize the dimetallacyclopropane ring even when π-donating N or S heteroatoms are present on the μ-C atom. Thus our recently synthesized sulfonium salt complex $[Fe_2\{\mu\text{-}C(CN)SMe_2\}(\mu\text{-}CO)(CO)_2Cp_2]SO_3CF_3$ (**15**) [19] containing both the CN group and the nucleofuge SMe_2 molecule as substituents at the μ-C, was supposed to be a suitable starting material for obtaining a great variety of heteroatom substituted Fischer type diiron bridging carbene complexes. Indeed, it has been possible to prepare a great variety of alkoxy, amino and phosphino substituted carbene derivatives [16, 20, 21]. Scheme 6 collects only the reactions with group 15 nucleophiles which give the opportunity to complete the study on the stability of the heteroatoms containing carbene products.

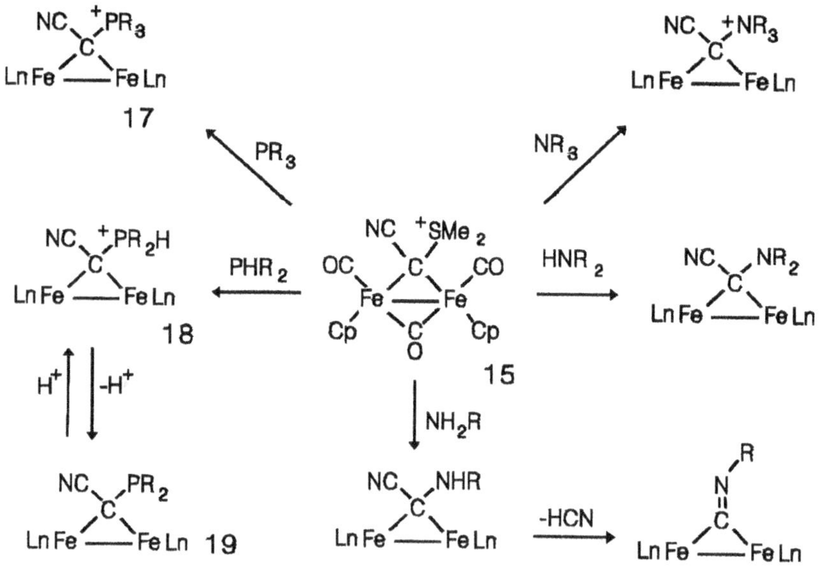

Scheme 6.

Tertiary, secondary and primary amines add at the SMe_2-protected μ-C atom to give a variety of ammonium salt, aminocarbene and isocyanide complexes. These latter are formed upon elimination of HCN from the aminocarbene $Fe_2\{\mu\text{-}C(CN)NHR\}(\mu\text{-}CO)(CO)_2Cp_2]$

intermediate. In the reaction with a weakly basic aniline, the supposed intermediate has been isolated and structurally characterized (see above). The reactions with secondary amines have revealed the importance of the steric hindrance of the nitrogen substituents in determining the position of the aminocarbene ligands $C(CN)NR_2$. As shown in Scheme 7, when R = Me the ligand is located in bridging position (Fig a); by increasing the steric hindrance to the Et group the IR analysis of the reaction products indicate the presence of an equilibrium mixture between the terminal and bridging complexes (Fig b),. Finally with the bulkiest isopropyl group the carbene ligand becomes terminally bonded (Fig c) both in solution and in the solid state as demonstrated by IR and X-ray structural analyses.

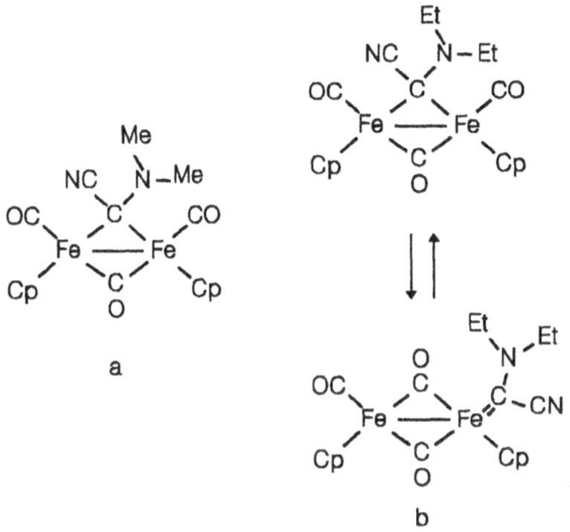

Scheme 7.

The reaction of the sulfonium salt **15** with ethylendiamine, forms the terminal aminocarbene complex **16** analogous to those obtained from $Fe_2(CO)_4Cp_2$ and electron rich olefins [22]. Its formation suggests that the presence of two nitrogen π-donating atoms favours the terminal over the expected bridging coordination position.

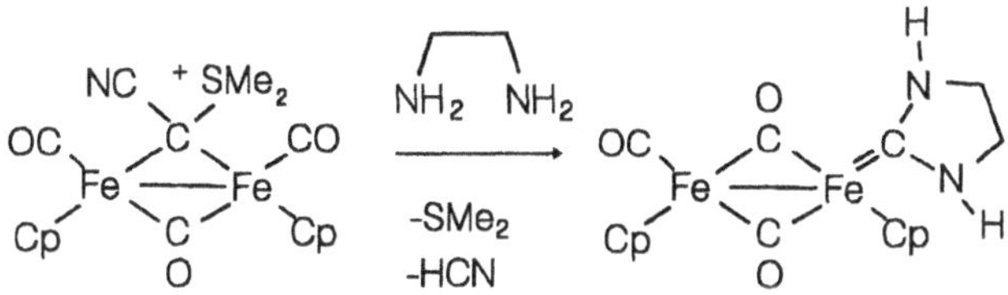

16

It has been reported that the addition of PR_3 to the alkylidyne carbon atom of the cations $[Fe_2(\mu\text{-}CX)(\mu\text{-}CO)(CO)_2Cp_2]^+$ (X=H, Me) [23] and $[MPt(\mu\text{-}CC_6H_4Me\text{-}4)(CO)_2(PR_3)Cp]^+$ (M = Mn, Re) [24] affords to the corresponding phosphonium ions. Likewise, the strong electrophilic nature of the pseudo-alkylidyne μ-carbon atom of the sulfonium salt complex permits the preparation of the phosphonium salts **17** and **18** (Scheme 6). When the reactions with the phosphines PHR_2 are performed in the presence of NEt_3 the first examples of phospinocarbenes $Fe_2[\mu\text{-}C(CN)PR_2](\mu\text{-}CO)(CO)_2Cp_2$ **19** [21] have been isolated.

Acknowledgments

Financial support from the MURST and CNR is gratefully acknowledged. The authors thanks Profs. Vincenzo G. Albano and Dario Braga who performed all the structural studies of the carbyne and carbene chemistry described.

References

1 K.H. Dotz, H. Fischer, P. Hofmann, F.R. Kreissl, U. Schubert, K. Weiss, *Transition Metal Carbene Complexes*, Ed. K.H.Dotz, Verlag Chemie (1983)
2 M.H.Quick, R.J. Angelici, *Inorg.Chem,* **20** (1981) 1123.
3 S. Willis, A.R. Manning, *J.Chem.Soc., Dalton Trans*, (1980) 186.
4 J.A.S. Howell, A.J. Rowan, *J.Chem.Soc., Dalton Trans*, (1980) 503.
5 E.D. Jemmis, B.V. Prasat, *Organometallics*, **11** (1992) 2528.
6 N.C. Schroeder, R. Funchess, R.A. Jacobson, R.J. Angelici, *Organometallics*, **8** (1989) 521.
7 L. Busetto, S. Bordoni, V. Zanotti, V.G. Albano, D. Braga, *Gazz.Chim.Ital.*, **118** (1988) 667.
8 J.A.K. Howard, J.C. Jeffery, M. Laguna, R. Navarro, F.G.A. Stone, *J.Chem.Soc., Dalton Trans*,(1981) 715.
9 L. Busetto, V. Zanotti, L. Norfo, A. Palazzi, V.G. Albano, D. Braga, *Organometallics*, in press.
10 J. Hackenbruch, W. Keim, M. Roper, H.Strutz, *J.Mol.Catalysis,* **26** (1984) 129.
11 L. Busetto, L. Carlucci, V. Zanotti, V.G. Albano, D. Braga, *J.Chem.Soc., Dalton Trans.* (1990) 243.
12 S.Willis, R.A. Manning, F.S. Stephens, *J.Chem.Soc., Dalton Trans.*,(1979) 23.
13 R.D. Adams, *Chem.Rev.*, **89** (1989) 1703.
14 J.H. Davis,Jr, C.M. Lukehart, L. Sacksteder, *Organometallics,* **6** (1987) 50.
15 R.D. Adams, J.E. Babin, *Organometallics,* **6** (1987) 1364.
16 V.G. Albano, S. Bordoni, D. Braga, L. Busetto, A. Palazzi, V. Zanotti, *Angew.Chem., Int.Ed.Engl.,* **7** (1991) 847.
17 L. Busetto, L. Carlucci, V. Zanotti, M. Monari, V.G. Albano, *J.Organomet.Chem.,* in press.
18 G. Cerichelli, G. Illuminati, G. Ortaggi, A.M. Giuliani, *J.Organomet.Chem.*, **127** (1977) 357.
19 L. Busetto, V. Zanotti, S. Bordoni, L. Carlucci, V.G. Albano, D. Braga, *J.Chem.Soc., Dalton Trans.*, (1990) 243.
20 L. Busetto, M.C. Cassani, V. Zanotti, V.G. Albano, D. Braga, *J.Organomet.Chem.*, **415** (1991) 395

21 L. Busetto, L. Carlucci, V. Zanotti, V.G. Albano, M. Monari, *Chem.Ber.*, **125** (1992) 1125.
22 M.F. Lappert, P.L. Pye, *J.Chem.Soc., Dalton Trans.*, (1977) 2172.
23 S.C. Kao, P.Y. Lu, R. Pettit, *Organometallics*, **1** (1982) 911.
24 J.C. Jeffery, R. Navarro, H. Razay, F.G.A. Stone, *J.Chem.Soc., Dalton Trans.*, (1981) 2471.

SYNTHESES AND REACTIONS OF HETERODINUCLEAR ALKOXYCARBYNE COMPLEXES

WILLIAM H. HERSH
Department of Chemistry and Biochemistry
Queens College of the City University of New York
Flushing, NY 11367-0904
USA

ABSTRACT. The alkoxycarbyne ligand is an isomer of the more common acyl and alkyl carbonyl moieties. We describe here the synthesis of two heterodinuclear methoxycarbyne complexes, and their decomposition to mononuclear methyl complexes. Reaction of $CpFe(CO)_2^- Na^+$ with $MeCpMn(CO)_2(THF)$ or of $MeCpFe(CO)_2^-$ with CpCr(CO)(NO)THF gives the heterodinuclear monoanions $[Cp(CO)Fe(\mu\text{-}CO)_2Mn(CO)MeCp]^- Na^+$ and $[MeCp(CO)Fe(\mu\text{-}CO)_2Cr(NO)Cp]^- Na^+$, respectively, in good yield. The X-ray crystal structure of the $Ph_3PCH_3^+$ salt of the FeCr anion confirms both the structure and the location of the NO ligand. Alkylation of the FeMn anion with $CH_3OSO_2CF_3$ and of the FeCr anion with $(CH_3)_3O^+ BF_4^-$ gives the methoxycarbynes **1** and **2**, respectively. Carbyne **1** is obtained in

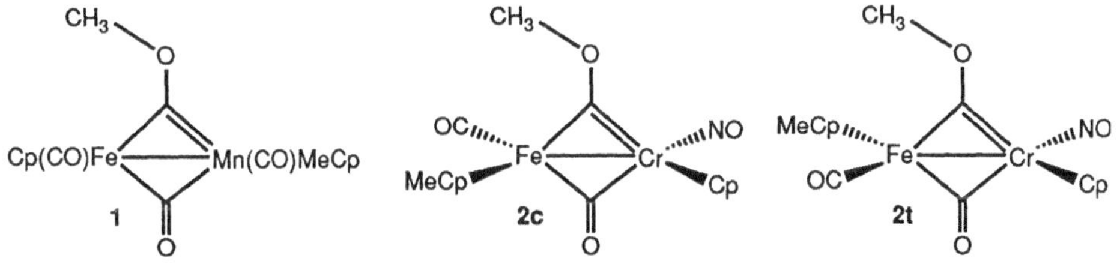

62% yield as a mixture of cis and trans isomers that interconvert rapidly at room temperature, while **2** is obtained as chromatographically separable cis (**2c**) and trans (**2t**) isomers, each in 10% yield.

First-order decomposition of both **1** and **2t** is observed in the absence and presence of PPh_3; kinetic analysis of the reactions of **2c** is complicated by partial conversion to **2t**. For both **1** and **2t**, parallel reactions occur, one that is independent of $[PPh_3]$ (k(75 °C) = 1×10^{-4} s^{-1} and k(50 °C) = 6.5×10^{-4} s^{-1} for **1** and **2t** respectively) and one that is first order in PPh_3 (k(75 °C) = 4.6×10^{-4} $M^{-1}s^{-1}$ and k(50 °C) = 6.6×10^{-4} $M^{-1}s^{-1}$ for **1** and **2t** respectively). Products in the absence of PPh_3 are $CpFe(CO)_2CH_3$, $[CpFe(CO)_2]_2$, $[CpFe(CO)]_4$, and $MeCpMn(CO)_3$ for **1**, and the analogous MeCp iron mono and dinuclear products and $CpCr(CO)_2NO$ for **2t**. In the presence of PPh_3, cleaner reactions to give $CpFe(CO)(PPh_3)CH_3$ as well as $MeCpMn(CO)_3$ occurred for **1**, while **2t** gave mixtures of $MeCpFe(CO)(L)CH_3$ and

F. R. Kreißl (ed.), Transition Metal Carbyne Complexes, 149–150.

$CpCr(CO)(NO)L$ (L = CO, PPh_3). The product of the bimolecular step was in both cases the $Ph_3PCH_3^+$ salt of the initial heterodinuclear monoanions, formed by S_N2 dealkylation of the carbynes. For **1**, the yields of the soluble mononuclear products decreased linearly with increasing [PPh_3] as expected, but for **2t** the yields dropped much faster than expected. For instance, for [PPh_3] = 0.06M, the rate was ~10% faster than in the absence of phosphine, but the yield of soluble Cr and Fe products was ~40% rather than the 90% expected if 10% of the carbyne was diverted to anion by the bimolecular pathway. The rate and yield data can be quantitatively fit to a mechanism that incorporates an intermediate that is trapped by PPh_3 - to give more [MeCp(CO)Fe(μ-CO)$_2$Cr(NO)Cp]$^-$ $Ph_3PCH_3^+$ - competitively with continuing on to mononuclear products, but since the rate does not level off at high [PPh_3], it appears that the intermediate irreversibly goes on to products regardless of phosphine concentration. A possible structure for the intermediate would be an unbridged methoxycarbyne, comparable to the intermediate that is presumably involved in cis/trans isomerization according to the widely-accepted Cotton mechanism for $[CpFe(CO)_2]_2$ cis/trans isomerization.

The apparent methyl migration reaction of **1** has been further probed to test for crossover. Reaction of **1** and its MeCp/CD_3 analogue **MeCp(CO)Fe(μ-COCD$_3$)(μ-CO)Mn(CO)MeCp** results in complete methyl scrambling in the products, giving $CpFe(CO)(PPh_3)CH_3$ (**3**) and $MeCpFe(CO)(PPh_3)CH_3$ (**3a**). The control experiment between **3** and **3a-CD$_3$** results in only very slow methyl exchange, but since exchange does eventually occur, the formation of an intermediate that might catalyze the exchange is possible. The "control crossover" between **1** and **3a-CD$_3$** confirms that possibility, since complete exchange was observed again. Thus, while we cannot prove that methyl migration in **1** is intramolecular, we can prove that it will occur among the products, mediated by some intermediate formed during thermal decomposition of **1**. The obvious choice of this intermediate is the 16-electron fragment $CpFe(CO)CH_3$, but kinetic experiments with phosphine exchange in **3** show that this choice is wrong. That is, **3** undergoes phosphine substitution by PPh_2Me at a rate *independent* of phosphine concentration, sufficient evidence for rate determining formation of $CpFe(CO)CH_3$. The rate constant at 65 °C is 6×10^{-5} s^{-1}, while the rate constant for decomposition of **1** at 65°C is essentially the same, ~4.5×10^{-5} s^{-1}. Thus, equivalent methyl scrambling would occur in both the crossover reaction of **3** and **3a-CD$_3$** and the reactions of **1** if $CpFe(CO)CH_3$ were responsible for the methyl exchange . We conclude that it is not responsible. A speculative proposal is that unbridging of the methoxycarbyne, proposed for cis/trans isomerization as well as for the structure of the Cr intermediate revealed by the product yield analysis for the reaction of **2t**, occurs and is followed by metal-metal bond cleavage to give the 16-electron intermediate $CpFe{\equiv}C\text{-}O\text{-}CH_3$, which either mediates methyl exchange itself, or is trapped by PPh_3 to give $Cp(PPh_3)Fe{\equiv}C\text{-}O\text{-}CH_3$ which could still be sufficiently reactive to initiate methyl exchange.

Acknowledgment. Financial support for this work from the National Science Foundation (CHE-9096105) is gratefully acknowledged. Arnold Rheingold (University of Delaware) carried out the X-ray structure of the FeCr anion, and Raymond Fong and Bing Wang carried out the FeMn and FeCr chemistry, respectively.

MULTICENTER LIGAND TRANSFORMATIONS OF TETRAMETHYLTHIOUREA ON RUTHENIUM CLUSTERS

Georg Süss-Fink
Institut de Chimie
Université de Neuchâtel
Avenue de Bellevaux 51
CH-2000 Neuchâtel
Switzerland

The reaction of tetramethylthiourea, $(Me_2N)_2CS$, with ruthenium carbonyl, $Ru_3(CO)_{12}$, has been studied under various conditions: A large variety of tri-, tetra-, penta-, and hexanuclear ruthenium clusters containing fragments of the thioureato molecule have been isolated and structurally characterized.

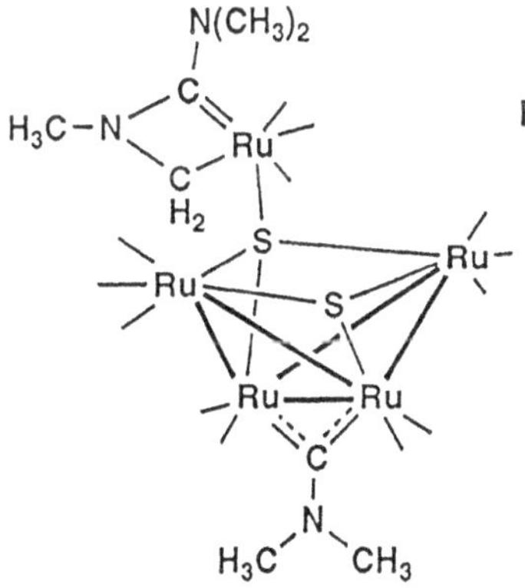

$Ru_4(CO)_{10}(\mu_2\text{-}CNMe_2)(\mu_3\text{-}S)[\mu_3\text{-}SRu(CO)_3(\eta^2\text{-}CH_2NMeCNMe_2)]$

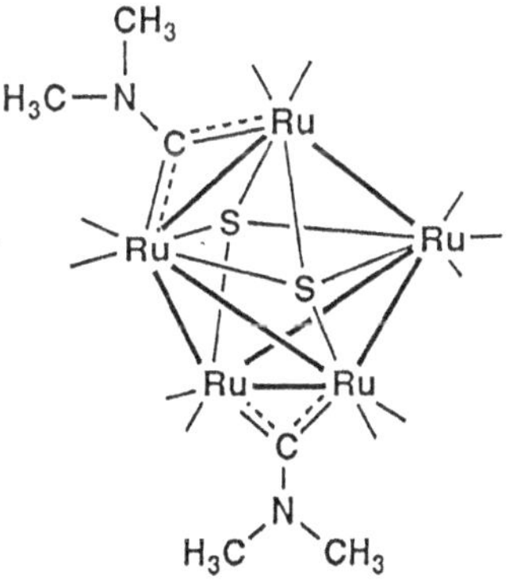

$Ru_5(CO)_{11}(\mu_2\text{-}CNMe_2)_2(\mu_4\text{-}S)_2$

F. R. Kreißl (ed.), Transition Metal Carbyne Complexes, 151–153.

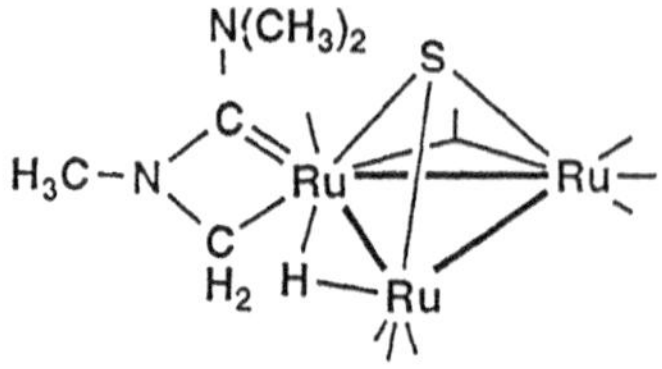

$(\mu_2\text{-H})Ru_3(CO)_7(\mu\text{-CO})(\mu_3\text{-S})(\eta^2\text{-}CH_2NMeCNMe_2)$

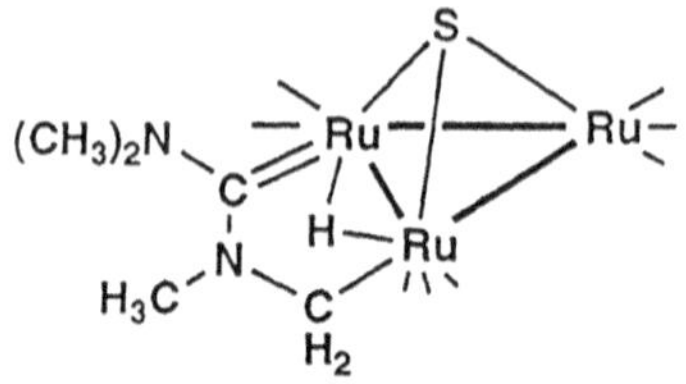

$(\mu_2\text{-H})Ru_3(CO)_8(\mu_3\text{-S})(\mu_2\text{-}\eta^2\text{-}CH_2NMeCNMe_2)$

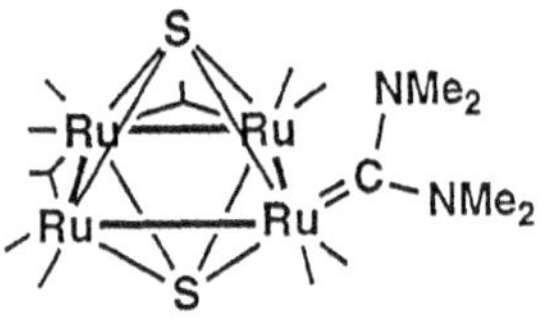

$Ru_4(CO)_8(\mu_2\text{-CO})_2(\mu_4\text{-S})_2[C(NMe_2)_2]$

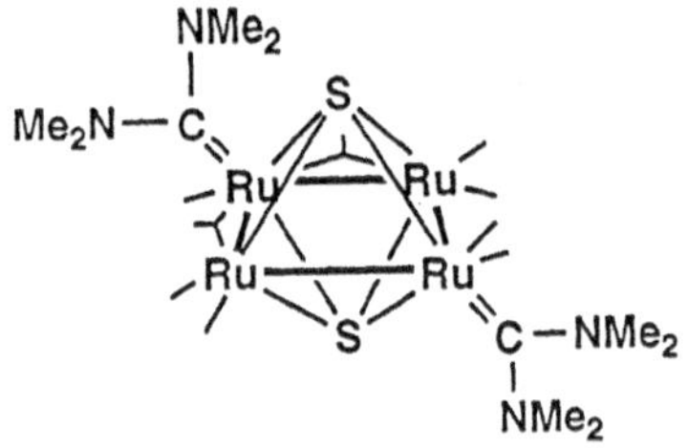

$Ru_4(CO)_7(\mu_2\text{-CO})_2(\mu_4\text{-S})_2[C(NMe_2)_2]_2$

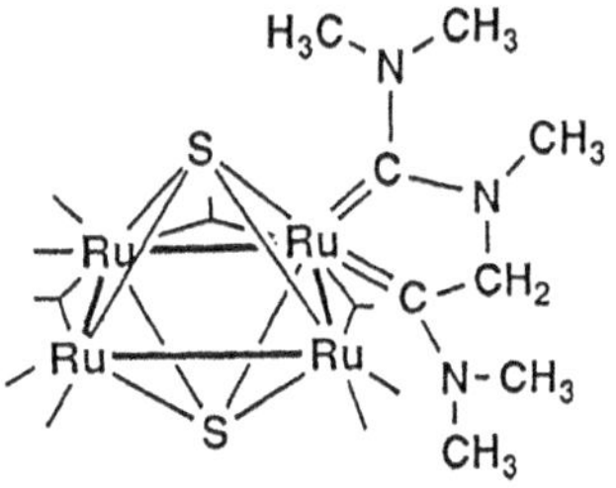

$Ru_4(CO)_6(\mu_2\text{-CO})_3(\mu_4\text{-S})_2[\eta^2\text{-}C(NMe_2)CH_2NMeC(NMe_2)]$

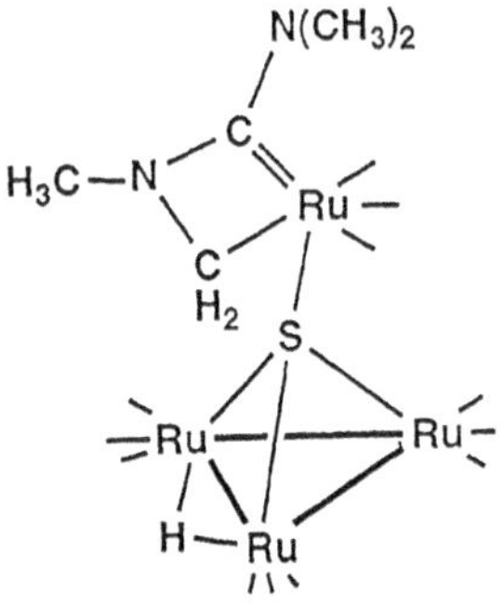

$(\mu_2\text{-H})Ru_3(CO)_9[\mu_3\text{-}SRu(CO)_3(\eta^2\text{-}CH_2NMeCNMe_2)]$

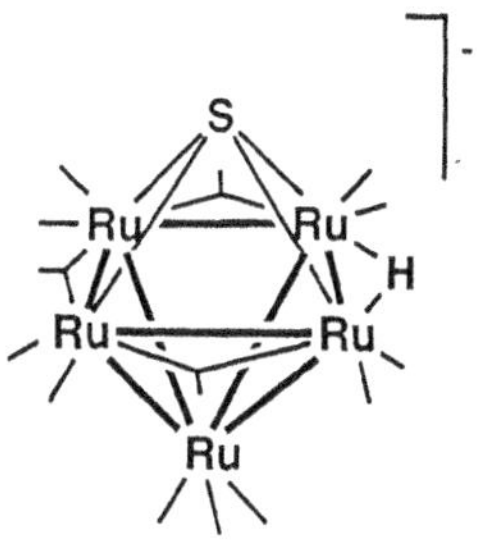

$[(Me_2N)_2CH][(\mu_2\text{-}H)Ru_5(CO)_{11}(\mu_2\text{-}CO)_3(\mu_4\text{-}S)]$

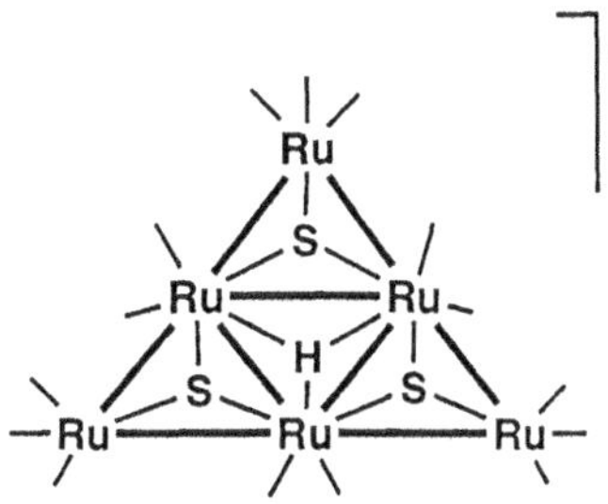

$[(Me_2N)_2CH][(\mu_3\text{-}H)Ru_6(CO)_{15}(\mu_3\text{-}S)_3]$

The fragmentation of the thioureato molecule into diaminocarbene and sulfur ligands, as well as consecutive transformations such as C-H activation and C-C coupling reactions in a multicenter coordination sphere are discussed.

CARBYNE COMPLEXES OF RUTHENIUM AND OSMIUM

W. R. ROPER
Department of Chemistry
The University of Auckland
Private Bag 92019, Auckland
New Zealand

ABSTRACT. The dichlorocarbene complexes $MCl_2(=CCl_2)(CO)(PPh_3)_2$ undergo a reaction with two equivalents of aryl lithium forming the five coordinate complexes $MCl(\equiv CR)(CO)(PPh_3)_2$, (M=Ru, R=Ph, C_6H_4-4-OMe; M=Os, R=Ph, *p*-tolyl, C_6H_4-4-NMe_2). This contribution will survey the many reactions of these carbyne complexes including (i) formation of cations by replacement of chloride with CO or CNR, (ii) protonation at the carbyne carbon (in some cases under photolytic conditions), (iii) coordination of the M≡C bond to Cu^I, Ag^I, Au^I, (iv) reaction with chalcogens yielding dihapto-chalcoacyl complexes, (v) addition of Cl_2 across M≡C, (vi) oxidation to octahedral carbyne complexes with I_2, (vii) oxidation to octahedral carbyne complexes with O_2, (viii) attack by nucleophiles at either the carbyne carbon or at a remote site depending on oxidation state. Structural data for a number of carbyne complexes and derived complexes will be presented and discussed. Reactivity as a function of oxidation state will also be considered. Finally, other possible methods for the synthesis of carbyne complexes of ruthenium and osmium will be presented. These include alkylation of chalcocarbonyl ligands in low oxidation state complexes and nitrogen loss from metallated diazoalkanes, $L_nMC(R)N_2$.

1. Introduction

Terminal carbyne complexes, $L_nM\equiv CR$, compounds with formal metal-carbon triple bonds, date from the 1973 report by E. O. Fischer et al. [1]. Development of this area since that time has produced examples involving transition metals from Group V-Group VIII and with a wide range of substituents on the carbyne carbon atom including H, alkyl, aryl, silyl, siloxy, amino, thioalkyl, selenoalkyl, telluroalkyl, and chloride. Supporting ligands include the common "organometallic" ligands carbonyl, isocyanide, cyclopentadienyl, pyrazolylborate, phosphine etc., as well as bulky alkyl, halide, alkoxide, amide, and thiolate [2]. This development has been driven by interest in the novel metal-carbon triple bond and the obvious implications for organic synthesis and for catalysis, particularly alkyne metathesis. Yet the total number of stable compounds involving metal-carbon triple bonds remains quite small and the synthetic methods available are limited in scope. This article focuses on carbyne complexes of just two metals, ruthenium and osmium. Special features associated with these compounds are (i) ruthenium and osmium represent the "right-hand boundary" of the Periodic Table map of stable carbyne complexes (a reported iridium example [3] is in equilibrium with

F. R. Kreißl (ed.), Transition Metal Carbyne Complexes, 155–168.

a hydrido,vinylidene complex and a nickel example [4] is thermally unstable in solution), (ii) the good stability permits structure determination by X-ray crystallography and an extensive examination of the reaction chemistry associated with these compounds, (iii) there are interesting parallels between the carbyne complexes and the corresponding nitrosyl-containing complexes, (iv) the compounds fall into two structural classes, five coordinate and six coordinate with structurally characterised examples of each class being either neutral or positively charged.

2. Synthesis of five coordinate carbyne complexes

Most transition metal carbyne complexes are prepared by transformation of a metal-bound carbene ligand into a carbyne ligand. The ruthenium and osmium compounds to be described are no exception in being derived from precursors with dichlorocarbene ligands. However, the reaction is unusual in involving also a reduction step. The required dichlorocarbene complexes are derived from the quite straightforward reaction between $MHCl(CO)(PPh_3)_3$ (M=Ru or Os) and $Hg(CCl_3)_2$ [5,6]. The products are stable, orange, crystalline materials formed in high yield.

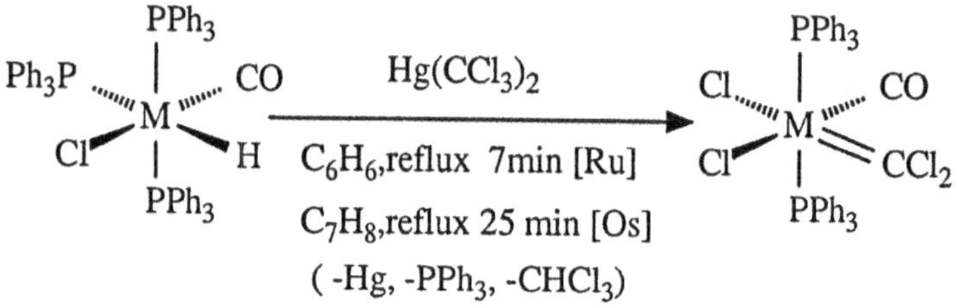

$RuCl_2(=CCl_2)(CO)(PPh_3)_2$ is formed in best yield when excess PPh_3 is present during the preparation. X-ray crystallography confirms the geometry depicted but disorder problems prevent the determination of reliable bond distances. The related iridium compound, $IrCl_3(=CCl_2)(PPh_3)_2$ [7] provided a good structure and revealed an iridium-carbon double bond of length 1.872(7)Å and "normal" carbon-chlorine distances of 1.721(5)Å. The chemistry of these compounds and of other dihalocarbene complexes has been reviewed [8]. Reaction between $MCl_2(=CCl_2)(CO)(PPh_3)_2$ and certain aryl lithium reagents leads directly to five coordinate carbyne complexes of ruthenium and osmium.

2.1 NEUTRAL FIVE COORDINATE COMPLEXES

Following Scheme 1, $MCl_2(=CCl_2)(CO)(PPh_3)_2$ is dissolved in a minimum of tetrahydrofuran and cooled to a temperature between -40° and -78°C. An amount of two equivalents of the lithium reagent is added dropwise over a period of a few minutes during which time the colour turns from orange to green and green material begins crystallising from the solution. Addition of diethylether completes crystallisation. Yields are in the range 80-90%. All the products (see Table 1. for a listing of products characterised from this reaction) have strong ν(CO) in the IR spectrum at very low frequencies and strong bands indicative of the presence of the carbyne ligand at ca. 1590 cm^{-1} and at 1330-1400 cm^{-1}(ν(M≡C) coupled with skeletal deformation modes of the aryl ring substituent [9]). For comparison, Table 1 also includes ν(CO) data for the corresponding nitrosyl complexes. A single crystal X-ray diffraction study of Os(≡C-*p*-tolyl)Cl(CO)$(PPh_3)_2$ confirmed the essentially trigonal bipyramidal geometry with an osmium-carbon triple bond distance of 1.78(2)Å [10]. The angle between the Os-Cl bond and the Os-carbyne bond is very large at 133° and the Os-Cl distance long at 2.507(4)Å. The chloride ligand proves to be labile and the next section deals with replacement of chloride by neutral ligands so forming cationic carbyne complexes.

Scheme 1

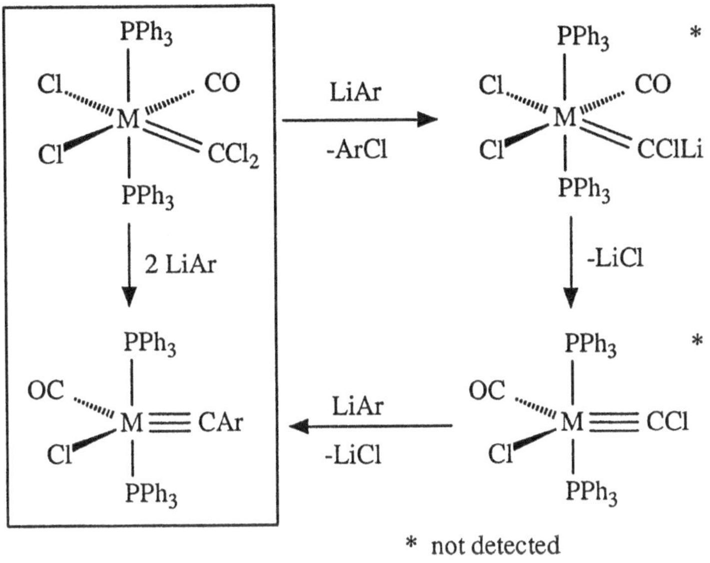

TABLE 1 IR Data[a] for 5-coordinate Carbyne Complexes[b]

Compound	ν(CO)	ν(carbyne)
Ru(≡CPh)Cl(CO)L_2	1875	1584 1328
Ru(≡CC_6H_4-OMe)Cl(CO)L_2	1880	1590 1340
Os(≡CPh)Cl(CO)L_2	1858	1584 1358
Os(≡C*p*-tolyl)Cl(CO)L_2	1864	1598 1359
Os(≡CC_6H_4-NMe_2)Cl(CO)L_2	1855	1595 1380
[Ru(≡CPh)$(CO)_2L_2]^+$	2020, 1960	1580 1348
[Ru(≡CC_6H_4-OMe)$(CO)_2L_2]^+$	2020, 1950	1585 1370
[Os(≡CPh)$(CO)_2L_2]^+$	2015, 1945	1582 1370
[Os(≡CC_6H_4NMe_2)$(CO)_2L_2]^+$	2015, 1970	1598 1405
[Os(≡C*p*-tolyl)$(CO)_2L_2]^+$	2010, 1944	1592 1375
[Os(≡C*p*-tolyl)(CO)L'$L_2]^+$	1934	1593 1374
Os(NO)Cl(CO)L_2	1910	
[Os(NO)$(CO)_2L_2]^+$	2055, 2000	

[a] in cm^{-1}, from references [11] and [12] [b] L=PPh_3, L'=CN*p*-tolyl

2.2 CATIONIC FIVE COORDINATE COMPLEXES

Treatment of dichloromethane/ethanol solutions of $Os(\equiv CR)Cl(CO)(PPh_3)_2$ containing CO with $AgClO_4$ leads to the formation of five coordinate cationic carbyne complexes.

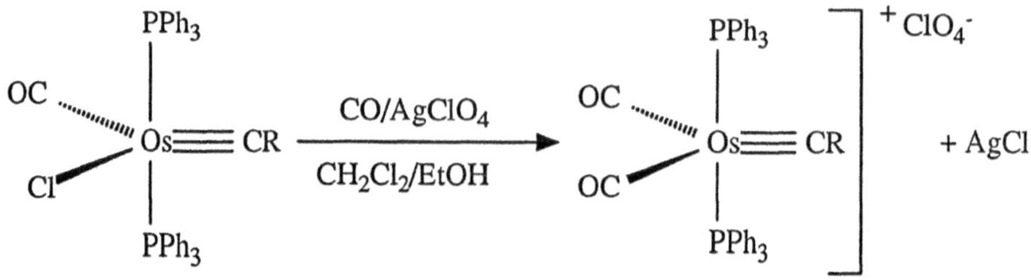

In the case of the ruthenium complexes the chloride is so labile that the presence of Ag^+ is unnecessary. In the osmium system replacement of CO by CNR (R=*p*-tolyl)

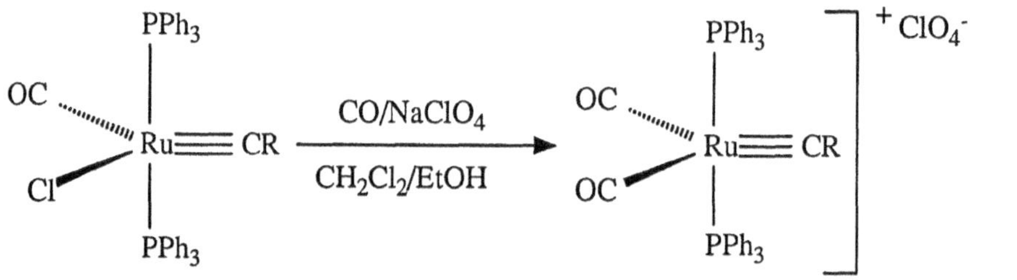

produces the corresponding salt $[Os(\equiv CR)(CO)(CNR)(PPh_3)_2]ClO_4$. IR data for these compounds are included in Table 1. All of these cationic compounds as solids show no sensitivity to moisture or oxygen and are therefore more easily handled than the rather sensitive neutral complexes.

3. Reactions of five coordinate carbyne complexes

The pattern of chemical reactions observed for these compounds clearly sets them apart from "Fischer-type" carbyne complexes of GroupVI e.g., $W(\equiv CR)X(CO)_4$. Whereas the "Fischer-type" complexes typically react with nucleophiles at the carbyne carbon all of the reactions observed for the five coordinate ruthenium and osmium complexes, including the cationic examples, are electrophilic additions to the $M\equiv C$ bond. The following sections deal individually with, protonation, addition of halides of the coinage metals, addition of chlorine and chalcogens, and finally an attempted nucleophilic addition where the nucleophile is directed to a remote site on the aryl ring of the carbyne substituent.

3.1 REACTION WITH ACIDS

$Os(\equiv CR)Cl(CO)(PPh_3)_2$ reacts rapidly with HCl producing the red carbene complexes $OsCl_2(=CHR)(CO)(PPh_3)_2$. The structure of one example, $OsCl_2(=CHPh)(CO)(PPh_3)_2$, was determined by X-ray crystallography revealing an Os=C double bond length of 1.94(1)Å and a very long *trans* Os-Cl distance of 2.550(3)Å [13]. One ruthenium example, $RuCl_2(=CHC_6H_4\text{-}4\text{-}OMe)(CO)(PPh_3)_2$, has also been characterised [12].

The situation with the cationic carbyne complexes of osmium is more complicated and may be followed in Scheme 2. The dicarbonyl cation, $[Os(\equiv CPh)(CO)_2(PPh_3)_2]^+$,

Scheme 2

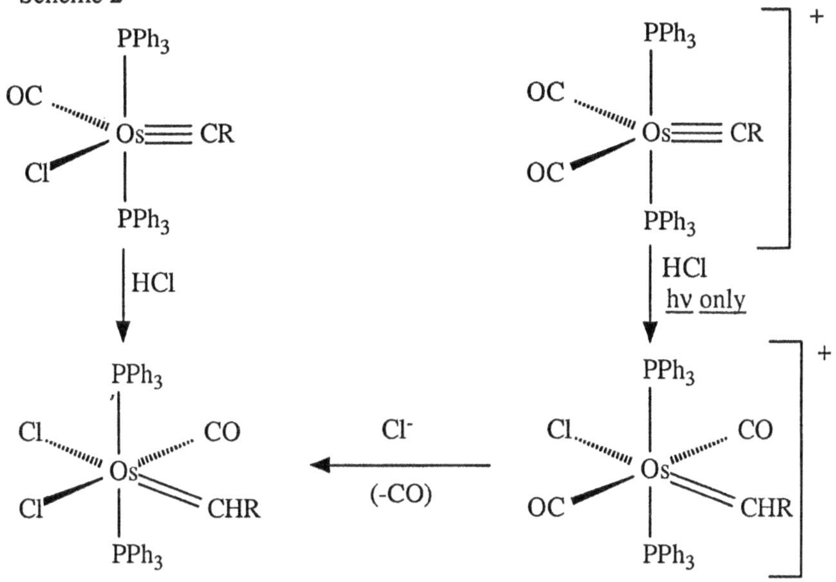

can be recovered unchanged after two hours in a CH_2Cl_2/EtOH solution containing conc. HCl. However, upon charge transfer (Os to carbyne) excitation (photolysis at wavelengths > 305 nm) HCl addition occurs rapidly, the final product being $OsCl_2(=CHPh)(CO)(PPh_3)_2$ [14]. It has been suggested that the relaxed CT state might be a square pyramidal Os(II) complex containing a bent carbyne ligand which carries a lone pair at the coordinating carbon atom. Addition of a proton to this carbon, in a step much faster than proton addition to the ground state molecule, and chloride to the metal followed by chloride displacement of CO gives the observed product. This suggestion was made on the basis of an analogy between the carbyne ligand (CR^+) and the nitrosyl ligand (NO^+). Bending of nitrosyl ligands has been used in the interpretation of the photochemistry of related nitrosyl complexes.

To further complicate the picture, the related isocyanide-substituted cation, $[Os(\equiv C$*p*-tolyl)(CN*p*-tolyl)$(CO)(PPh_3)_2]^+$, undergoes a fast thermal reaction with HCl as depicted in Scheme 3. The cationic carbene complex produced is converted by treatment with $LiEt_3BH$ to the neutral substituted-benzyl complex labelled **B** in Scheme 3. The overall two-step transformation of **A** into **B** is a good example of the conversion of a metal-carbon triple bond to a metal-carbon single bond by sequential addition of "H^+" followed by "H^-" [11].

Weaker acids also add to $Os(\equiv CR)Cl(CO)(PPh_3)_2$, for example, acetic acid with $Os(\equiv C$*p*-tolyl)$Cl(CO)(PPh_3)_2$ gives the cationic carbene complex with a *dihapto* acetato-ligand, $[Os(\eta^2\text{-}O_2CCH_3)(=CH$*p*-tolyl)$(CO)(PPh_3)_2]^+$ [12].

3.2 REACTION WITH COPPER(I), SILVER(I). AND GOLD(I)

The metalla-alkyne nature of the metal carbyne bond is demonstrated by the reaction between $Os(\equiv C$*p*-tolyl)$Cl(CO)(PPh_3)_2$ and Ph_3PAuCl. Triphenylphosphine is displaced and the pink solid, Os(=C[AuCl]p-tolyl)Cl(CO)(PPh_3)$_2$ is formed. Analogous

Scheme 3 (R = p-tolyl)

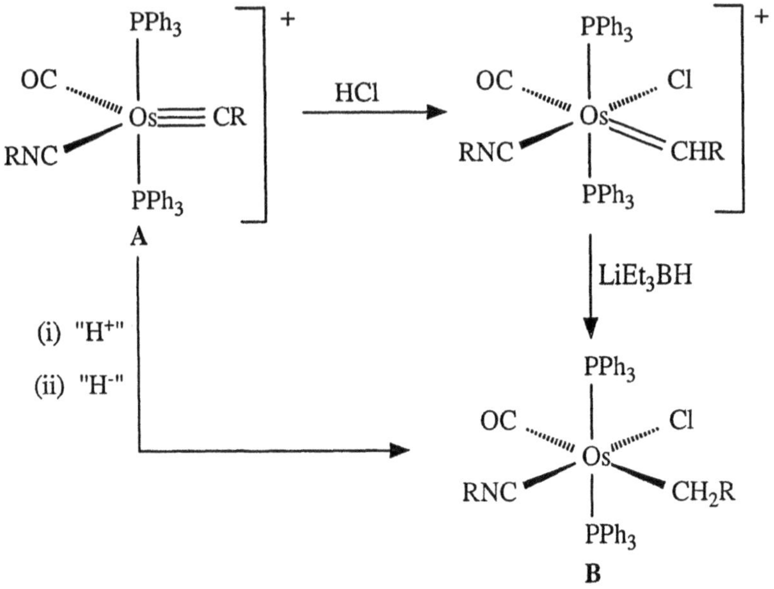

reactions occur, albeit more slowly, when a benzene solution of Os(≡Cp-tolyl)Cl(CO)(PPh_3)$_2$ is treated with a suspension of AgCl or CuI. The AgCl adduct is a violet colour and X-ray crystal structure determination confirms the dimetallacyclopropene formulation with the osmium-carbon distance being elongated from 1.78(2)Å to 1.839(5)Å [10]. Even a cationic carbyne complex will bind to Ag^+ as is illustrated in Scheme 4. One ruthenium-containing dimetallacyclopropene has been characterised from reaction between Ru(≡CPh)Cl(CO)(PPh_3)$_2$ and [Et_4N][AuI_2] giving

Ru(=C[AuI]Ph)Cl(CO)(PPh_3)$_2$.

3.3 REACTION WITH Cl_2

Treatment with chlorine usually cleaves carbyne ligands from metals but in the case of the ruthenium and osmium carbyne complexes addition of one equivalent of Cl_2 (dissolved in CCl_4) to benzene solutions of the carbyne complexes gives the orange-red, moisture-sensitive, but otherwise quite stable mono-chlorocarbene complexes (see Scheme 5). Although Scheme 5 depicts osmium complexes $RuCl_2$(=CClPh)(CO)(PPh_3)$_2$ and several derivatives have been characterised [12].The hydrolysis reaction is multistep involving formation of an acyl derivative which by reverse migration gives the observed aryl dicarbonyl complex. The chlorine substituent on the carbene carbon atom is readily replaced in reaction with $LiEt_3BH$, primary amines, and chalcogenide anions thus providing convenient routes to a wide range of different carbene ligands. The chalcoacyl ligands adopt an η^2-attachment which has been confirmed for a thioacyl derivative by crystal structure determination [15].

Scheme 4 (R = *p*-tolyl)

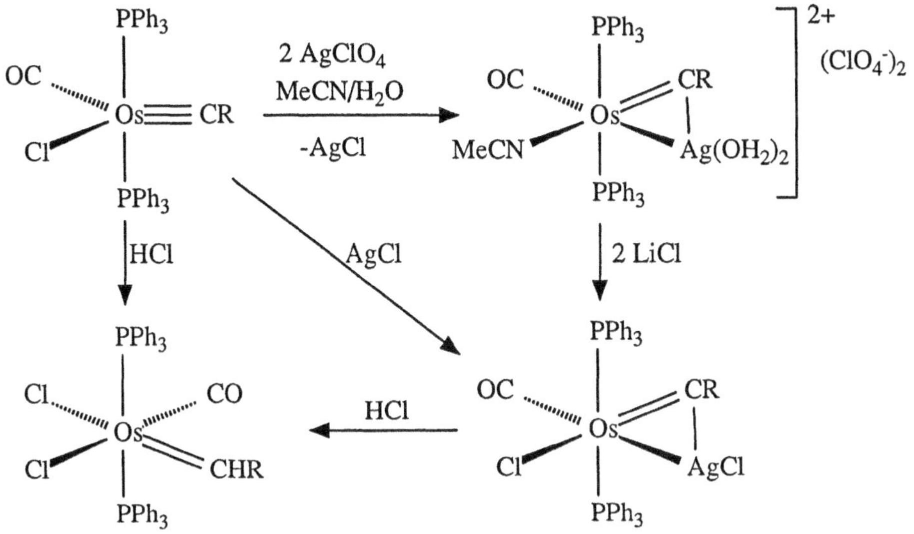

3.4 REACTION WITH CHALCOGENS

The η^2-chalcoacyl derivatives of the previous section are also accessible by direct reaction between the carbyne complex and elemental sulfur, selenium, or tellurium, once again emphasising the metalla-alkyne nature of these carbyne complexes.

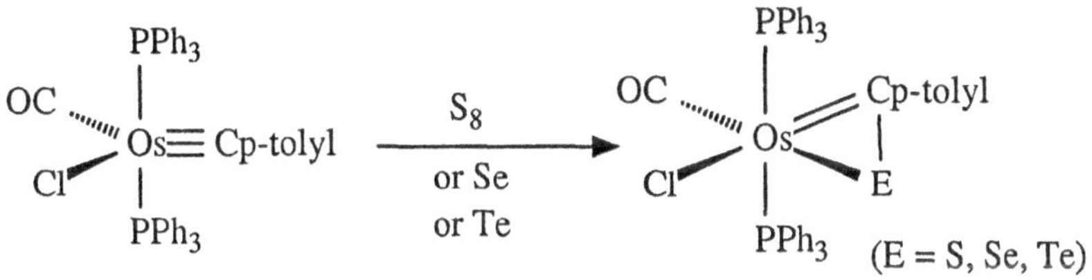

3.5 REACTION OF CATIONIC CARBYNE COMPLEXES WITH $LiEt_3BH$

The reluctance of the carbyne carbon to react with nucleophiles is revealed by the reaction with $LiEt_3BH$ (see Scheme 6). Here the most electrophilic site is not the carbyne carbon but the *para* position of the aryl ring in the carbyne substituent. Both ruthenium and osmium five coordinate, cationic, carbyne complexes undergo this reaction. The structure of a representative example, the osmium compound derived from the *p*-tolyl carbyne complex, has been determined by X-ray crystallography [16]. The unusual vinylidene complex reacts with HCl to produce a substituted benzyl derivative. The reaction may proceed through the intermediate σ-vinyl complex depicted in Scheme 6 although there is also the possibility that the vinylidene compound is in equilibrium with the carbene tautomer as shown below.

Scheme 5 (E = S, Se, Te)

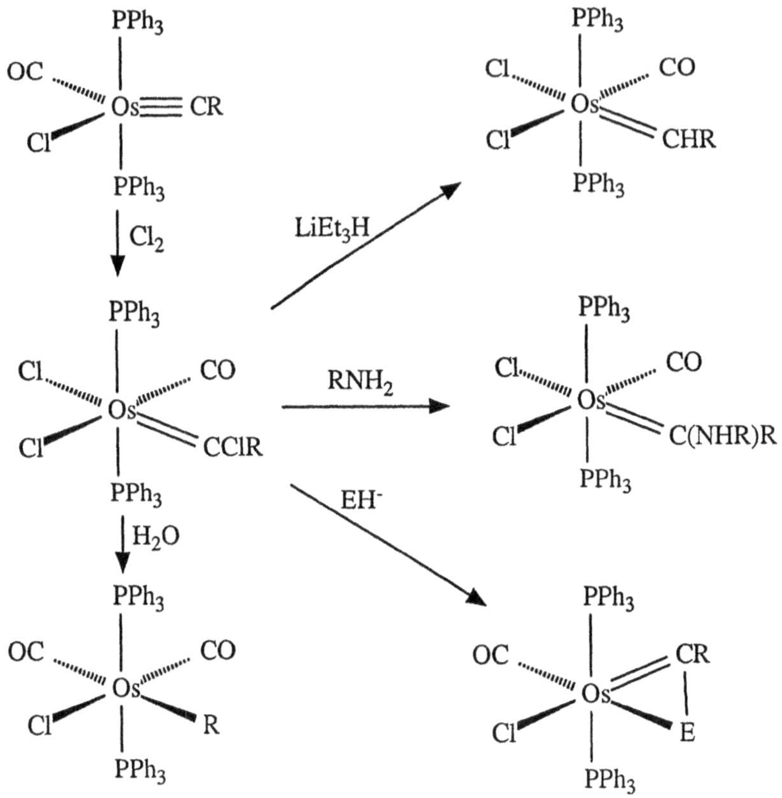

4 Synthesis of octahedral carbyne complexes

Consideration of the octahedral carbyne complexes of tungsten, W(≡CR)XL_4, suggests that there should exist a parallel set of osmium(II) compounds i.e., Os(≡CR)X_3L_2. Indeed, attention has already been drawn to the useful analogy between the carbyne and nitrosyl ligands and for the nitrosyl ligand compounds of both types Os(NO)XL_3 and Os(NO)X_3L_2 are well-known. Accordingly, various ways of oxidizing the five coordinate (formally zero oxidation state) complexes were considered. As described in section 3.3, chlorine adds across the M≡C bond forming monochlorocarbene complexes of type MCl_2(=CClR)(CO)$(PPh_3)_2$. In principle these compounds should be suitable precursors of octahedral carbyne complexes by chloride removal from the carbene carbon atom. This has not yet been achieved. However, reaction of the five coordinate carbyne complexes with iodine leads *directly* to cationic carbyne complexes and this is discussed in section 4.1. One other effective means of oxidizing the five coordinate carbyne complexes, which follows closely the known chemistry of the related nitrosyl complexes, is through reaction with molecular oxygen. This is detailed in section 4.2.

Scheme 6

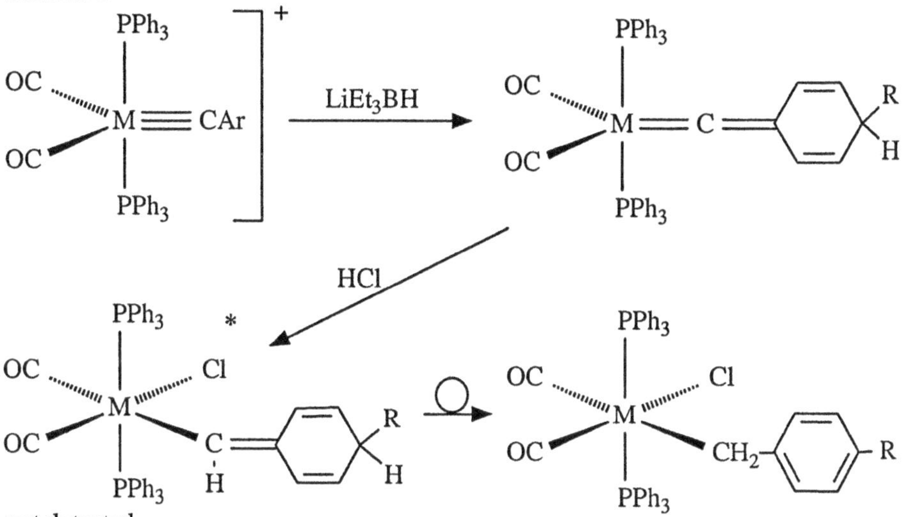

* not detected

4.1 REACTION OF FIVE COORDINATE CARBYNE COMPLEXES WITH I_2

Addition of iodine to Os(≡CR)Cl(CO)$(PPh_3)_2$ yields a number of compounds which have not been identified. However addition of iodine to Ru(≡CPh)Cl(CO)$(PPh_3)_2$ at low temperatures yields a single product formulated as [Ru(≡CPh)ClI(CO)$(PPh_3)_2$]I on the basis of IR data and further reactions of the compound. The compound is moisture-sensitive and hydrolyses to the dicarbonyl RuPhCl$(CO)_2(PPh_3)_2$. This reaction must involve OH^- attack at the carbyne carbon producing an intermediate acyl complex. When [Ru(≡CPh)ClI(CO)$(PPh_3)_2$]I is heated in an inert solvent CO is lost and the stable neutral carbyne complex Ru(≡CPh)Cl$I_2(PPh_3)_2$ is produced. IR data for this and other octahedral carbyne complexes are collected in Table 2. Iodine addition to the cationic complex [Ru(≡CPh)$(CO)_2(PPh_3)_2$]ClO_4 gives another cationic complex, [Ru(≡CPh)I_2(CO)$(PPh_3)_2$]ClO_4.

4.2 REACTION OF FIVE COORDINATE CARBYNE COMPLEXES WITH O_2

The nitrosyl complex, OsCl(NO)(CO)$(PPh_3)_2$, in the presence of oxygen is converted into the novel peroxycarbonyl complex Os(C[O]OO)Cl(NO)$(PPh_3)_2$ [13]. In a similar reaction Os(≡CR)Cl(CO)$(PPh_3)_2$ is converted into Os(C[O]OO)(≡CR)Cl$(PPh_3)_2$ [17]. The reaction is most successful when R = *p*-dimethylaminophenyl and when conducted at temperatures below 0°C. The product is formed in 72% yield as golden-yellow crystals. The bands associated with the peroxycarbonyl ligand in the IR spectrum at 1680 and 1060 cm^{-1} can be compared with those for the nitrosyl complex Os(C[O]OO)Cl(NO)$(PPh_3)_2$ at 1710 and 1025 cm^{-1}. For the carbyne complex with R=phenyl the peroxycarbonyl derivative is produced in only about 15% yield the major product being OsPhCl(CO)$(PPh_3)_2$. The ruthenium compound,

$Ru(\equiv CPh)Cl(CO)(PPh_3)_2$ with oxygen gives $RuPhCl(CO)_2(PPh_3)_2$ as the only identifiable product. Selected reactions of the peroxycarbonyl-carbyne complex are shown in Scheme 7. The peroxycarbonyl ligand is removed by treatment with HCl (note that in these octahedral carbyne complexes the carbyne ligand is unreactive towards acids) and both neutral and cationic derivatives are accessible. Two of these, $Os(\equiv CC_6H_4\text{-}4\text{-}NMe_2)Cl_2(NCS)(PPh_3)_2$, and $[Os(\equiv CC_6H_4\text{-}4\text{-}NMe_2)Cl_2(CN\textit{p}\text{-tolyl})(PPh_3)_2]ClO_4$ have been fully characterised by X-ray crystal structure determination giving Os≡C distances of 1.75(1)Å and 1.78(1)Å respectively [17]. A significant feature of both structures is a bending of the Os-Cl bonds away from the carbyne ligand. For one chlorine in the cationic complex the angle between the Os-Cl and Os≡C bonds is increased from the ideal octahedral angle of 90° to 104.4(5)°.

TABLE 2 IR Data[a] for Octahedral Carbyne Complexes[b]

Compound	ν(carbyne) cm^{-1}
$Ru(\equiv CPh)ClI_2(PPh_3)_2$	1582, 1385
$[Ru(\equiv CPh)I_2(CO)(PPh_3)_2]^+$	1580, 1318, 1280
$Os(C[O]OO)(\equiv CPh)Cl(PPh_3)_2$	1585, 1395
$Os(C[O]OO)(\equiv CC_6H_4\text{-}4\text{-}NMe_2)Cl(PPh_3)_2$	1598, 1405
$Os(\equiv CPh)Cl_3(PPh_3)_2$	1585, 1410
$Os(\equiv CC_6H_4\text{-}4\text{-}NMe_2)ClI_2(PPh_3)_2$	1597, 1420
$Os(\equiv CC_6H_4\text{-}4\text{-}NMe_2)Cl_2(NCS)(PPh_3)_2$	1600, 1410
$[Os(\equiv CC_6H_4\text{-}4\text{-}NMe_2)Cl_2L(PPh_3)_2]^+$	1600, 1412

[a] from references [12] and [17] [b] L = CN*p*-tolyl

5 Reactions of six coordinate carbyne complexes

The formal similarity of these ruthenium and osmium compounds to the octahedral Group VI carbyne compounds would lead to an expectation of reactivity towards nucleophiles. In fact the neutral octahedral complexes prove to be rather unreactive compounds but the cationic complexes do react with nucleophiles. Mention has already been made of the hydrolysis of $[Ru(\equiv CPh)ClI(CO)(PPh_3)_2]I$ to $RuPhCl(CO)_2(PPh_3)_2$. A very clear-cut example is provided by the reaction of $[Os(\equiv CR)Cl_2(CNR')(PPh_3)_2]ClO_4$ (R=*p*-dimethylaminophenyl, R'=*p*-tolyl) with NaSH.

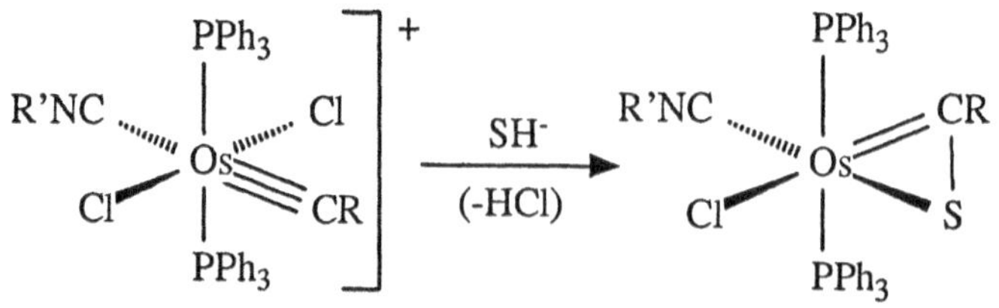

Scheme 7 (R = *p*-dimethylaminophenyl, R' = *p*-tolyl)

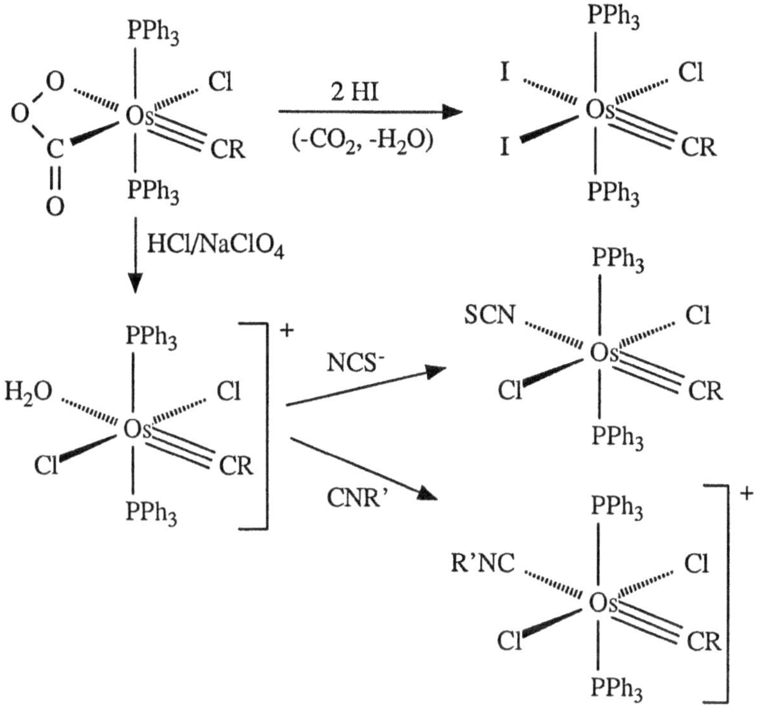

6 Methylation of a tellurocarbonyl ligand

Electrophilic addition to the sulfur atom of terminal thiocarbonyl ligands is a well-established route to thiocarbyne complexes of Group VI metals [2, 18, 19]. A requirement for the success of this approach is that the thiocarbonyl ligand be bound to a very electron-rich metal centre, preferably in an anionic complex. A methyltellurocarbyne complex of tungsten, $[L(CO)_2W{\equiv}CTeMe]$ ($L{=}HB[3,5\text{-}N_2C_3Me_2H]_3^-$) was obtained by treatment of the anionic electron-rich tellurocarbonyl complex, $[L(CO)_2WCTe]^-$ with MeI [20]. Osmium provided the first example of a tellurocarbonyl ligand from reaction between $OsCl_2(CCl_2)(CO)(PPh_3)_2$ and $TeI I^-$ which gave $OsCl_2(CTe)(CO)(PPh_3)_2$ [21]. This octahedral osmium(II) complex shows no interaction with electrophiles at tellurium. It is possible to reduce $OsCl_2(CTe)(CO)(PPh_3)_2$ to the green, zero oxidation state, tellurocarbonyl complex, $Os(CTe)(CO)_2(PPh_3)_2$, by the sequence shown in Scheme 8 [22]. This five coordinate complex shows a band in the IR spectrum at 997 cm^{-1} which can be associated with $\nu(CTe)$. Comparison with the $\nu(CTe)$ value for $OsCl_2(CTe)(CO)(PPh_3)_2$ at 1046 cm^{-1} suggests that the CTe ligand in $Os(CTe)(CO)_2(PPh_3)_2$ should be susceptible to electrophilic addition at Te. When $Os(CTe)(CO)_2(PPh_3)_2$ is treated with MeI/EtOH an orange solution is produced and the cationic carbyne complex can be crystallised as the orange perchlorate salt $[Os({\equiv}CTeMe)(CO)_2(PPh_3)_2]ClO_4$.

$$(CO)_2(PPh_3)_2Os{=}CTe \xrightarrow[\text{LiClO}_4]{\text{MeI/EtOH}} [(CO)_2(PPh_3)_2Os{\equiv}CTeMe]ClO_4$$

Scheme 8

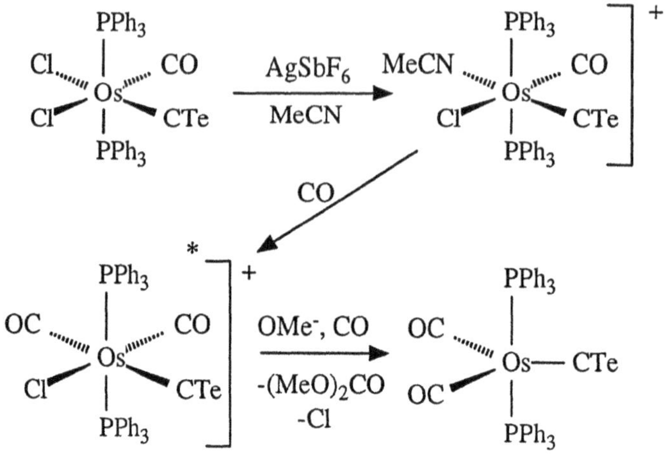

* not isolated

$[Os(\equiv CTeMe)(CO)_2(PPh_3)_2]^+$ has ν(CO) bands in the IR spectrum at 2011 and 1951 cm^{-1}. Comparison with ν(CO) values for other five coordinate cationic carbyne species given in Table 1 shows a close correspondence. An X-ray crystal structure determination confirmed the essentially trigonal bipyramidal geometry and gave an Os≡C bond distance of 1.841(16)Å [13].

7 α-Diazoalkyl complexes of osmium as carbyne complex precursors

Nitrogen loss from α-diazoalkyl transition metal complexes offers a potential route to carbyne complexes.

$$L_nM\text{-}\underset{\underset{N_2}{||}}{C}\text{-}R \xrightarrow[-N_2]{} L_nM\text{-}\ddot{C}\text{-}R \longleftrightarrow L_nM{\equiv}CR$$

Reported examples of this approach include the preparation of bridging-carbyne complexes of manganese from [2+2] cyclisation of intermediate mononuclear carbyne complexes [23] and the thermolysis or photolysis of $(PEt_3)_2Pd[C(N_2)Ph]_2$ which produced diphenylacetylene possibly *via* a carbyne coupling reaction [24]. An osmium α-diazoalkyl complex results from the oxidative addition of $Hg(C[N_2]CO_2Et)_2$ to $OsCl(NO)(PPh_3)_3$ [25]. Reaction with one equivalent of I_2 removes the diazoalkyl group from mercury leaving $Os(C[N_2CO_2Et)Cl(HgI)(NO)(PPh_3)_2$. The further reactions of the compound may be followed in Scheme 9. The conversion of compound **C** to compound **D** may be understood by the following sequence, acid cleavage of the Os-Hg bond gives an osmium-hydride, nitrogen loss followed by hydride migration to the α-carbon atom, and finally protonation at the resulting carbene centre gives the observed compound **D.** Conversion of compound **E** to compound **F** may proceed *via* an intermediate which can be written as $[Os(\equiv CCO_2Et)Cl(NO)(PPh_3)_2]^+$ (an analogue of $[Os(\equiv CR)(CO)_2(PPh_3)_2]^+$) which undergoes insertion into a C-H bond of a phenyl ring of one triphenylphosphine ligand. The structure of compound **F** was verified by X-ray crystal structure determination. While this limited study has not produced a stable carbyne complex the results are sufficiently promising to warrant further investigation

Scheme 9

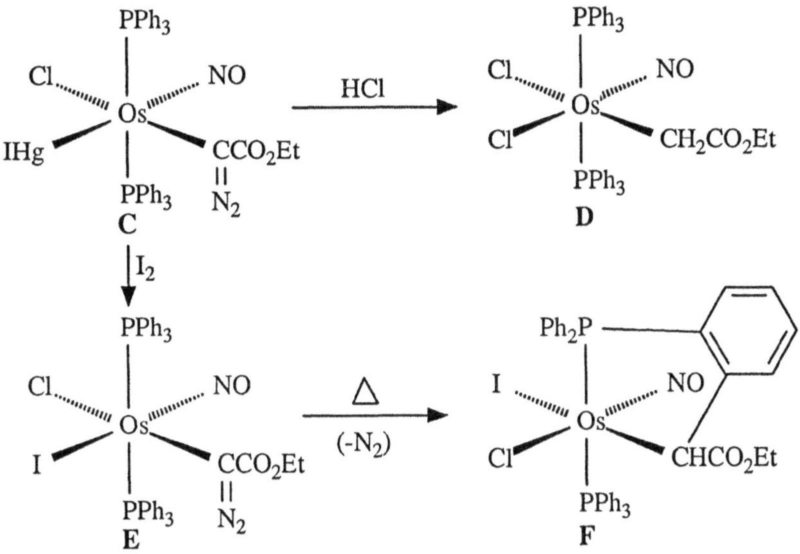

of metallated diazoalkanes as carbyne complex precursors.

8 Concluding remarks

The metal-carbon triple bond chemistry of ruthenium and osmium described in this article bears a close resemblance to the metal-carbon double bond chemistry of these elements as exemplified by the methylene complexes [26]. In both systems two structural classes are found, five coordinate (trigonal bipyramidal, formally zero oxidation state) and six coordinate (octahedral, formally +2 oxidation state). In both systems the five coordinate compounds exhibit multiple metal-carbon bonds which are rather non-polar and typically undergo addition reactions with electrophilic reagents. On the other hand the six coordinate compounds, both M=C and M≡C, begin to show electrophilic character at the carbon centres especially in cationic complexes. Further development of the carbyne chemistry of ruthenium and osmium will depend upon the discovery of new synthetic methods allowing the preparation of a broader range of compounds with widely differing carbyne substituents.

Acknowledgments

Support from the N.Z. Universities Grants Committee and the University of Auckland Research Committee is gratefully acknowledged. The author thanks all his coworkers named in the references but particularly Drs. James Wright and Tony Wright who were principally responsible for developing the carbyne complex chemistry of ruthenium and osmium.

References

1 E. O. Fischer, G. Kreis, C. G. Kreiter, J. Müller, G. Huttner, H Lorenz, *Angew. Chem.* **85** (1973) 618; *Angew. Chem. Int. Ed. Engl.* **12** (1973) 564.
2 *Carbyne Complexes*: H. Fischer, P. Hofmann, F. R. Kreissl, R. R. Schrock, U. Schubert, K. Weiss, (eds.) Weinheim, Verlag Chemie, 1988.
3 A. Höhn, H. Werner, *Angew. Chem.* **98** (1986) 745; *Angew. Chem. Int. Ed. Engl.* **25** (1986) 737.
4 E. O. Fischer, J. Schneider, *J. Organomet. Chem.* **295** (1985) C29.
5 W. R. Roper, A. H. Wright, *J. Organomet. Chem.* **233** (1982) C59.
6 G. R. Clark, K. Marsden, W. R. Roper, L. J. Wright, *J. Amer. Chem. Soc.* **102** (1980) 6570.
7 G. R. Clark, W. R. Roper, A. H. Wright, *J. Organomet. Chem.* **236** (1982) C7.
8 P. J. Brothers, W. R. Roper, *Chem. Rev.* **88** (1988) 1293.
9 N. Q. Dao, E. O. Fischer, C. Kappenstein, *Nouv. J. Chim.* **4** (1980) 85.
10 G. R. Clark, C. M. Cochrane, K. Marsden, W. R. Roper, L. J. Wright, *J. Organomet. Chem.* **315** (1986) 211.
11 L. J. Wright, Ph. D. Thesis, University of Auckland, 1980.
12 A. H. Wright, Ph. D. Thesis, University of Auckland, 1983.
13 W. R. Roper, *J. Organomet. Chem.* **300** (1986) 167.
14 A. Vogler, J. Kisslinger, W. R. Roper, *Z. Naturforsch.* **38b** (1983) 1506.
15 G. R. Clark, T. J. Collins, K. Marsden, W. R. Roper, *J. Organomet. Chem.* **259** (1983) 215.
16 W. R. Roper, J. M. Waters, L. J. Wright, F. Van Meurs, *J. Organomet. Chem.* **201** (1980) C27.
17 G. R. Clark, N. R. Edmonds, R. A. Pauptit, W. R. Roper, J. M. Waters, A. H. Wright, *J. Organomet. Chem.* **244** (1983) C57.
18 H. P. Kim, R. J. Angelici, *Advances Organomet. Chem.* **27** (1987) 51.
19 A. Mayr, H. Hoffmeister, *Advances Organomet. Chem.* **32** (1991) 227.
20 T. Desmond, F. J. Lalor, G. Ferguson, M. Parvez, *J. Chem. Soc. Chem. Commun.* (1984) 75.
21 G. R. Clark, K. Marsden, W. R. Roper, L. J. Wright, *J. Amer. Chem. Soc.* **102** (1980) 1206; G. R. Clark, K. Marsden, C. E. F. Rickard, W. R. Roper, L. J. Wright, *J. Organomet. Chem.* **338** (1988) 393.
22 A. F. Hill, M. Sc. Thesis, University of Auckland, 1983.
23 W. A. Herrmann, *Angew. Chem. Int Ed. Engl.* **13** (1974) 812.
24 S-I. Murahashi, Y. Kitani, T. Uno, T. Hosokawa, K. Miki, T. Yonezawa, N. Kasai, *Organometallics,* **5** (1986) 356.
25 M. A. Gallop, T. C. Jones, C. E. F. Rickard, W. R. Roper, *J. Chem. Soc. Chem. Commun.* (1984) 1002.
26 W. R. Roper, *NATO ASI Ser., Ser. C (Adv. Met. Carbene Chem.)* **269** (1989) 27.

TOWARDS THE SYNTHESIS OF CARBINE COMPLEXES OF GOLD AND COPPER: NEW CARBENE COMPLEXES

H.G. RAUBENHEIMER, S. CRONJE, R. OTTE, W. VAN ZYL, I. TALJAARD and P. OLIVIER

Rand Afrikaans University
Department of Chemistry
Johannesburg 2000
South Africa

ABSTRACT. Lithiumthiazoles react by substitution or addition with gold and copper chlorides and subsequent protonation or alkylation affords stable mono- or bis(carbene) complexes. Complications which occur during these syntheses include homoleptic rearrangement, dissociative polymerization and carbon-carbon ligand coupling.

1. Introduction

Carbine complexes of gold and copper are unknown. As a preliminary to the synthesis of such compounds, we studied the affinity of the metals towards the formation of carbene complexes.[1] Since the standard Fischer-type precursor ligands, i.e. acyls, thioacyls or even imidoyls are not available for them, we made use of their natural tendency to form 'ate' complexes and employed lithiumthiazoles for this purpose. Subsequent protonation or alkylation of one or both of the nitrogen atoms afforded the coordinated carbene ligands. Carbene complexes derived from thiazolyl precursors have been prepared long ago by Stone and co-workers.[2] Our results are related to the recent work of Bonati et al. who prepared gold derivatives of imidazoles and, unexpectedly, also obtained a biscarbene complex.[3]

2. Results and Discussion

2.1. THE 'ATE' METHOD

In the following discussion 'thiazole' indicates benzothiazole and/or 4-methylthiazole. When differentiation is necessary derivatives of the former will be indicated by a and the latter by b.

The lithiated thiazole reacted with [ClAutht] (tht =

F. R. Kreißl (ed.), *Transition Metal Carbyne Complexes*, 169–173.

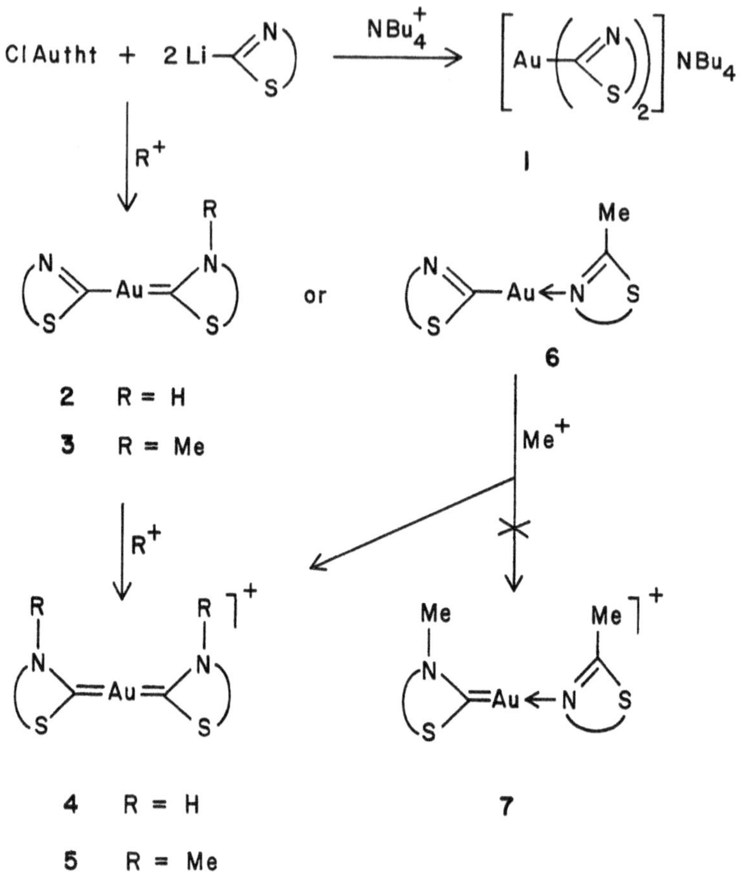

Scheme 1.

tetrahydrothiophene) to form an aurate which could be either isolated as its tetrabutyl ammonium salt or protonated or alkylated to give mono-and bis(carbene) complexes 2-5 (Scheme 1). In the attempted preparation of 3b, compound 6b also formed which, on further alkylation, did not give 7b, but underwent a homoleptic rearrangement to 5b.

2.2. OTHER CATIONIC MONO(CARBENE) COMPLEXES

The compound [Ph_3PAuCl] served as precursor for the synthesis of complexes of types 11 and 12. (Scheme 2). The substitution products 8 were however unstable at room temperature and polymerized to insoluble 9 (n probably 3[3]). These polymeric complexes served as starting material for the formation of the carbene(chloro) complexes 10. Compound 12b could not be isolated because a homoleptic rearrangement, once again, afforded 5b.

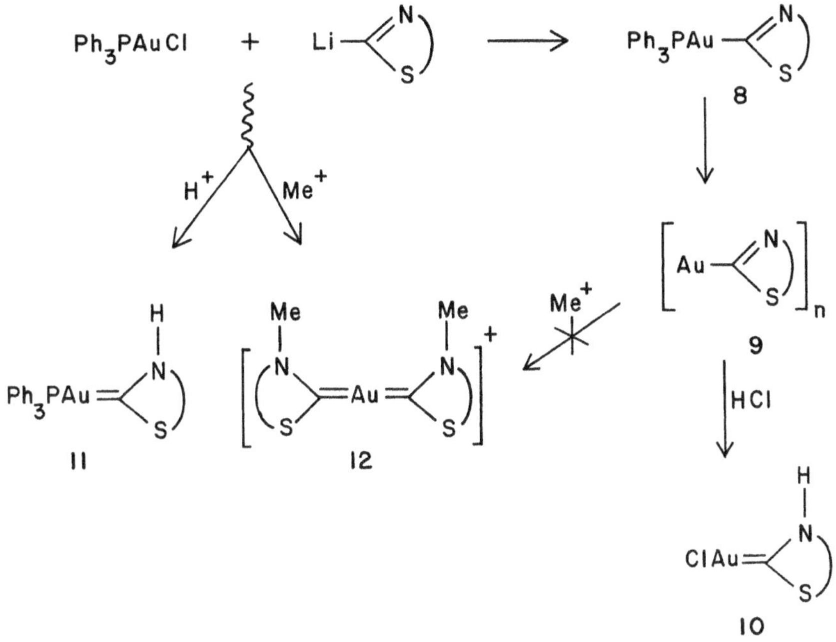

Scheme 2.

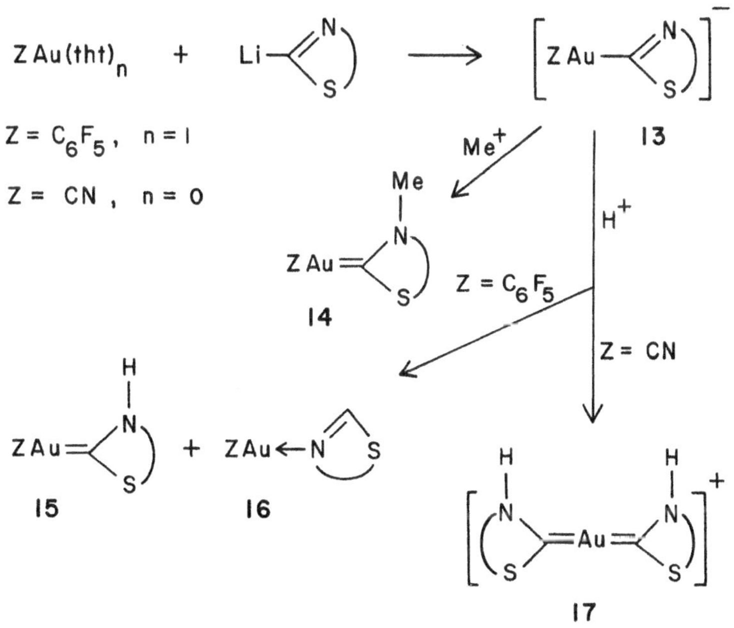

Scheme 3.

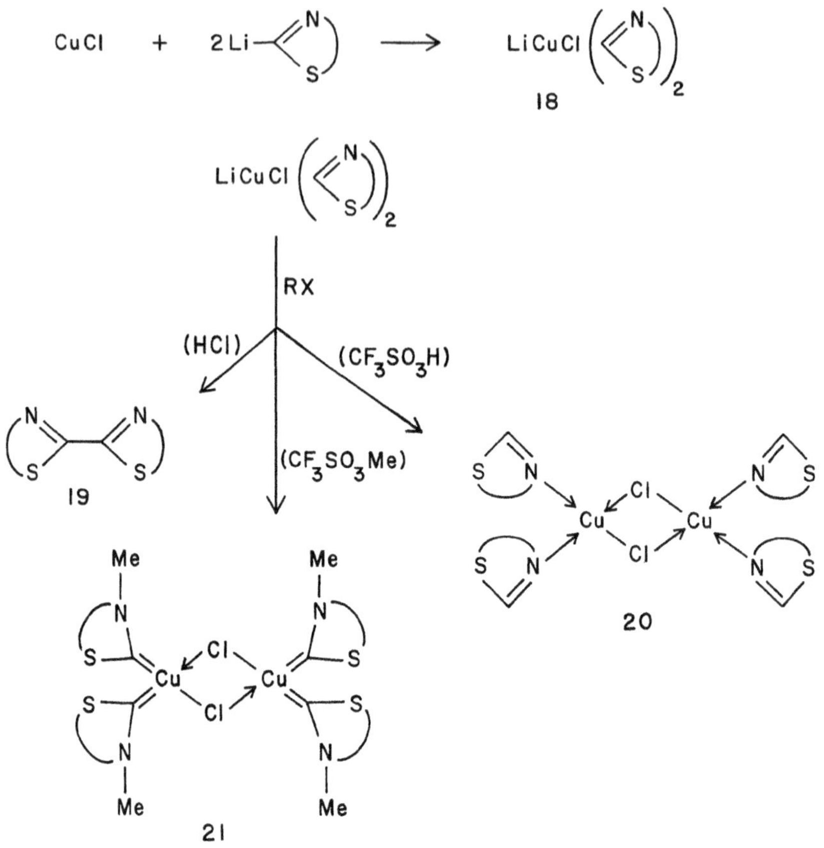

Scheme 4.

2.3. OTHER NEUTRAL CARBENE COMPLEXES

With the exception of the *in situ* acidification of the aurate **13b**, which did not afford **14b** but rather a coordinated 4-methylthiazole in **16b**, all the reactions of [C_6F_5Autht] (in Scheme 3) were straightforward. The cyanide-containing carbene complexes **14** were easy to isolate and characterize, but upon *in situ* protonation of **13a** (Z=CN; this **13b** was not prepared) a homoleptic rearrangement was responsible for the formation of **17a**.

Although structural data cannot be discussed here, suffice it to say that in many of the new compounds strong gold-gold interactions are apparent leading to metal-metal bonds of less than 3.2 Å.

2.4. NEUTRAL COPPER(I) CARBENE COMPLEXES

The reactions in Scheme 4 show that only alkylation of the cuprate salts 18 gave the first isolable carbene complexes of copper. The structure proposed for 21 is based on an X-ray study of 20b, elemental analysis of 21 (indicating Cu:Cl, 1:1), and NMR studies. The unexpected coupling and C-protonation reactions which afforded 19 and 20 showed reaction dependency on the type of acid used for acidification.

We have also undertaken a study involving the synthesis of diaminocarbene complexes of copper and we succeeded in preparing and characterizing a neutral chloride-bridged carbene complex derived in a similar manner as before from N-methylimidazolyllithium, CuCl and CF_3SO_3Me. It decomposes, however, much easier than its thiazolinylidene counterparts which are already very air sensitive.

References

1 H.G. Raubenheimer, F. Scott, M. Roos and R. Otte, *J. Chem. Soc., Chem. Commun.*, (1990), 1722.

2 M. Green, F.G.A. Stone and M. Underhill, *J. Chem. Soc., Dalton Trans.*, (1975), 939.

3 F. Bonati, A. Burini and B.R. Pietroni, *J. Organomet. Chem.*, 475, (1989) 147.

DOUBLE AND TRIPLE BONDS TO f-ELEMENTS: STRUCTURE AND CHEMISTRY OF ACTINIDE COMPLEXES OF MULTIELECTRON PAIR DONOR LIGANDS

John W. Gilje and Roger E. Cramer
Chemistry Department
2545 The Mall
University of Hawaii
Honolulu, HI 96822
U.S.A.

ABSTRACT. F-element-ligand multiple bonds are not well studied. In the course of our studies of phosphoylide-early actinide chemistry we obtained $Cp_3U{=}CHPR_3$. Based on very good x-ray data, the first neutron diffraction structure of an organo-uranium complex, and theoretical results, we have assigned multiple bond character to the uranium-carbon bond in this molecule. Besides being the first reasonably well characterized f-element-carbon multiple bond, $Cp_3U{=}CHPR_3$ is one of the first members of a class of compounds in which an electron rich, dicarbanion donor is stabilized by an electron withdrawing substituent.

While we have not yet discovered further uranium-carbon multiple bonds, Cp_3UNPPh_3, $Cp^*{}_2UCl(NSPh_2)$, and $Cp^*{}_2U(NSPh_2)_2$, which contain $[NPPh_3]^-$ and $[NSPh_2]^-$ ligands whose electronic structures are similar to $[CHPR_3]^-$, have been prepared and structurally characterized. All contain short, nearly triple U-N bonds.

Reflecting differences between the various classes of metal-carbon multiple bonds, the chemistry of $Cp_3U{=}CHPR_3$ contrasts sharply to that of transition metal-carbon double bonds. Most notable is an extensive insertion chemistry for $Cp_3U{=}CHPR_3$. In addition to reactions with a variety of small, unsaturated molecules, $Cp_3U{=}CHPR_3$ reacts with terminal metal-carbonyls to produce melallaphosphoniumenolates in which the carbonyl residue is activated by tight association to uranium. Further reactions include C-O bond cleavage, isomerization, C-P bond rupture, and carbonyl coupling reactions.

Introduction

In contrast to the d-block, where there are many examples of metal-ligand multiple bonds,[1] f-element-ligand multiple bond chemistry is in its infancy. In part this reflects the entire area of molecular f-element chemistry which is much less developed than that of the transition metals, and, in part, it is the result of difficulties in evaluating the nature of the bonding in heavy metals which are highly ionic, do not follow simple electron counting rules (e.g. the 18

F. R. Kreißl (ed.), Transition Metal Carbyne Complexes, 175–188.

electron rule), and where high quality theoretical calculations are difficult because many electrons and relativistic effects must be considered. Even the much studied AnO_2^{n+} (n = 1 and 2) "yl" ions, which are ubiquitous in high valent Am, U, Np, and Pu chemistry, remain controversial, although most would agree that their extremely short and strong An-O bonds reflect An-O multiple bonding.[2]

If the donor orbitals of a ligand are at significantly lower energy than the acceptor orbitals on a metal, the ligand can usually be described as an anion. This is an appropriate approximation for many complexes of the transition metals with nitrogen and oxygen ligands, but for carbon donors it is only appropriate with the more electropositive transition metals.[1] This relationship is congruent with the chemist's intuition that metal-carbon bonds are different from those between metals and oxygen or nitrogen. However, with the quite electropositive f-elements all metal-ligand bonds will possess a large ionic component and differences in the ionicity of the metal ligand interaction between oxygen, nitrogen, and carbon donors may decrease in f-element complexes. This has been a basic question which we have attempted to address with our studies of the early actinides.

An-C Multiple Bonds

As part of our studies[3,4,5,6,7] on early actinide-phosphoylide chemistry we have prepared $Cp_3U{=}CHPR_3$:[8]

$$Cp_3UCl + Li(CH_2)_2PR_3 \text{ ----> } Cp_3U{=}CHP(CH_3)R_2 + LiCl$$

and have determined the structures of $Cp_3U{=}CHPMe_2Ph$[9,10] and $Cp_3U{=}CHPMe_3$ [11] at room temperature by x-ray diffraction and that of $Cp_3U{=}CHPMe_3$ by neutron diffraction at 20K.[12] All the structures are very similar with the U=C bonds close to 2.29 Å and the U-C-P angles slightly greater than 140°. Compared with other organouranium complexes $Cp_3U{=}CHPR_3$ contains long U-C(Cp) distances which are comparable to those in UCp_4.[13] This indicates that $Cp_3U{=}CHPR_3$ is sterically quite crowded. (A similar conclusion can be drawn from short intramolecular H-H contacts in $Cp_3U{=}CHPMe_3$.[12]) Thus, it is significant that the shortest U-C distances known, e.g. 2.293(1) Å from the neutron structure of $Cp_3U{=}CHPMe_3$, also occur in these crowded molecules.

The ^{13}C-H coupling constants within the α CH group of $Cp_3U{=}CHPMePh_2$, 113 Hz, of $Cp_3U{=}CHPMe_2Ph$, 100 Hz, and of $Cp_3U{=}CHPMe_3$, 95 Hz are low.[38] Particularly the 95 Hz coupling, which is in the range for electron deficient alkylidene complexes with agostic metal-hydrogen interactions, suggested that agostic interactions might contribute to the short U-C bonds in these molecules. However, the accurate location of the hydrogen atoms in the neutron structure of $Cp_3U{=}CHPMe_3$ shows, in spite of a large U-C-P angle, a normal U-C-H angle, 110.6(2)°, and a U-H separation, 2.859(3) Å, which precludes a U - H_α interaction.[12]

A logical explanation for the short uranium-carbon bond and wide U-C-P angle involves multiple U=C bonding. Two limiting resonance forms can be written for the $[CHPR_3]^-$ ligand:

$[\colon\!CH\text{-}PR_3]^-$ <---> $[\colon CH\text{=}PR_3]^-$
A **B**

High quality ab initio molecular orbital calculations indicate that **A** best describes in the ion.[14,15] In **A** the carbon atom carries two lone pairs of electrons and a formal charge of two, while the phosphonium substituent, with it formal positive charge, reduces the overall charge on the anion to uninegative. The m.o. calculations show the HOMO is essentially a nitrogen p-orbital and the HOMO-1 is a nitrogen sp hybrid. While both contain some hyperconjugative interaction with P-H antibonding orbitals both are mainly localized on nitrogen and have the correct spatial arrangement to serve as π (HOMO) and σ (HOMO-1) donor orbitals.[15] Thus, $[\colon\!CHPR_3]^-$ should be pictured as a dicarbanion which can serve as a four electron donor ligand. The bonding in $Cp_3U{=}CHPR_3$ can be described in terms of:

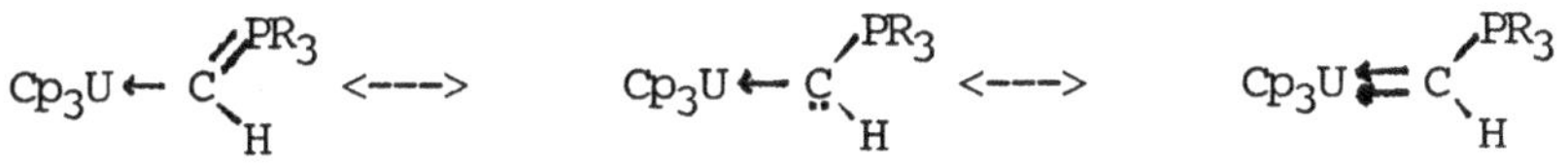

a hybrid of these:

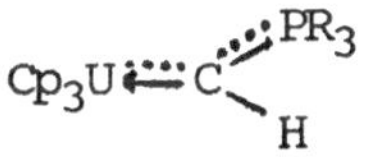

pictures the short uranium-carbon bond as arising from a 2-electron carbon-uranium σ bond combined with a 2-electron three-center π bond over the U-C-P fragment. Reflecting the electropositive nature of uranium, the U-C bond will be will be highly polar and an ionic model which involves the interaction of $[Cp_3U]^+$ with the dicarbanion $[\colon\!CHPR_3]^-$ may be appropriate in many discussions. However, it is important to recognize that four electrons are involved in the uranium-carbon bond of $Cp_3U{=}CHPR_3$ regardless of whether a polar covalent or purely ionic bonding is assumed.

EHMO calculations[16] using a uranium-carbon bonds distance of 2.40 Å (the U-C separation expected in Cp_3UCH_3) yield U-C overlap populations of 0.40 in Cp_3UCH_3 and 0.57 in $Cp_3U{=}CHPR_3$. When calculations are repeated on $Cp_3U{=}CHPR_3$ at its observed U-C distance of 2.29 Å, the overlap population rises to 0.61, with a π component of 0.19. These data support the notion of a uranium-carbon multiple bond character in $Cp_3U{=}CHPR_3$.

Superficially, $Cp_3U{=}CHPR_3$ bears resemblance to Schrock type alkylidnes which also are sometimes formulated as complexes of dicarbanions, $[\ CHR]^{2-}$. Although such analogies can be drawn, $Cp_3U{=}CHPR_3$ and related compounds actually should occupy a separate place within the framework of metal-carbon multiple bonding. In this respect a subdivision of compounds containing metal-carbon double bonds into three classes is convenient.

In the first class are the Fischer Carbenes Complexes in which an electron poor alpha carbon atom in a C(XR)R' ligand is coordinated to an electron-rich metal and is stabilized by an electron donating heteroatom, X. Multiple metal-carbon bonding arises from σ donation from ligand to metal accompanied by back-donation of π electrons from the metal to the electron deficient carbene ligand. Even though the carbenoid center is further stabilized by π-electron donation from an electron rich substituent such as O or N, it remains electrophilic.

The second class are the alkylidene complexes, where no heteroatom substituents are attached to the α carbon atom. The metal involved in this class of bond can either be low valent, e.g. $(OC)_5WCPh_2$, in which case metal carbon bonding is similar to that in Fischer complexes, or can be in a higher oxidation state. In the latter case, a metal-carbon double bond formally results from the donation of both σ and π electrons from the ligand to metal. Complexes of this type usually contain a group 5 or 6 transition metal in which there is a good match in energy and size between the metal and carbon σ and π orbitals. Consequently, considerable metal metal-carbon double bonding occurs and no stabilizing groups are required on the α carbon atom.

The uranium-carbon bond in $Cp_3U{=}CHPR_3$ is in a third class in which a negative charge on the α carbon atom is stabilized by an electron-withdrawing heteroatom subsituent. Because extensive ligand to metal charge delocalization is unlikely in compounds of very electropositive metals, this third type of metal carbon multiple bonding may be the most prevalent among the f-elements.

There Is No $Cp_3Th{=}CHPR_3$

Organo U(IV) and Th(IV) chemistries are usually very similar. Consequently we were surprised to observe that the reaction of Cp_3ThCl with $Li(CH_2)_2PR_2$ does not produce $Cp_3Th{=}CHPMeR_2$.[17] Rather, in the 1:1 reaction of Cp_3ThCl with $Li(CH_2)_2PR_2$ the only organo Th complexes detected by nmr and mass spectroscopy are $Cp_2Th[(CH_2)_2PR_2]_2$ and unreacted Cp_3ThCl. The x-ray structure of $Cp_2Th[(CH_2)_2PR_2]_2$ demonstrates that, if the Cp group is considered to occupy a single coordination site, the Th is approximately octahedral with the two $[(CH_2)_2PR_2]^-$ ligands coordinating in a bidentate fashion. The Th-C σ bonds, 2.79 Å and 2.68 Å, in this complex are among the longest know. We are uncertain why the U and Th chemistry differs, but the consequence is that $[CHPR_3]^-$ has, of yet, only been observed to coordinate uranium.

Ylide Complexes With Cp^*_2U

Two types of $Cp_2(X)M$(ylide), M = Ti, Zr, Hf, complexes have been reported. One type contains the monodentate $[CHPPh_3]^-$ ligand and includes $Cp_2(Cl)Zr[CHPMe_3]$,[18] $Cp_2(R)Zr[CHPPh_3]$ (R = Ph,[19] Et[20]), $Cp_2(Cl)M[CHPPh_3]$ (M = Zr, Hf),[21] and $Cp_2(Cl)M[CHP(NR_2)_2R']$ (M = Ti, Zr, Hf; R = Me, Et; R' = NEt_2, Me).[22,23] Metal carbon multiple bonding has been suggested in several of these compounds.[23] In the second type of $Cp_2(X)M$ complex, the ylide is chelating: $Cp_2(H)Zr[(CH_2)_2PMe_2]$,[18] $Cp_2(Cl)Zr[(CH_2)_2PMe_2]$,[18] and $Cp_2(Cl)Zr[(CH_2)_2P(NEt_2)_2]$.[24] A most interesting observation is the spontaneous conversion of some of this second group into complexes of the first type upon mild heating,[18] and implies that monodentate $[CHPR_3]$ is stabilized by coordination to $Cp_2(X)M$. In view of structural and chemical similarities which are well recognized between $Cp_2(X)M$ and $Cp^*_2(X)An$ systems,[25] the behavior of the group 4 transition metals gave us great hope that $Cp^*_2(X)An[CHPR_3]$ complexes could be prepared and that they would provide further information on actinide-carbon multiple bonding.

In fact $Cp^*_2AnCl_2$ reacts cleanly with $Li(CH_2)_2PR_2$:[26]

$$Cp^*_2AnCl_2 + Li(CH_2)_2PR_2 \longrightarrow Cp^*_2(Cl)An[(CH_2)_2PR_2] + LiCl$$

to yield only $Cp^*_2(Cl)An[(CH_2)_2PR_2]$ (An = U, Th; $R_2 = Me_2$, MePh, Ph_2), in which the ylide ligands are chelating. Further, these complexes are thermally stable and do not convert into $Cp^*_2(Cl)An[CHP(Me)R_2]$ upon heating. While the molecules are fluxional on the nmr time scale, a detailed analysis of the dnmr spectra of $Cp^*_2(Cl)Th[(CH_2)_2PPh_2]$ and $Cp^*_2(Cl)Th[(CH_2)_2PMePh]$ shows that the fluxional process involves dissociation of a methylene group from the metal followed by rotation about the remaining $An-CH_2$ and CH_2-P bonds:

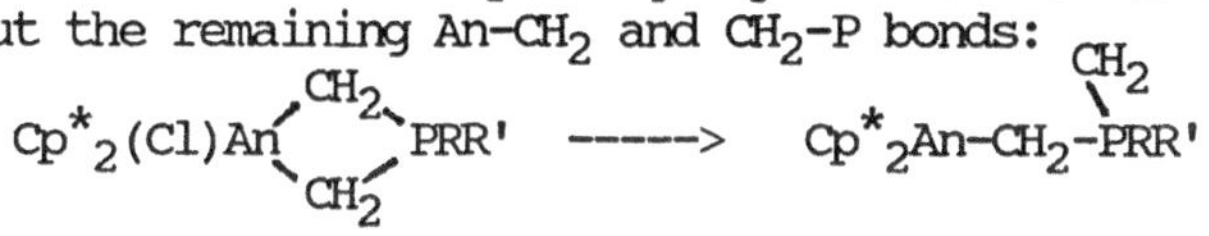

The high temperature spectrum of $Cp^*_2(Cl)Th[(CH_2)_2PMePh]$, in which both Cp^* and both CH_2 groups are equivalent, but the $P-CH_3$ protons do not exchange with the CH_2 protons, rules out mechanisms in which a methylene group become a CH_3:

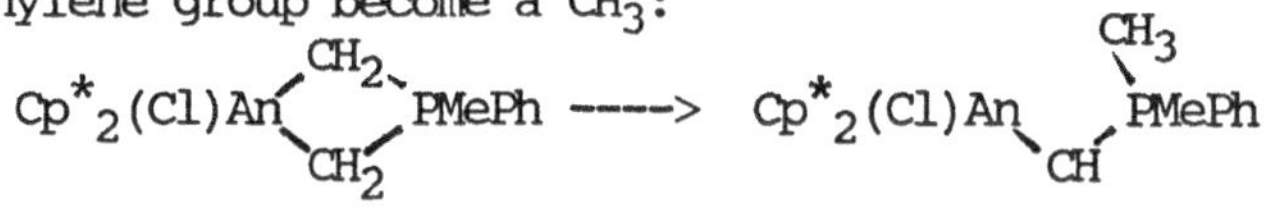

Thus, $Cp^*_2(Cl)An[CHP(Me)R_2]$ is eliminated even as an intermediate in the nmr exchange process. We must deduce from these data that the ylide ligand is more prone to serve as a monodentate $[CHPR_3]^-$ ligand with transition metal $Cp_2(X)M$ than with $Cp^*_2(X)An$ fragments.

At this point we were forced to conclude that An=C multiple bonds are rare species and we turned our attention to the bonds between actinides and other multielectron pair donor ligands.

Actinide-Nitrogen Multiple Bonds

Imido ligands, NR^{2-}, can donate as many as three electron pairs upon coordination to a metal and are good ligands toward electron-poor, high-valent transition metals, where they often form metal-nitrogen triple bonds. At present, actinide imides are largely unkown. We reported the first of these: $Cp_3UNC(Me)CHPMePh_2$ in 1984,[27] Cp_3UNR,[28] $[(Me_3Si)_2N]_3UNR$,[29] and $[(Me_3Si)_2N]_3U(F)R$[30] (R=Ph and $SiMe_3$) have subsequently been described by other groups. Because of our work with the phosphoylides our attention was drawn to $HNPR_3$ and $[NPR_3]^-$, which are isoelectronic with $[CHPR_3]^-$, and the very similar $HNSR_2$ and $[NSR_2]^-$. With the transition metals there is a small, but significant chemistry of $[NPR_3]^-$,[31] a very few complexes of $HNPR_3$[32,33] and $[NSR_2]^-$ [34,35] and $HNSR_2$ is unstudied.

Transmetallation of Cp_3AnCl with $LiNPPh_3$ and $LiNSPh_2$ produces $Cp_3AnNPPh_3$[36] (An = U, Th) and Cp_3UNSPh_2,[37] respectively:

$$Cp_3AnCl + LiNPPh_3 \longrightarrow Cp_3AnNPPh_3 + LiCl$$
$$Cp_3UCl + LiNSPh_2 \longrightarrow Cp_3UNSPh_2 + LiCl$$

Cp_3UNSPh_2 and Cp_3UNSPh_2 also form in acid-base reactions:[37]

$$Cp_3U{=}CHPR_3 + HNPPh_3 \longrightarrow Cp_3UNPPh_3 + H_2C{=}PR_3$$
$$Cp_3U{=}CHPR_3 + HNSPh_2 \longrightarrow Cp_3UNSPh_2 + H_2C{=}PR_3$$

as does a bis $[NPPh_2]^-$ complex, $Cp_2U(NPPh_3)_2$:

$$Cp_3UNPR_3 + HNPPh_3 \longrightarrow Cp_2U(NPPh_3)_2 + CpH$$

Unfortunately, we have not yet been able to obtain diffraction quality crystals of Cp_3UNSR_2, or $Cp_2U(NPPh_3)_2$. However, we have obtained an x-ray structure of Cp_3UNPPh_3, which facilitates its comparison with $Cp_3U{=}CHPR_3$. The geometry about both the phosphorus and uranium is the usual distorted tetrahedron. The average U-C(Cp) distance, 2.78(2) Å, is the same as in $Cp_3U{=}CHPR_3$, and the $[NPPh_3]^-$ ligand does not differ significantly from those in transition metal-phosphine imide complexes.[31] The most interesting feature of the structure is the U-N bond length of 2.07(2) Å. Some U-N and U-C bond lengths are compared in Table 1. Here the uranium oxidation state, the ligand charge, and the number of ligand lone pairs have been assigned from the point of view of an ionic model, i.e. by considering heterolytic cleavage of the metal-ligand bond. In terms of a strictly ionic model the distance to uranium of equally charged ligands should be about the same, with the U-N separations being, perhaps, a few percent shorter than the U-C distances. However, data from Table 1 show poor correlation between oxidation state, ligand charge, and bond length. Rather there is an obvious correlation between bond length and the number of lone pairs that are available for metal-ligand bonding.

TABLE 1. Some Uranium-Carbon and Uranium-Nitrogen Bond Distances

Compd	U Ox. State	Ligand Charge	Ligand Lone Pairs	U-N or U-C, Å
Cp_3UNPh[28]	V	2-	3	2.019(6)
Cp_3UNPPh_3	IV	1-	3	2.07(2)
Cp_3UNPh_2[38]	IV	1-	2	2.29(2)
$Cp_3U{=}CHPMe_3$	IV	1-	2	2.274(8)
$Cp_3UC_4H_9$[39]	IV	1-	1	2.43(2)

If electron delocalization does not occur within the R group, the bonding in transition-metal imide complexes can be described in terms of the resonance forms:

$$M{-}MR \longleftrightarrow M{=}NR \longleftrightarrow M{\equiv}NR$$

When the electron count about the metal allows donation of three electron pairs from nitrogen, the contribution from A and B is usually minor and the metal-nitrogen bond order is nearly 3. In such complexes the metal-nitrogen triple bond has been found to be 0.41 ± 0.02 Å longer than the appropriate Pauling metallic radius.[40] While the metallic radius of four valent uranium has not been tabulated by Pauling, 1.60 Å can be estimated by subtracting the difference of ionic radii between U(IV) and Th(IV), 0.05 Å, from the metallic radius of Th(IV), 1.65 Å. Subtraction of 1.60 Å from the U-N distance, 2.02 Å, in Cp_3UNPh yields 0.42 Å, which argrees with the value for the transition-metal imides which possess triple metal nitrogen bonds

Table 2 summarizes the metal nitrogen bond distances in structurally characterized transition metal-NPR_3 complexes in which a triple metal-nitrogen bond would not force a violation of the 18 electron rule. The differences between M-N distances and metallic radii for these complexes are tend to be somewhat larger than 0.41 Å expected for a triple bond.[40]

TABLE 2. Selected Data for Phosphine Imine Complexes

Cmpd	M-N	Met. Rad.	diff	P-N	M-N-P
Cp_3UNPPh_3	2.07(2)Å	1.60Å	0.47Å	1.61(2)Å	172(1)°
$[Cl_4NbNPPh_3]_2$[41]	1.776(8)Å	1.342Å	0.44Å	1.637(9)Å	171.1(6)°
$[Cl_4TaNPPh_3]_2$[42]	1.801(8)Å	1.343Å	0.46Å	1.593(9)Å	176.8(7)°
$(PhS)_4ReNPPh_3$[43]	1.743(7)Å	1.283Å	0.46Å	1.634(9)Å	163.1(6)°
$[Cl_4MoNPPh_3]_2$[44]	1.723(4)Å	1.296Å	0.43Å	1.656(4)Å	168.4(3)°
$[Cl_4(py)MoNPPh_3]$[45]	1.719(9)Å	1.296Å	0.42Å	1.653(9)Å	176.6(6)°

Consistent with the idea that there is some charge delocalization from nitrogen to phosphorus and in view of the nearly linear U-N-P unit, the following resonance structures are probably most important in

Cp_3UNPPh_3:

$$Cp_3U{\equiv}N\text{-}PPh_3 \text{ <---> } Cp_3U{=}N{=}PPh_3$$

and the overall conclusion, which is supported by EHMO calculations,[36] is that the metal-nitrogen bond order in the Cp_3UNPPh_3 is high but somewhat less than in Cp_3UNR.

Cp^*_2U Complexes with NPR_3 and NSR_2.

While we have so far been unable to form Cp^*_2U complexes which contain the $CHPR_3$ ligand, we have been able to obtain and structurally characterize $Cp^*_2UCl_2(HNPPh_3)$ and $Cp^*_2UCl_2(HNSPh_2)$ by direct reaction of $Cp^*_2UCl_2$ with the free ligands:[46,37]

$$Cp^*_2UCl_2 + HNPPh_3 \text{ ----> } Cp^*_2UCl_2(HNPPh_3)$$
$$+ HNSPh_2 \text{ ----> } Cp^*_2UCl_2(HNSPh_2)$$

The first of these is a rare example of a coordinated, intact $HNPR_3$ ligand and the second is the first metal complex of $HNSR_2$. Both are are thermally stable for several days at 100°, even in the presence of excess $HNPPh_3$ or $HNSPh_2$.

Both $Cp^*_2UCl_2(HNPPh_3)$ and $Cp^*_2UCl_2(HNSPh_2)$ are of the Cp_2MX_2Y type with uranium being bound by two pentamethylcyclopentadienide rings, two chlorides and the nitrogen of the imine ligand. The U-Cl distances, 2.66 - 2.73 Å, in both are longer than the usual U-Cl distances of about 2.56 - 2.60 Å, and the Cl-U-Cl angles, which are among the widest known, probably reflect crowding in the equatorial girdle. The 2.43(1) Å U-N bond distances in both compounds are the shortest yet found in complexes of uranium with an uncharged nitrogen donor ligand, indicates that the $HNPPh_3$ and $HNSPh_2$ ligands are tightly bound to the metal. We have previously noted that U-N bond distances, which range between 2.02-2.68 Å, appear to reflect both the ligand charge and the number of donor electrons on the ligand.[36,38] High quality ab initio calculations[47,15] indicate a high negative charge on nitrogen and the importance **C**:

$$\underset{\mathbf{C}}{H_n\overset{+}{E}\text{-}\overset{-}{N}H} \text{ <---> } \underset{\mathbf{D}}{H_nE{=}NH} \qquad (H_nE = H_3P,\ H_2S)$$

Both the charge on nitrogen and/or the donation of two electron pairs from nitrogen could shorten the U-N bond in $Cp^*_2UCl_2(HNPPh_3)$ and $Cp^*_2UCl_2(HNSPh_2)$.

In fact the interactions between $Cp^*_2UCl_2$ and $HNPPh_3$ or $HNSPh_2$ may be complex. In the x-ray structure of $Cp^*_2UCl_2(HNPPh_3)$ we were able to locate and refine the hydrogen which is attached to nitrogen. Even though there are large uncertainties in its position, the U-H distance, 2.2(2) Å, is short, similar to the terminal Th-H distance, 2.29(3) Å, in $(Cp^*ThH_2)_2$.[48] In addition, the N-H vector points in the general direction of one of the chloro ligands and both the Cl-H, 2.3(2) Å, and Cl-N, 3.06(1) Å, distances, are considerably shorter than the sum of the van der Waals radii of Cl (1.8 Å), H (1.2 Å), and N (1.5 Å). The N-H i.r. frequency is decreases upon coordination and is consistent with hydrogen bonding to Cl. and/or an interaction of the N-H bond with the electron deficient uranium center. As we have not yet attempted a theoretical analysis, we hesitate to draw more specific conclusions on the nature of U-N bonding in these molecules.

Cp^*_2U complexes of $[NSPh_2]^-$ are easily formed by several routes.[37] Perhaps the most convenient are reactions of $LiNSPh_2$ with $Cp^*_2UCl_2$:

$$Cp^*_2UCl_2 + LiNSPh_2 \longrightarrow Cp^*_2UCl(NSPh_2)$$
$$+ 2LiNSPh_2 \longrightarrow Cp^*_2U(NSPh_2)_2$$

Alternatively:

$$Cp^*_2U(Cl)[(CH_2)_2PR_2] + HNSPh_2 \longrightarrow Cp^*_2UCl(NSPh_2) + CH_2{=}P(Me)Ph_2$$
$$+ 2HNSPh_2 \longrightarrow Cp^*_2U(NSPh_2)_2 + [Me_2PR_2]Cl$$
$$Cp^*_2UCl_2(HNSPh_2) + LiNSPh_2 + HNSPh_2 \longrightarrow Cp^*_2U(NSPh_2)_2 + LiCl + [H_2NSPh_2]Cl$$

Crystals of both complexes were obtained. Analogous $[NPPh_3]^-$ derivatives can also be obtained, but we were unsuccessful in growing crystals suitable for structure determinations.

$Cp^*_2UCl(NSPh_2)$ and $Cp^*_2U(NSPh_2)_2$ have typical bent metallocene structures. In both, the most notable feature is the short U-N bond: 2.10(1) Å in $Cp^*_2UCl(NSPh_2)$ and 2.138(5) Å in $Cp^*_2U(NSPh_2)_2$. While we have not yet attempted a thorough theoretical analysis of the bonding in $Cp^*_2UCl(NSPh_2)$ and $Cp^*_2U(NSPh_2)_2$, the frontier orbitals on a Cp^*_2An group are more extensive than those normally available on a transition metal Cp_2M fragment. Acceptor orbitals perpendicular to (and possibly in) the equatorial plane extend the ability of the Cp^*_2An moiety to function as a π acceptor.[49,50] Consequently, orbital constraints should not prevent the donation of lone pairs from $[NSPh_2]^-$ ligands. Indeed, the U-N bond lengths in both complexes are near those in Cp_3U-imide complexes.

REACTIONS OF $Cp_3U{=}CHPR_3$

Initial studies demonstrated that the alpha carbon atom in $Cp_3U{=}CHPR_3$ is nucleophilic and reacts with both Bronsted and Lewis acids.[38,51,52] With internal alkenes and alkynes no reactions take place,[52] but a rich insertion chemistry is encountered with polar unsaturated molecules. With carbon monoxide,[53] nitrile,[27] and isonitrile:[54]

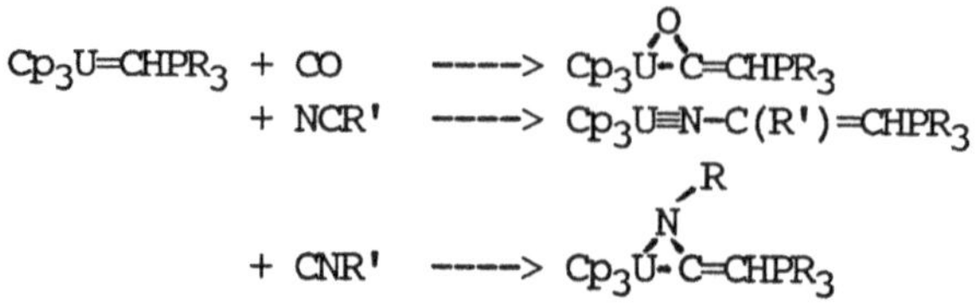

insertion occurs into the U=C bond to form products in which the hetereoatom is tightly bonded to uranium and all four electrons from the uranium-carbon bond have been used to form a new unsaturated carbon carbon bond.

We have investigated some reactions with more complex unsaturated systems. A number of hetereocumulenes, including RN=C=O, CO_2, RHC=N-N=CHR, and R-N=N-R, react. With phenylisocyanate insertion:[55]

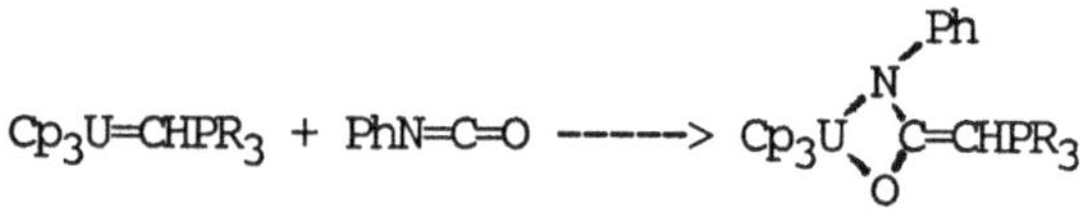

is followed by a further reaction:

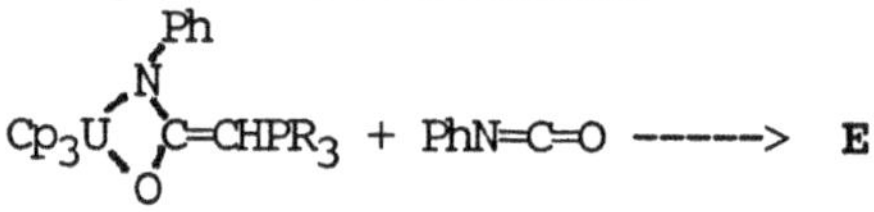

E, while not yet well characterized, is an efficient catalyst for the cyclodimerization of Ph-N=C=O into a uretidinedione:[60]

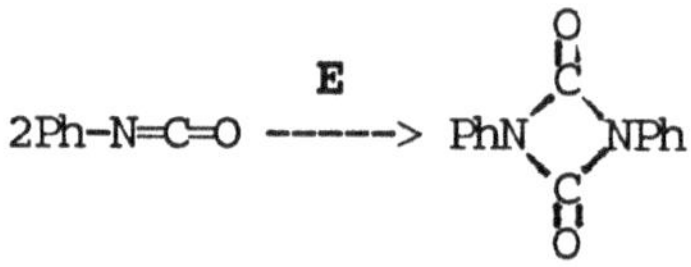

Carbon dioxide also reacts and while this bright yellow complex is still being characterized, ir and nmr spectra indicate a Cp_3U complex, and suggest CO_2 oligimerization.[52]

Some of the most interesting chemistry which we have encountered involves the reactions of $Cp_3U{=}CHPR_3$ with metal carbonyls. The typical reaction with terminal carbonyls is insertion:

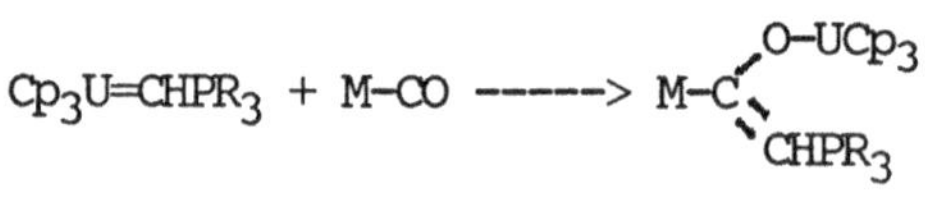

Structures have been obtained for M = $Cp(OC)_2Mn$-,[56] Cp(OC)Co-,[57] $(OC)_5W$-,[58] and $Cp_2(OC)_3Ru_2$.[59] The metrical parameters within the various $-C(OUCp_3){=}CHPR_3$ ligands are identical within experimental uncertainties and are consistent with the resonance form shown above, which is a metal enolate not a Fischer Carbene Complex. The uranium is tightly coordinated to oxygen and all four electrons from the uranium-carbon bond are utilized to form a new carbon-carbon "enolate" multiple bond.[38,56] While NMR data reveals that the $M\text{-}C(OUCp_3)\text{=}CHPR_3$ unit is usually the initial product with other metal carbonyls, the insertion product is often unstable.[51,60,7,61] In fact, even the $M\text{-}C(OUCp_3)\text{=}CHPR_3$ complexes which can be isolated show futher reactivity.

Mild heating of $Cp(OC)_2Mn\text{-}C(OUCp_3)\text{=}CHPR_3$[62] causes C-O bond cleavage:

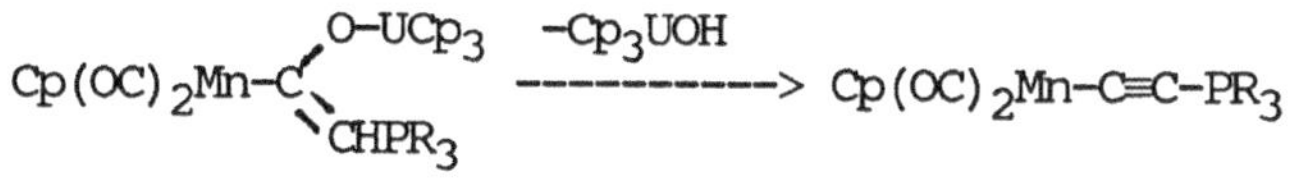

but this is not the only reaction of the $M\text{-}C(OUCp_3)\text{=}CHPR_3$ unit. $Cp(OC)Co\text{-}C(OUCp_3)\text{=}CHPR_3$ undergoes both C-O and C-P cleavage:[57]

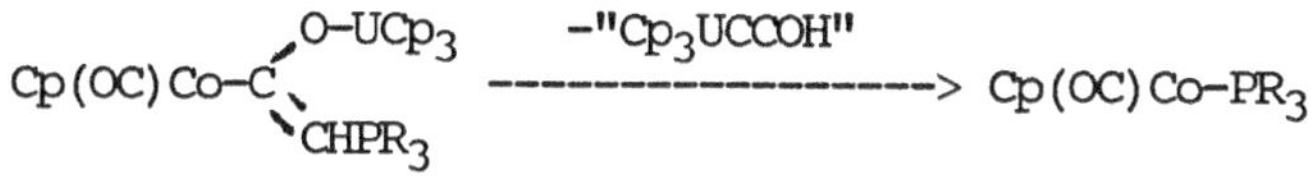

$(OC)_5W\text{-}C(OUCp_3)\text{=}CHPR_2(CH_3)$ undergoes an unusual isomerization:[58]

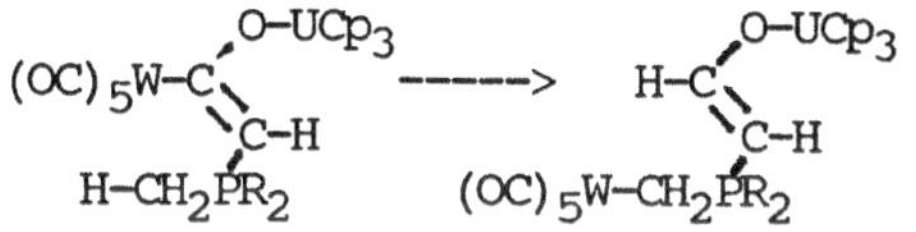

$(OC)_5Mo\text{-}C(OUCp_3)\text{=}CHPR_3$ is similar to $(OC)_5W\text{-}C(OUCp_3)\text{=}CHPR_3$, but with $(OC)_5Cr\text{-}C(OUCp_3)\text{=}CHPR_3$ isomerization is greatly depressed and C-O cleavage is the major reaction.[61]

While the reaction of $[CpM(CO)_2]_2$, M = Ru or Fe, with $Cp_3U\text{=}CHPMeRR'$ produces $Cp(OC)M(\mu\text{-}CO)_2M(Cp)[C(OUCp_3)\text{=}CHPRR']$:[59]

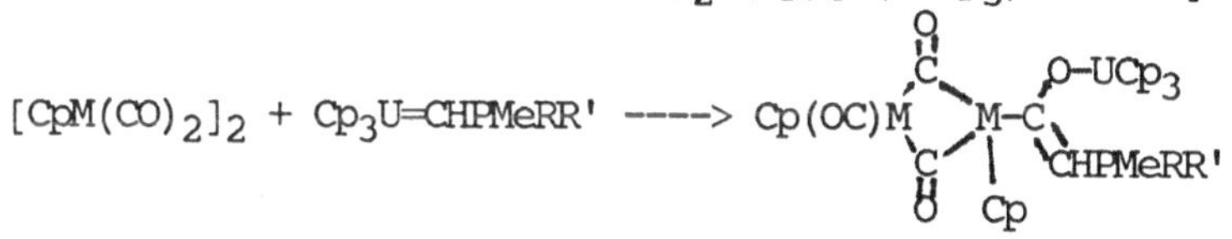

the nature of R and R' and M influences the stability of the products. Thus, $Cp_3U\text{=}CHPMe_3$ and $Cp_3U\text{=}CHPMe_2Ph$ react with $[CpRu(CO)_2]_2$ at room temperature to form products which are stable for long periods in both the solid state and solution. At least in our hands, an analogous product cannot be isolated from reactions of $Cp_3U\text{=}CHPMePh_2$. Rather, nmr spectra show that if the insertion product forms, it decomposes within a few minutes to as of yet uncharacterized materials. Qualitatively, then, the stability of the insertion product appears to decrease with increasing phenyl substitution on the phosphorus.

Similarly, $Cp(OC)Fe(\mu\text{-}CO)_2Fe(Cp)[C(OUCp_3)\text{=}CHPMe_3]$ is the most stable of the insertion products obtained from reactions with

$[CpFe(CO)_2]_2$. From $[CpFe(CO)_2]_2$ and $Cp_3U{=}CHPMe_2Ph$ the insertion product can be isolated. However, it slowly decomposes at room temperature even in the solid state. Finally, the insertion product could be observed in the nmr spectra of freshly made solutions of $Cp_3U{=}CHPMePh_2$ and $[CpFe(CO)_2]_2$, but was never obtained in pure form. In all cases the insertion products from the $[CpFe(CO)_2]_2$ reactions undergo an unusual carbonyl coupling reaction to yield:

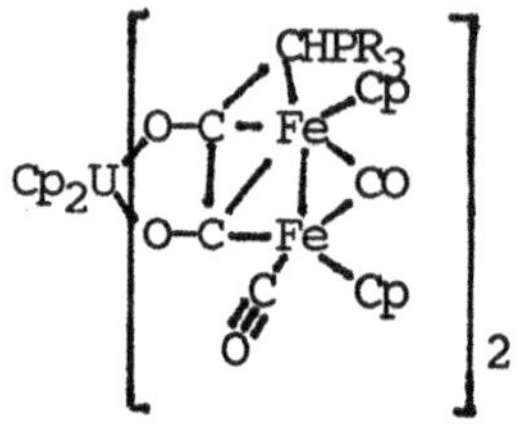

which contains an allyl group which has formed from a bridging and a terminal carbonyl, originally present in $[CpFe(CO)_2]_2$, and a carbon atom from the ylide moiety in $Cp_3U{=}CHPR_3$. The allyl is π bonded to one Fe and σ bonded to the other.[63,59]

Reactions which occur with other dinuclear carbonyls are not yet well characterized. $Mn_2(CO)_{10}$ forms a product which does not contain an $Mn{-}C(OUCp_3){=}CHPR_3$ unit. In contrast $Re_2(CO)_9[C(OUCp_3){=}CHPR_3]$ and $Re_2(CO)_8[C(OUCp_3){=}CHPR_3]_2$ has been characterized by nmr when $Re_2(CO)_{10}$ reacts with $Cp_3U{=}CHPR_3$.

Obviously, there are subtle effects which influence the course of the reactions of $Cp_3U{=}CHPR_3$ with metal carbonyls, and one of our current goals is to systematize the chemistry of the $M{-}C(OUCp_3){=}CHPR_3$ moiety. In a formal sense all of these reactions are examples of carbon monoxide activation via coordination of both C and O. The products of the C-O cleavage reactions are novel acetylides, $-C{\equiv}C{-}PR_3$, the isomerization produces an unusual enolate, and the carbonyl coupling reaction has formed an allyl moiety. We are continuing our investigations of this fascinating series of molecules.

References

1. Nugent, W. A.; Mayer, J. M. Metal-Ligand Multiple Bonds, John Wiley & Sons, New York, 1988.
2. Pyykkö, P. Inorg. Chim. Acta **1987**, 139, 243-245 and references therein.
3. Cramer, R. E.; Maynard, R. B.; Gilje, J. W. J. Am. Chem. Soc. **1978**, 100, 5562-5564.
4. Cramer, R. E.; Maynard, R. B.; Gilje, J. W. Inorg. Chem. **1980**, 19, 2564-2569.
5. Cramer, R. E.; Maynard, R. B.; Gilje, J. W. Inorg. Chem. **1981**, 20, 2466-2470.
6. Cramer, R. E.; Mori, A. L.; Maynard, R. B.; Gilje, J. W.; Tatsumi, K.; Nakamura, A. J. Am. Chem. Soc. **1984**, 106, 5920-5926.
7. Gilje, J. W.; Cramer, R. E.; Bruck, M. A.; Higa, K. T.; Panchanatheswaran, K. Inorg. Chim. Acta **1985**, 110 139-143.

8. Cramer, R. E.; Jeong, J. H.; Maynard, R. B.; Gilje, J. W. Inorg. Synth. **1990**, 27, 177-181.
9. Cramer, R. E.; Maynard, R. B.; Paw, J. C.; Gilje, J. W. J. Am. Chem. Soc. **1981**, 103, 3589-3590.
10. Cramer, R. E.; Maynard, R. B.; Paw, J. C.; Gilje, J. W. Organometallics **1983**, 2, 1336-1340.
11. Cramer, R. E.; Bruck, M. A.; Edelmann, F.; Afzal, D.; Gilje, J. W.; Schmidbaur, H. Chem. Ber. **1988**, 121, 417-420.
12. Stevens, R. C.; Bau, R.; Cramer, R. E.; Afzal, D.; Gilje, J. W.; Koetzle, T. F. Organometallics **1990**, 9, 694-697.
13. Raymond, K. N.; Eigenbrot, C. W., Jr. Acc. Chem. Res. **1980**, 13, 276-283.
14. McDowell, R. S.; Streitwieser, A., Jr. J. Am. Chem. Soc. **1984**, 106, 4047-8.
15. Head, J. D., University of Hawaii, Unpublished Results.
16. Tatsumi, K.; Nakamura, A. J. Organomet. Chem. **1984**, 272, 141-154.
17. Chun, T. C.; Hitt, J., Unpublished Results.
18. Schwartz, J.; Gell, K. I. Inorg. Chem. **1980**, 19, 3207-3211.
19. Erker, G.; Czisch, P.; Mynott, R.; Tsay, Y. H.; Kruger, C. Organometallics **1985**, 4, 1310-1312.
20. Erker, G.; Czisch, P.; Kruger, C.; Wallis, J. M. Organometallics **1985**, 4, 2059-2060.
21. Baldwin, J. C.; Keder, N. L.; Strouse, C. E.; Kaska, W. C. Z. Naturforsch. **1980**, 35B, 1289-1297.
22. Schmidbaur, H.; Pichl, R. Z. Naturforsch. **1985**, 40B, 352-356.
23. Schmidbaur, H.; Pichl, R.; Müller, G. Angew. Chem. Int. Ed. Engl. **1986**, 25, 574-575.
24. Schmidbaur, H.; Pichl, R.; Müller, G. Chem. Ber. **1987** 120, 39-44.
25. Cardin, D. J.; Lappert, M. F.; Raston, C. L. Chemistry of Organo-Zirconium and -Hafnium Compounds, Ellis Horwood Limited, West Sussex, England, 1986, Chapter 10.
26. Cramer, R. E.; Roth, S.; Edelmann, F.; Bruck, M. A.; Cohn, K. C.; Gilje, J. W. Organometallics **1989**, 8, 1192-1199.
27. Cramer, R. E.; Panchanatheswaran, P.; Gilje, J. W. J. Am. Chem. Soc. **1984**, 106, 1853-1854.
28. Brennan, J. G.; Andersen, R. A. J. Am. Chem. Soc. **1985**, 107, 514-516.
29. Zalkin, A.; Brennan, J. G.; Andersen, R. A. Acta Cryst. **1988**, C44, 1553-1554.
30. Burns, C. J.; Smith, W. H.; Huffman, J. C.; Sattleberger, A. C. J. Am. Chem. Soc. **1990**, 112, 3237-3239.
31. Dehnicke, K.; Strähle, J. Polyhedron **1989**, 8, 707-726.
32. Schrumpf, F.; Roesky, H. W.; Noltemeyer, M. Z. Naturforsch. **1990**, 45B, 1600-1602.
33. Bock, H.; tom Dieck, H. Z.Naturforsch. **1966**, 21b, 739-746.
34. Roesky, H. W.; Zimmer, M.; Schmidt, H. G.; Noltemeyer, M. Z. Naturforsch. **1988**, 43B, 1490-1494.
35. Roesky, H. W.; Zimmer, H.; Noltemeyer, M.; Sheldrick, G. M. Chem. Ber. **1988**, 121, 1377-1379.
36. Cramer, R. E.; Edelmann, F.; Mori, A. L.; Roth, S.; Gilje, J. W.;

Tatsumi, K.; Nakamura, A. Organometallics **1988**, 7, 841-849.
37. Ariyaratne, K. A. N. S., Ph. D. Thesis, University of Hawaii, 1992.
38. Cramer, R. E.; Engelhardt, U.; Higa, K. T.; Gilje, J. W. Organometallics **1987**, 6, 41-45.
39. Perego, G.; Cesari, M.; Farina, F.; Lugli, G. Acta Crystallogr., Sect. B **1976**, B32, 3034-3039.
40. Nugent, W. A.; Haymore, B. I. Coord. Chem Rev. **1980**, 31, 123-175.
41. Bezler, H.; Strähle, J. Z. Naturforsch. **1979**, 34b, 1199-1202.
42. Bezler, H.; Strähle, J. Z. Naturforsch. **1983**, 38b, 317-320.
43. Dilworth, J. R.; Neaves, B. D.; Hutchinson, J. P.; Zubieta, J. A. Inorg. Chim. Acta **1982**, 65, L223-L224.
44. Hösler, K.; Weller, F.; Dehnicke, K. Z. Naturforsch. **1987**, 42B, 1563-1566.
45. Völp, K.; Weller, F.; Dehnicke, K. Z. Naturforsch. **1987**, 42B, 947-950.
46. Cramer, R. E.; Roth, S.; Gilje, J. W. Organometallics **1989**, 8, 2327-2330.
47. Sudhakar, P.V.; Lammertsma, K. J. Am. Chem. Soc. **1991**, 113, 1899-1906.
48. Bruno, J. W.; Smith, G. M.; Marks, T. J.; Fair, C. K.; Schultz, A. J.; Williams, J. M. J. Am. Chem. Soc. **1986**, 108, 40-56.
49. Tatsumi, K.; Nakamura, A. J. Am. Chem. Soc. **1987**, 109, 3195-3206.
50. Bursten, B. E.; Strittmatter, R. J. Angew. Chem. Int. Ed. Engl. **1991**, 30, 1069-1085.
51. K. T. Higa, Ph.D. Thesis, University of Hawaii, 1984.
52. Panchanatheswaran, K., Ph. D. Thesis, University of Hawaii, 1984.
53. Cramer, R. E.; Maynard, R. B.; Paw, J. C.; Gilje, J. W. Organometallics **1982**, 1, 869-871.
54. Cramer, R. E.; Panchanatheswaran, K.; Gilje, J. W. Angew. Chem. Int. Ed. Engl. **1984**, 23, 912-913.
55. Cramer, R. E.; Jeong, J. H.; Gilje, J. W. Organometallics **1987**, 5, 2010-2012.
56. Cramer, R. E.; Higa, K. T.; Gilje, J. W. J. Am. Chem. Soc. **1984**, 106, 7245-7247.
57. Cramer, R. E.; Jeong, J. H.; Richmann, P. R.; Gilje, J. W. Organometallics **1990**, 9, 1141-1146.
58. Cramer, R. E.; Jeong, J. H.; Gilje, J. W. Organometallics **1986**, 5, 2555-2557.
59. Afzal, D., Unpublished Results.
60. Jeong, J. W., Ph.D. Thesis, University of Hawaii, 1987.
61. Gilje, J. W.; Cramer, R. E. Inorg. Chim. Acta **1987**, 139, 177-181.
62. Cramer, R. E.; Higa, K. T.; Gilje, J. W. Organometallics **1985**, 4, 1496-1499.
63. Cramer, R. E.; Higa, K. T.; Pruskin, S. L.; Gilje, J. W. J. Am. Chem. Soc. **1983**, 105, 6749-6750.

Metal Carbenes and Metal Carbynes as Precursors for a Rational Synthesis of Carbido and Hydrocarbon Bridged Complexes

Wolfgang Beck*, Josef Breimair, Peter Fritz, Wolfgang Knauer, Torsten Weidmann
Institut für Anorganische Chemie der Universität München,
Meiserstr. 1, 8000 München 2, Germany.

Abstract.
Addition of anionic Fischer carbene complexes, of their CS_2 adducts and of an anionic thiocarbyne complex to cationic organometallic Lewis acids and to coordinated, unsaturated hydrocarbons gives novel hydrocarbon bridged heterodimetallic compounds. Oxidative addition of halocarbyne complexes to zerovalent platinum and palladium complexes provides a synthesis of carbido bridged complexes.

1. Introduction

Nucleophilic attack on coordinated unsaturated hydrocarbons is one of the fundamental and particularly well studied reactions in Organometallic Chemistry. The addition of carbonylmetallates instead of common nucleophiles provides a directed synthesis of hydrocarbon bridged complexes. Carbonylmetallates (particularly $Re(CO)_5^-$, $Os(CO)_4^{2-}$) add to π-bonded olefin, acetylene, allyl, diene, trimethylenemethane, dienyl, benzene, triene and cycloheptatrienyl ligands in cationic complexes and give hydrocarbon bridged bi- and trimetallic, homo- or heteronuclear complexes [1].

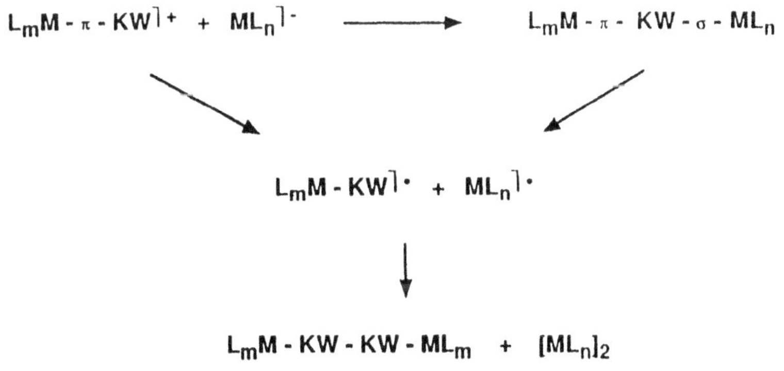

KW = unsaturated hydrocarbon

F. R. Kreißl (ed.), Transition Metal Carbyne Complexes, 189–199.

In competition to nucleophilic attack electron transfer from the carbonylmetallate to the cationic complex may occur [2] to give 17 and 19 electron complexes which couple with formation of metal-metal and C-C bonds.

Another directed route to hydrocarbon bridged complexes is C-C coupling by addition of anionic π-hydrocarbon complexes to coordinated hydrocarbons of cationic complexes [3,4].

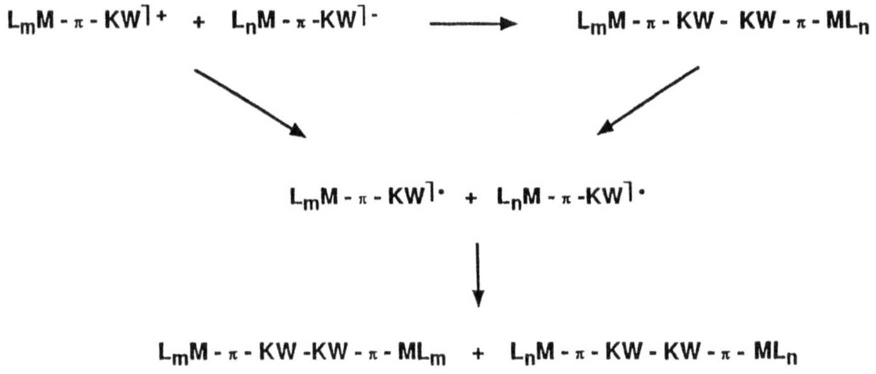

In continuation of these studies we have used anionic Fischer carbene complexes [5] and the anionic thiocarbyne complex HB(3,5-dimethylpyrazolyl)$_3$(OC)$_2$MO≡C-S$^-$ [6] as nucleophiles. The latter reaction represents a third method for a rational synthesis of hydrocarbon bridged complexes according to the scheme:

M-KW-Nu]$^-$ + KW-M]$^+$ ⟶ M-KW-Nu-KW-M

Nu = nucleophile, e.g. sulfur.

We also have found new examples of rare μ-carbido complexes M=C=M.

2. Addition of Anionic Fischer Carbene Complexes to Coordinated unsaturated Hydrocarbons and to Cationic Organometallic Lewis Acids

The acidity of hydrogen atoms in α-position to the carbene atom of Fischer carbene complexes has been demonstrated by Kreiter [7]. The CH_2 group of the complexes $(OC)_5M{=}C(OMe)CH_2^-$ (M = Cr,Mo) can be alkylated and silylated [8]. The addition of $(OC)_5M{=}C(OMe)CH_2^-$ which shows a versatile and rich chemistry [9a,b] to the cationic complexes $(OC)_5Re(C_2H_4)^+$ and $(OC)_3M'(\eta^7\text{-}C_7H_7)^+$ (M' = Cr,Mo) affords the heterobimetallic hydrocarbon bridged complexes $(OC)_5M{=}(COMe)CH_2CH_2CH_2Re(CO)_5$ and $(OC)_5M{=}(COMe)CH_2C_7H_7\text{-}\eta^6\text{-}M'(CO)_3$ which structures have been determined by X-ray diffraction [10].

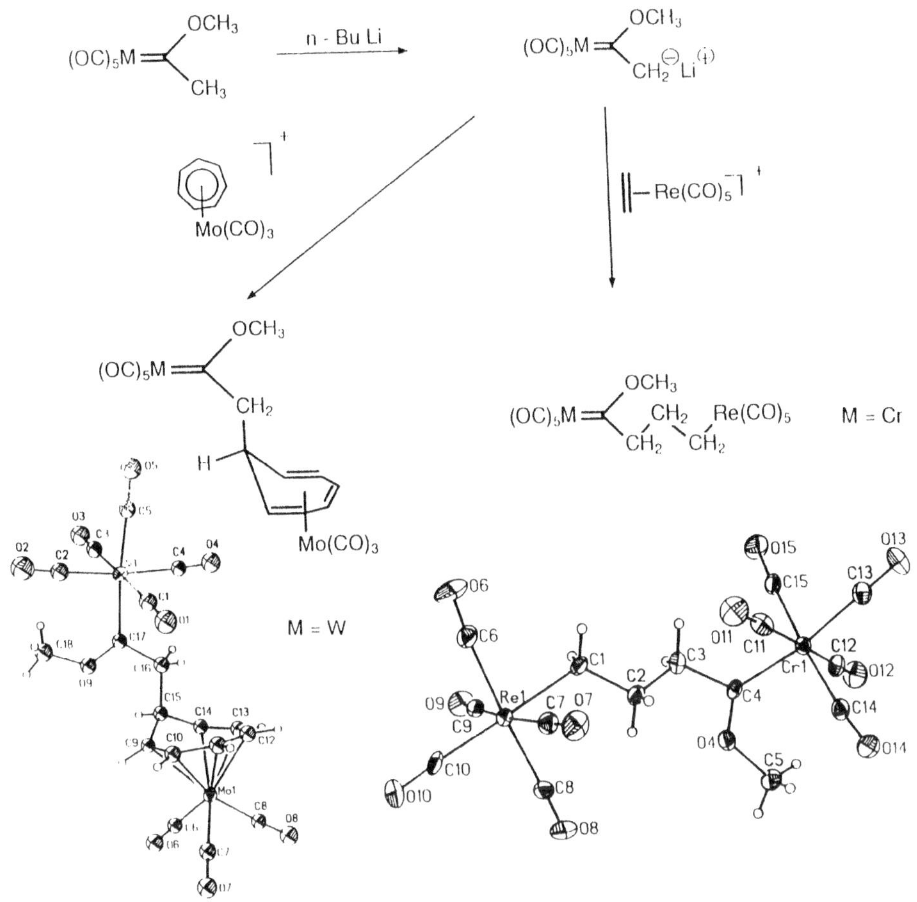

Aumann has recently studied the addition of the uncoordinated tropylium ion to these anionic Fischer carbene complexes [9c]. Similarly, the addition of the cationic allyl complex $Cp(OC)(ON)Mo(\eta^3\text{-}C_3H_5)^+$ gives new hydrocarbon bridged complexes.

It has been shown by Raubenheimer [11] that carbondisulfide adds to mono- and dianionic Fischer carbene complexes. We have used these anions as nucleophiles; they react with cationic Organometallic Lewis Acids [12] and with cationic ethylene and allyl complexes to give a series of novel ligand bridged dimetallic compounds:

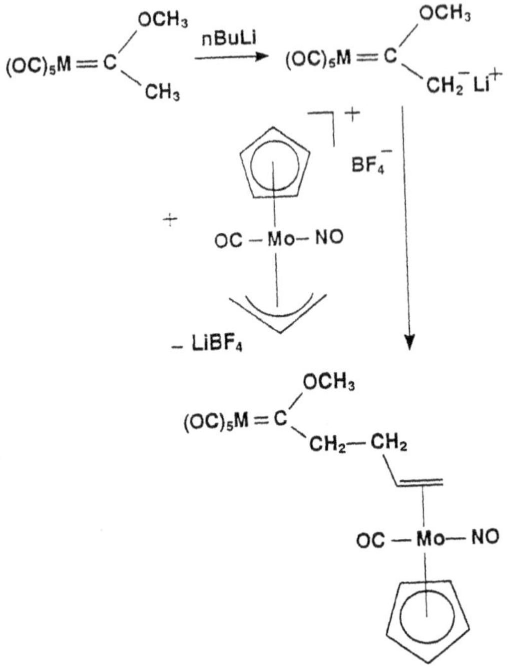
OCH3
nBuLi
(OC)5M = C
CH3
CH2⁻ Li+
BF4⁻
OC – Mo– NO
– LiBF4
CH2— CH2

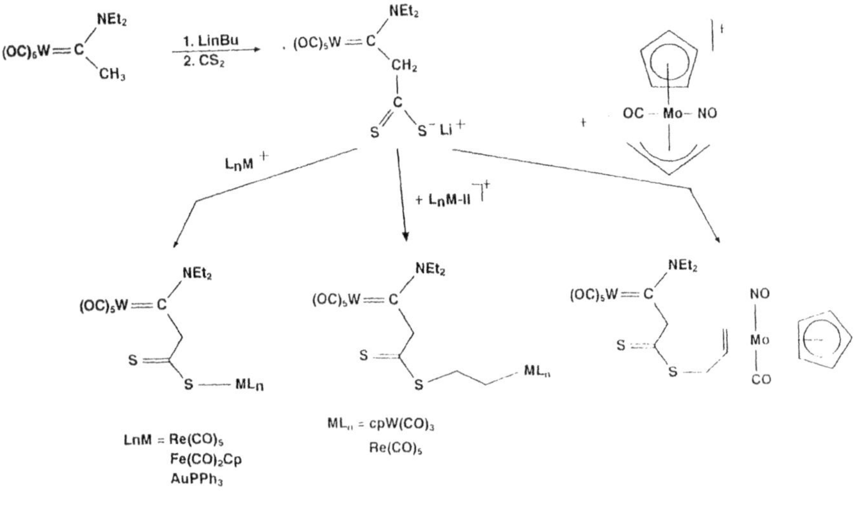
NEt2
(OC)5W = C
CH3
1. LinBu
2. CS2
CH2
S⁻ Li+
OC – Mo– NO
LnM +
+ LnM-II
MLn
LnM = Re(CO)5
Fe(CO)2Cp
AuPPh3
MLn = cpW(CO)3
Re(CO)5
NO
Mo
CO

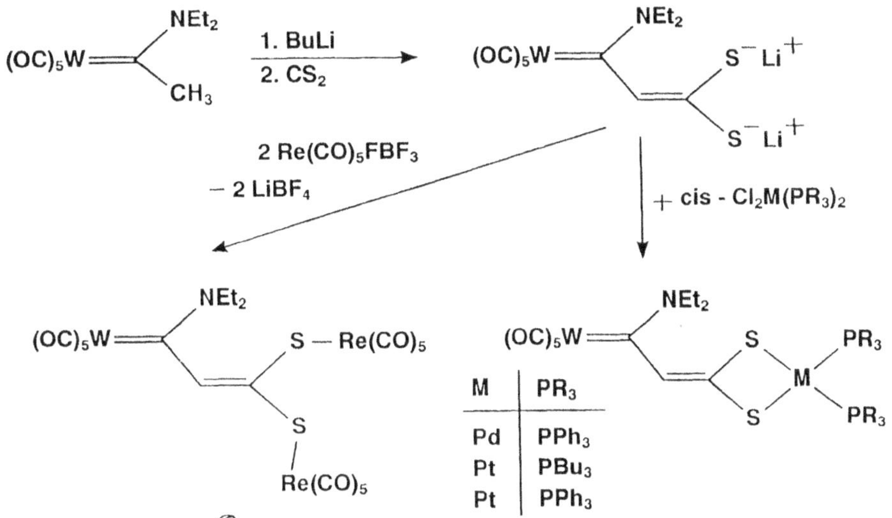

M	PR_3
Pd	PPh_3
Pt	PBu_3
Pt	PPh_3

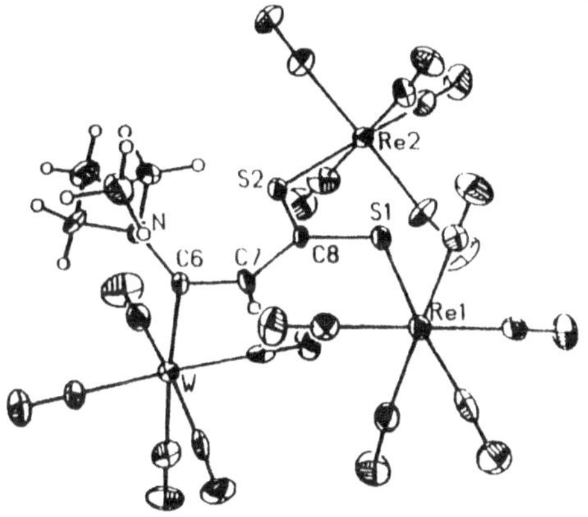

3. Addition of the Anionic Thiocarbyne Complex $HB(pz)_3(OC)_2Mo{\equiv}C\text{-}S^-$ [6] to Cationic Organometallic Lewis Acids and to Coordinated Unsaturated Hydrocarbons

These reactions proceed with formation of a series of rare [13] thiocarbonyl bridged complexes and of novel hydrocarbon bridged complexes as it is shown in the following schemes:

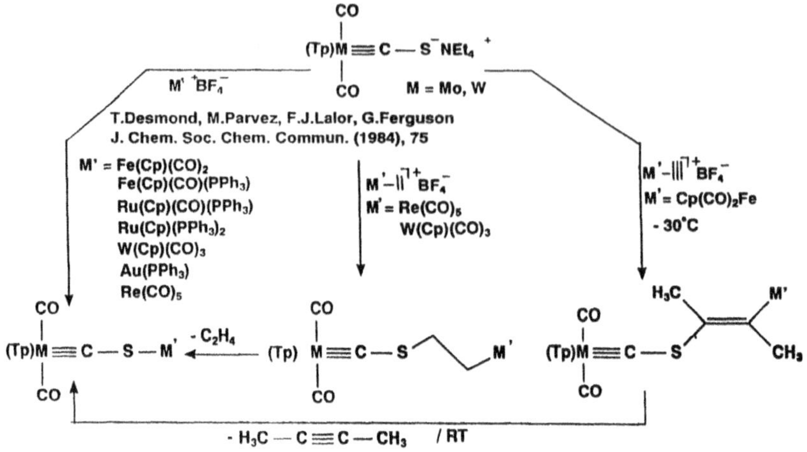

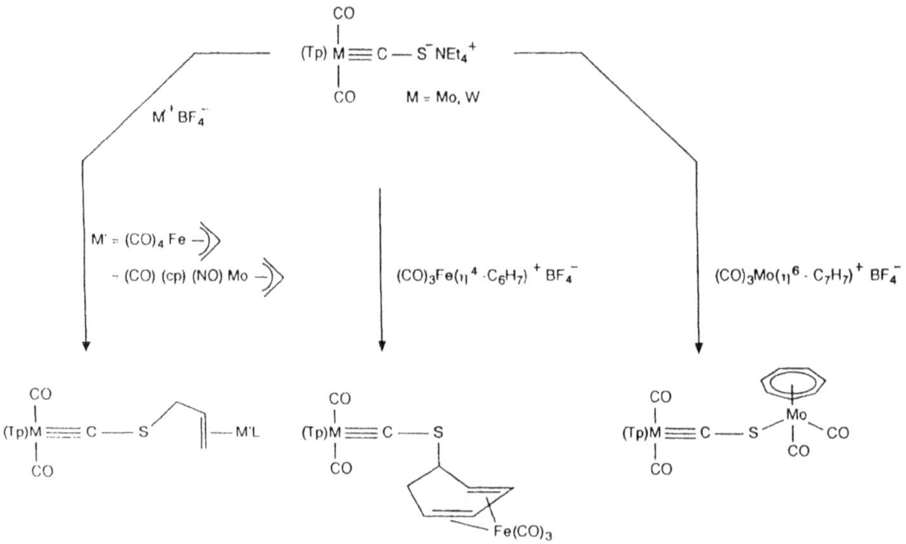

4. New μ_2-Carbido Metal Complexes M=C=M

Bimetallic carbido bridged complexes (M=C=M) or metallated carbyne complexes (M≡C-M) with a "naked" carbon [14] atom between two metal atoms are rare. Until now only the tetraphenylporphinato [15] and phthalocyanato iron complexes [16], the W≡C-Ru complex which is formed by an interesting methatesis reaction from $(Me_3CO)_3W{\equiv}W(OCMe_3)_3$ and $Cp(OC)_2Ru\text{-}C{\equiv}C\text{-}Me$ [17] and the Mo≡C-Fe compound (from $HB(pyrazolyl)_3Mo{\equiv}C\text{-}Cl$ and $Cp(OC)_2Fe^-$) [18] are known.

$(TPP)Fe{=}C{=}Fe(TPP)$	**Mansuy et al. 1981**
$(Me_3CO)_3W{\equiv}C{-}Ru(CO)_2Cp$	**Selegue et al. 1987**
$(pc)Fe{=}C{=}Fe(pc)$	**Ercolani et al. 1988**
$(TPP)Fe{\equiv}C{-}Re_2(CO)_9$	**Beck, Knauer 1990**
$HB(pz)_3Mo{\equiv}C{-}Fe(CO)_2Cp$	**Templeton 1991**

We have used Mansuy's dichlorcarbene iron complex [19] as starting material which reacts with the carbonyl metallates $Re(CO)_5^-$, $Cr(CO)_5^{2-}$ and $Fe(CO)_4^{2-}$ to give heterobimetallic μ-carbido bridged compounds.

$$(P)Fe{=}CCl_2 + 2M(CO)_5^- \xrightarrow[-2Cl^-]{-CO} (P)Fe{\equiv}C{-}M(CO)_4{-}M(CO)_5$$

M = Mn, Re

$$(P)Fe{=}CCl_2 + Cr(CO)_5^{2-} \xrightarrow[-2Cl^-]{} (P)Fe{=}C{=}Cr(CO)_5$$

$$(P)Fe{=}CCl_2 + Fe(CO)_4^{2-} \xrightarrow[-2Cl^-]{} (P)Fe{=}C{=}Fe(CO)_4$$

(P) = TPP and 4 - phenyl substituted
(COMe, i-Pr, CN) porphyrinate

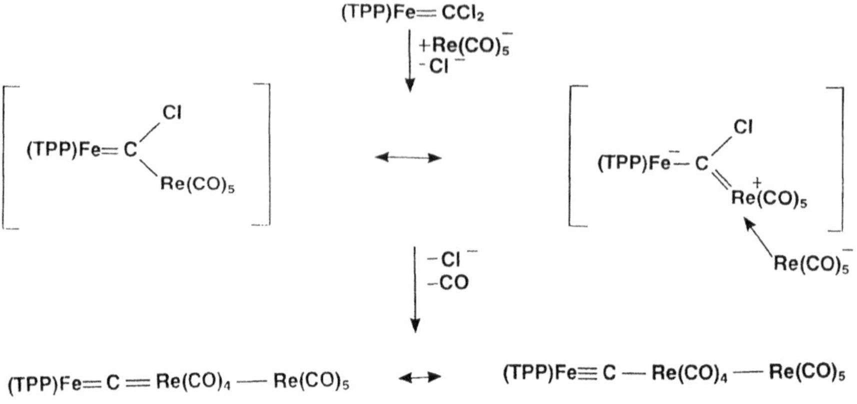

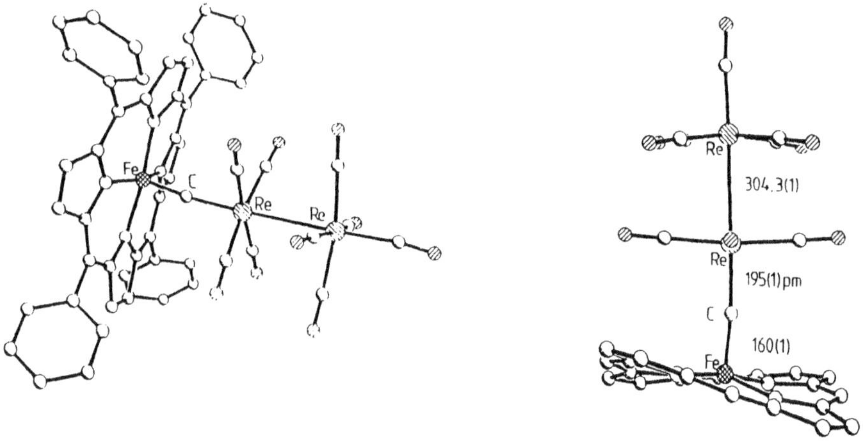

The reaction of (TPP)Fe=CCl_2 with pentacarbonyl rhenate may proceed by substitution of one chloride by $Re(CO)_5^-$ and addition of another $Re(CO)_5^-$ anion to a (CCl)-bridged intermediate. The iron carbon bond [1.605(13)A] of the complex (TPP)Fe≡C-$Re_2(CO)_9$ appears to be the shortest metal carbon bond so far reported [20].

Another route to μ-Carbido complexes which we have found is oxidative addition of Lalor's halocarbyne complexes [21] to zero valent triphenylphosphine compounds of nickel, palladium and platinum. In these reactions the heterodimetalla cyclopropenes could be isolated which isomerize at higher temperatures to the dimetalla allene complexes. The first adducts of arylcarbyne complexes to zerovalent platinum complexes (arylcarbyne as analogue of ace-

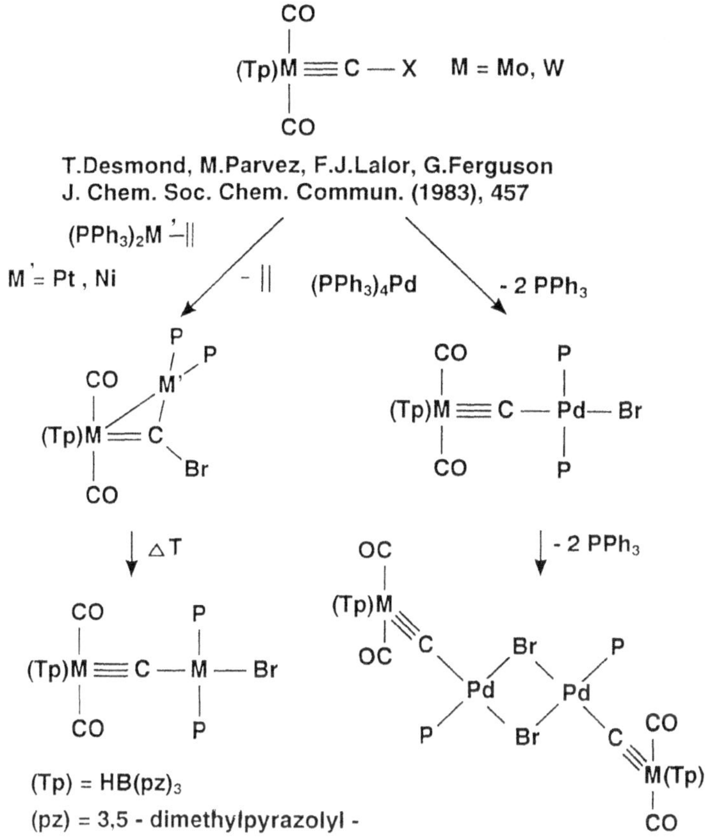

tylene) have been obtained by Stone and coworkers [22] and many examples of these complexes have been studied in his group [23].

Generous support by Deutsche Forschungsgemeinschaft and Fonds der Chemischen Industrie is gratefully acknowledged. We thank Dr. Ch. Robl, Dr. K. Sünkel and Dr. B. Wagner for X-ray structural studies.

References

1 W. Beck, Polyhedron 7 (1988) 2255; W. Beck, B. Niemer, J. Breimair and J. Heidrich, J.Organomet. Chem. 372 (1989) 79; B. Niemer, J. Breimair, T. Völkl, B. Wagner, K. Polborn and W. Beck, Chem. Ber. 124 (1991) 2237; B. Niemer, T. Weidmann and W. Beck, Z.Naturforsch. Teil B, 47 (1992) 509 and earlier publication in the series "Hydrocarbon Bridged Metal Complexes" by W. Beck and coworkers.

2 R. E. Lehmann and J.K. Kochi, Organometallics 10 (1991) 190; H.-J. Müller, U. Nagel, M. Steimann and W. Beck, Chem. Ber. 122 (1989) 1387.

3 G. Deganello, T. Boschi and L. Toniolo, J.Organomet. Chem. 97 (1975) C46; M. Moll, P. Würstl, H. Behrens and P. Merbach, Z. Naturforsch., Teil B 33 (1978) 1304; M. Airoldi, G. Deganello, G. Dia, P. Saccone and J. Takats, Inorg. Chim. Acta 41 (1980) 171.

4 M. Wieser, K. Sünkel, C. Robl and W. Beck, Chem. Ber. 125 (1992) 1369.

5 E.O. Fischer and A. Maasböl, Chem. Ber. 100 (1967) 2445.

6 T. Desmond, F.J. Lalor, G. Ferguson and M. Parvez, J. Chem. Soc. Chem. Commun. (1984) 75.

7 C.G. Kreiter, Angew. Chem. Int. Ed. Engl. 7 (1968) 390; C.P. Casey and R.L. Anderson, J.Am.Chem.Soc. 96 (1974) 1230.

8 Y.-Ch. Xu and W.D. Wulff, J. Org. Chem. 52 (1987) 3263; D.W. Macomber, P. Madhukar and R.D. Rogers, Organometallics 8 (1989) 1275; D.W. Macamber and P. Madhukar, J. Organomet. Chem. 433 (1992) 279 and references cited therein.

9a C.P. Casey, CHEMTECH (1979) 378; C.P. Casey and W.R. Brunsvold, Inorg. Chem. 16 (1977) 391.

9b R. Aumann and H. Heinen, Chem. Ber. 120 (1987) 537 and other publications by R. Aumann on "Organic Syntheses with Transition Metal Complexes".

9c R. Aumann, Chem. Ber. 125 (1992) 1141; R. Aumann and M. Runge, Chem. Ber. 125 (1992) 259.

10 J. Breimair, T. Weidmann, B. Wagner and W. Beck, Chem. Ber. 124 (1991) 2431.

11 H.G. Raubenheimer, G.J. Krüger and H.W. Viljoen, J. Chem. Soc., Dalton Trans. (1985) 1963.

12 W. Beck and K. Sünkel, Chem. Rev. 88 (1988) 1405.

13 B.D. Dombek and R.J. Angelici, J.Am.Chem. Soc. 96 (1974) 7568; S. Lotz, R.R. Pille and P.H. Van Rooyen, Inorg. Chem. 25 (1986) 3053.

14 W.A. Herrmann, Angew. Chem. Int. Ed. Engl. 25 (1986) 56.

15 D. Mansuy, J.-P. Lecomte, J.-C. Chottard and J.-F. Bartoli, Inorg. Chem. 20 (1981) 3119; V.L. Goedken, M.R. Deakin and L.A. Bottomley, J. Chem. Soc. Chem. Commun. (1982) 607.

16 C. Ercolani, M. Gardini, V.L. Goedken, G. Pennesi, G. Rossi, U. Russo and P. Zanonato, Inorg. Chem. 28 (1989) 3097.

17 S.L. Latesky and J.P. Selegue, J. Am. Chem. Soc. 109 (1987) 4731; G.A. Koutsantonis and J.P. Selegue, J. Am. Chem. Soc. 113 (1991) 2316.

18 M. Etienne, P.S. White and J.L. Templeton, J. Am. Chem. Soc. 113 (1991) 2324.

19 D. Mansuy, M. Lange, J.-C. Chottard, P. Guerin, P. Morliere, D. Brault and M. Rougee, J. Chem. Soc. Chem. Commun. 1977, 648; D. Mansuy, M. Lange, J.-C. Chottard, J.F. Bartoli, B. Chevrier and R. Weiss, Angew. Chem. Int. Ed. Engl. 17 (1978) 781; D. Mansuy, Pure Appl. Chem. 52 (1980) 681; review on dihalocarbene complexes: P.J. Brothers and W.R. Roper, Chem. Rev. 88 (1988) 1293.

20 W. Beck, W. Knauer and C. Robl, Angew. Chem. Int. Ed. Engl. 29 (1990) 318.

21 T. Desmond, F.J. Lalor, G. Ferguson and M. Parvez, J. Chem. Soc. Chem. Commun. (1983) 457.

22 T.V. Ashworth, J.A.K. Howard and F.G.A. Stone, J. Chem. Soc. Chem. Commun. (1979) 42.

23 F.G.A. Stone, Angew. Chem. Int. Ed. Engl. 23 (1984) 89; M. Green, J.A.K. Howard, A.P. James, C.M. Nunn and F.G.A. Stone, J.Chem. Soc. Dalton Trans. (1986) 187; J.E. Goldberg, D.F. Mullica, E.L. Sappenfield and F.G.A. Stone, J. Chem. Soc., Dalton Trans. (1992) 2495.

Note: Recently, Geoffroy and coworkers have reported closely related additions of anionic Fischer carbene complexes to coordinated olefin and cyclohexadienyl ligands (Inorg.Chim.Acta 198-200 (1992) 601).

SOME CHEMISTRY OF $Tp'(CO)_2W{\equiv}C\text{-}H$, A SIMPLE TERMINAL CARBYNE

G. M. Jamison, P. S. White, D. L. Harris and J. L. Templeton*

Department of Chemistry
University of North Carolina, Chapel Hill
Chapel Hill, North Carolina 27599-3290

ABSTRACT: Hydridocarbynes with the formula $Tp'(CO)_2M{\equiv}C\text{-}H$ (M = Mo, W; Tp' = $[HB(3,5\text{-}Me_2C_3HN_2)_3]^-$) have been prepared by fluorodesilylation of carbyne precursors, $Tp'(CO)_2M{\equiv}C\text{-}SiMe_2Ph$. Both molybdenum and tungsten terminal carbyne monomers undergo dimerization reactions to form unusual vinylidene-bridged products, $Tp'(CO)_2M(\mu\text{-}\eta^2{:}\eta^2\text{-}CCH_2)M(CO)_2Tp'$. Spectroscopic properties of these complexes are described herein as is the solid state x-ray structure of the tungsten dimer. Electrochemical data and a qualitative molecular orbital description of the dimeric products are also presented.

INTRODUCTION

Discovery of the first mononuclear carbyne complexes by E. O. Fischer nearly twenty years ago[1] has been followed by numerous fundamental studies of the multiple bonding of these simple monovalent carbon nuclei to a single metal center.[2] Typically intermolecular coupling of two mononuclear carbyne ligands results in the formation of bimetallic bridging alkyne complexes (eq. 1); this transformation was observed by Stone via proton-induced dimerization of $Cp(CO)_2W{\equiv}C\text{-}C_6H_4CH_3$,[3] and also by catalytic activation toward coupling by $Cp_2Cr_2(CO)_6$.[4] Similar reactivity

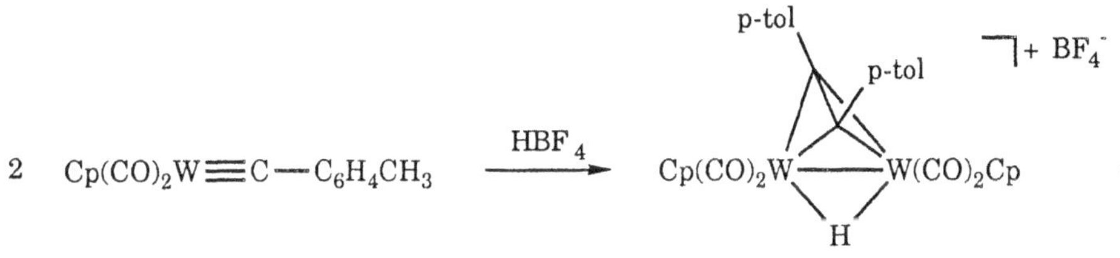

(1)

pathways have also been detected in high-valent methylidyne complexes.[5] In these examples the alkyne bridges perpendicularly across the metal-

F. R. Kreißl (ed.), Transition Metal Carbyne Complexes, 201–218.

metal bond, as a four electron donor, with both carbons interacting equally with both metal nuclei (this is in contrast to the two electron donor σ, σ bridging alkyne mode observed in certain instances).[6] This type of complex is also available from reaction of free organic alkynes with triply-bonded $L_nM{\equiv}ML_n$ in the presence of donor ligands as detailed by Chisholm et. al. (eq. 2).[5]

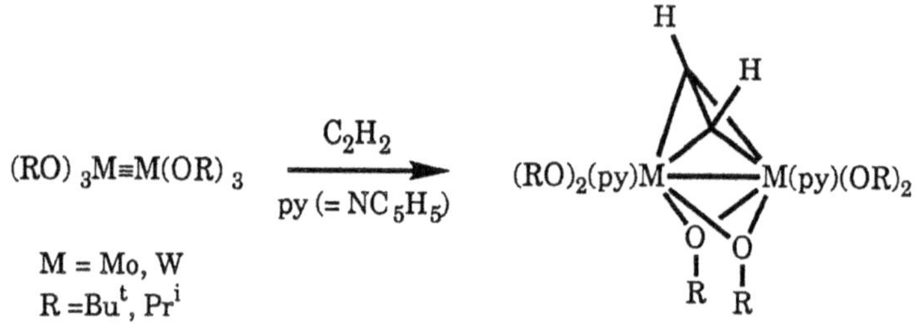

(2)

1

A number of homo- and heterobimetallic complexes bearing bridging vinylidenes, CCR_2, have appeared over the last decade.[7] The majority of these compounds bear the vinylidene bridge symmetrically across the M-M axis via metal-vinylidene σ bonds (the σ, σ ($2e^-$) bridging mode; see Figure 1); a notable exception is the σ, η^2 ($4e^-$) bridging mode found in $Cp(CO)_2Mo$[8] and CpRhL (L = $Pr^i{}_3P$)[9] systems in which the CCR_2 moiety donates four electrons to the bimetallic substrate in a "side-on" fashion.

Figure 1

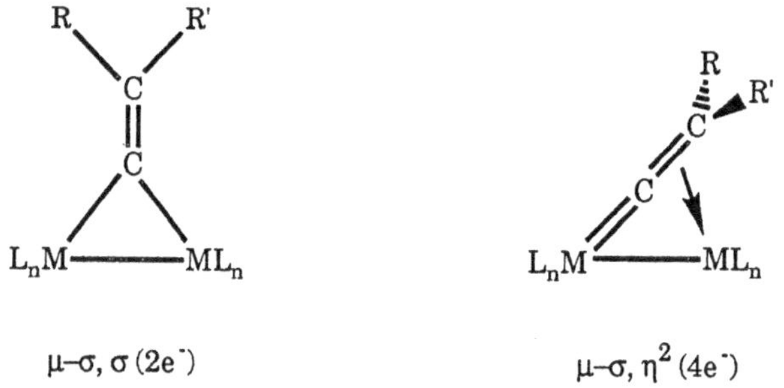

μ–σ, σ ($2e^-$) μ–σ, η^2 ($4e^-$)

L_nM = $Cp(CO)_2Mo$ or $Cp(Pr^i{}_3P)Rh$

Herein we present details of the formation of stable terminal mononuclear methylidyne complexes of molybdenum and tungsten bearing the Tp' ligand (**3a** and **3b**, Tp' = $HB[3,5\text{-}Me_2C_3HN_2]_3^-$).[10] These unusual Fischer carbynes are accessible through reaction at the carbyne ligand of their respective silylcarbyne precursors **2a** and **2b**. Furthermore, these methylidynes dimerize to generate unusual homobimetallic products bearing a formal bridging vinylidene moiety in a rare μ-η^2:η^2-CCH_2 bridging mode (eq. 3-5).

$$M(CO)_6 \xrightarrow[\text{ii.}\ \text{iii.}]{\text{i.}} Tp'(CO)_2M{\equiv}C\text{-}SiMe_2Ph \qquad (3)$$

2a: M = Mo
2b: M = W

i. $Me_2PhSiLi$, 10:1 Et_2O:THF, 0 °C.
ii. $(F_3CC(O))_2O$, -78 °C.
iii. KTp'/MeOH, -78 °C → 20 °C, 6hr.

$$Tp'(CO)_2M{\equiv}C\text{-}SiMe_2Ph \xrightarrow[\text{THF}]{Bu_4NF} Tp'(CO)_2M{\equiv}C\text{-}H \qquad (4)$$

2a, 2b -78 °C → 25 °C

3a: M = Mo
3b: M = W

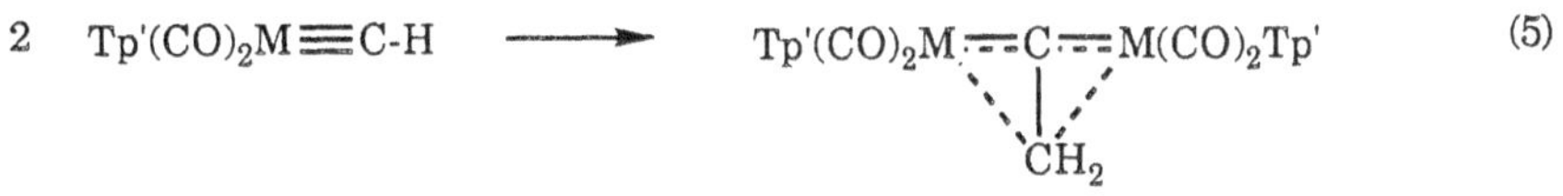

4a: M = Mo
4b: M = W

RESULTS AND DISCUSSION

Synthesis and Characterization of Molybdenum and Tungsten Silylcarbynes.

Coordination of the bulky Tp' ligand (Tp' = $HB[Me_2C_3HN_2]_3^-$) in organometallic complexes, including carbynes, has been shown to inhibit aggregation and affect reactivity pathways by virtue of the large size of this tridentate ligand (cone angle ~225°).[11] Incorporation of Tp' into the ligand sphere can be accomplished by displacement of three mutually cis terminal carbonyls from metal hexacarbonyls prior to generation of the carbyne ligand.[12] The Tp' ligand can also be incorporated into the ligand sphere of more highly functionalized organometallic substrates in much the same way that Cp and its substituted derivatives are employed to generate $C_5H_nR_{(5-n)}$-containing carbynes and related complexes.[13]

By employing the dimethylphenylsilyl anion as a nucleophile to initiate Mayr's multistep Fischer carbyne synthesis,[14] and later utilizing the Tp' anion as a capping tridentate ligand, new silylcarbyne complexes Tp'$(CO)_2$M≡C-$SiMe_2$Ph (**2a**, M = Mo; **2b**, M = W) can be obtained (eq. 3). Isolated yields of the new carbynes are low; purification involves separation from unreacted metal hexacarbonyl and salts by repeated alumina chromatography. Nonetheless, multigram quantities of **2a** and **2b** are available by this method.

Silylcarbynes **2a** and **2b** are characterized by two strong stretching frequencies in their infrared spectra (at 1997 and 1911 cm^{-1} for **2a**; 1982 and 1889 cm^{-1} for **2b**). ^1H NMR spectroscopy indicates C_s molecular symmetry, as a 2:1 out-of-plane: in-plane pattern is observed for those signals associated with the three pyrazole rings of the Tp' ligand. The symmetry plane contains the M≡C bond and one Tp' pyrazole ring and bisects the two terminal carbonyls. Additional resonances are easily assigned to the silicon substituents.

Low field ^{13}C NMR resonances assignable to the carbyne C_α nuclei of **2a** (360.4 ppm) and **2b** (339.0 ppm, $^1J_{W-C}$ = 160 Hz; ^{183}W 14% abundant, I = 1/2) verify the triple bond which links the $CSiR_3$ unit to the metal center.[2] A single terminal carbonyl resonance in each spectrum reflects the C_s symmetry of these complexes.

Synthesis and Characterization of Tungsten Hydridocarbyne (3b).

We have improved upon the isolation of milligram quantities of the parent hydridocarbyne complex **3b** from hydrido(phosphonium)carbene precursors.[11f] Tungsten hydridocarbynes trans-X$(PMe_3)_4$W≡C-H (X = Cl, I, OTf, BH_4) have been reported by Schrock (eq. 6).[15] The $(t\text{-}BuO)_3$W≡C-H hydridocarbyne has been observed spectroscopically by Chisholm (eq. 7).[5a]

Preparation of tungsten carbyne **3b** from silylcarbyne **2b** yields gram-scale quantities of the methylidyne product. Reaction of silylcarbyne **2b** with a slight excess of Bu_4NF in wet THF at low temperature results in quantitative conversion to hydridocarbyne **3b** as monitored by IR spectroscopy (eq. 4). Solvent evaporation followed by alumina chromatography yields **3b** as a bright yellow powder. A molecular weight of 552, as determined by vapor pressure osmometry in CH_2Cl_2, confirms the monomeric nature of the hydridocarbyne product.

$$WCl_2(PMe_3)_4 \xrightarrow[\text{tmeda}]{2\ AlMe_3,\ PhCH_3} Cl(PMe_3)_4W{\equiv}C\text{-}H \quad (6)$$

$$W_2(Bu^tO)_6(\mu\text{-}C_2H_2) \rightleftharpoons 2\ (Bu^tO)_3W{\equiv}C\text{-}H \quad (7)$$

Hydridocarbyne **3b** displays a dicarbonyl pattern in its IR spectrum (ν_{CO} = 1986 and 1891 cm^{-1}). In addition, a low-field ^1H NMR singlet at 8.22 ppm flanked by satellites due to tungsten coupling ($^2J_{W\text{-}H}$ = 83 Hz) is assigned to the proton of the hydridocarbyne ligand. The low-field chemical shift and large $^2J_{W\text{-}H}$ value are diagnostic for the η^1-CH ligand. There is obviously a highly efficient coupling mechanism operating to generate a two-bond $J_{W\text{-}H}$ value of 83 Hz. The magnitude of this coupling exceeds many $^1J_{W\text{-}H}$ values.[16, 17]

A characteristic carbyne C_α resonance at 280 ppm ($^1J_{W\text{-}C}$ = 192 Hz, $^1J_{C\text{-}H}$ = 142 Hz) in the ^{13}C NMR spectrum unambiguously identifies the η^1-carbyne ligand.[2, 15a, c, d] In contrast to large $^1J_{C\text{-}H}$ (~250 Hz) which characterize terminal alkynes,[18] the organotungsten terminus of the triple bond in **3b** perturbs the coupling mechanism so as to give a remarkably low $^1J_{C\text{-}H}$ value of 142 Hz.

Formation and Characterization of Tp'(CO)$_2$M(μ-η^2:η^2-CCH$_2$)M(CO)$_2$Tp' (4a and 4b).

In our attempts to isolate molybdenum methylidyne **3a** through fluorodesilylation of **2a**, formation of the parent carbyne **3a** was observed by IR spectroscopy as the reaction mixture was warmed to 0 °C (ν_{CO} = 2001 and 1913 cm^{-1}). Further warming to room temperature resulted in a color change from yellow to dark red, with simultaneous generation of a new four-band pattern (ν_{CO} = 1990, 1934, 1901 and 1855 cm^{-1}; eq. 5). Isolation of the new product by alumina chromatography gave **4a** as an olive-green powder. FABMS of the new product gave a molecular ion at M = 923, approximately twice the molecular weight of monomeric Tp'(CO)$_2$Mo≡C-H (MW = 462).

Similarly, in attempting to obtain crystalline samples of the monomeric tungsten carbyne **3b**, solutions of the hydridocarbyne in numerous solvents slowly changed from yellow to olive green. IR spectroscopy indicated a transformation of the monomer to a new species with carbonyl stretching bands at 1974, 1916, 1880 and 1831 cm^{-1}. Isolation by alumina chromatography yielded **4b** as an olive-green powder whose FABMS gave a molecular ion at 1099 amu. The material could be recrystallized to give dark red crystals of the new complex, one of which was subjected to x-ray structural analysis.

Molecular Structure of Tp'(CO)$_2$W(μ-η^2:η^2-CCH$_2$)W(CO)$_2$Tp' (4b).

An ORTEP drawing of **4b** is shown in Figure 2; selected bond lengths and angles are listed in Table I. The molecule is a bimetallic complex in which two pseudo-octahedral Tp'(CO)$_2$W fragments are bridged by a formal vinylidene fragment, CCH$_2$.

Figure 2. ORTEP Representation of **4b**.

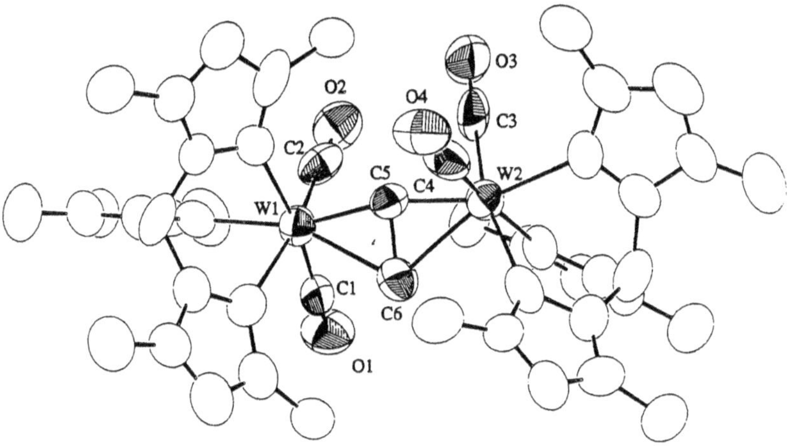

Table I. Selected bond lengths (Å), bond angles (deg) and torsion angles for Tp'(CO)$_2$W(μ-η^2:η^2-CCH$_2$)W(CO)$_2$Tp' (4b).

W(1)-W(2)	3.96 (1)	W(1)-C(6)	2.36 (2)
W(1)-C(5)	2.00 (2)	W(2)-C(6)	2.34 (2)
W(2)-C(5)	1.98 (2)	C(5)-C(6)	1.51 (2)
W(1)-C(5)-W(2)	166.6 (9)	W(1)-C(6)-W(2)	114.5 (8)
W(1)-C(5)-C(6)	83 (1)	W(1)-C(6)-C(5)	57.4 (9)
W(2)-C(5)-C(6)	83 (1)	W(2)-C(6)-C(5)	57.1 (9)
C(6)-W(1)-C(5)-W(2)	0.4 (5)		

The geometry of the W$_2$C$_2$ core of **4b** is unique for bimetallic vinylidene complexes, and has precedent in only a handful of examples in which bimetallic alkenes bridge symmetrically to both metal termini in a μ-η^2:η^2 manner.[19] The vinylidene ligand is perpendicular to the W(1)-W(2) axis, and the carbide-like vinylidene terminus, C(5), lies almost directly between the two tungsten nuclei (W(1)-C(5)-W(2) = 166.6 (9)°) and sets the two metals 3.96 (1) Å from one another. The carbide atom is tightly bound to both tungsten nuclei with W(1)-C(5) = 2.00 (2) Å, W(2)-C(5) = 1.98 (2) Å. These distances approximate those of W=C double bonds.[20] Bonding of the two tungsten nuclei to methylene carbon C(6) is best described in terms of weak single bonds, with W(1)-C(6) and W(2)-C(6) interatomic distances of 2.36 (2) Å and 2.34 (2) Å, respectively. A W(1)-C(6)-W(2) bond angle of 114.5 (8)° couples with the two attatched hydrogens to indicate a distorted tetrahedral geometry at C(6). Another notable structural feature is the essential planarity of the four membered ring defined by W(1)-C(5)-W(2)-C(6) (torsion angle of 0.4 (5)°).

A short interatomic C(5)-C(6) distance (1.51 (2) Å) constitutes a definitive C-C single bond. Taken together with the metal-vinylidene distances described above, the existence of a C-C single bond effectively renders each vinylidene carbon pentavalent. Obviously, the bonding scenario here is unusual and involves nonclassical bonding in the W_2C_2 core (vide infra).

Electrochemical Investigations of Vinylidenes 4a and 4b.

In order to gain some insight into the reactivity of bridging vinylidenes **4a** and **4b**, electrochemical studies have been carrried out. The cyclic voltammograms reveal that both the molybdenum and tungsten vinylidenes are readily oxidized. Molybdenum derivative **4a** exhibits a quasi-reversible redox couple at $E_{1/2}$ = +0.64 V vs. SCE (ΔE_p = 270 mV at 100 mV/sec). The accompanying cathodic response indicates that the oxidized species is somewhat unstable, as $i_p^c/i_p^a = 0.7$.

Tungsten complex **4b** also generates a quasi-reversible redox couple upon oxidation ($E_{1/2}$ = +0.44 V vs. SCE; ΔE_p = 425 mV at 100 mV/sec). The couple appears to be chemically reversible, with $i_p^c/i_p^a = 0.9$; the more electron-rich tungsten analogue is more easily oxidized than its molybdenum congener.

Variable Temperature NMR Behavior of Bridging Vinylidenes 4a and 4b.

At room temperature, dimolybdenum complex **4a** generates a simple ^1H NMR spectrum. Aromatic pyrazole resonances appear as two sharp singlets at 5.88 and 5.75 ppm, integrating for 4H and 2H, respectively. Upfield are singlet resonances assignable to the Tp' methyl substituents at 2.43 (6H), 2.40 (12H), 2.35 (br, 12H) and 2.26 (6H) ppm. An additional singlet appears at 2.44 ppm integrating for 2H; this signal is assigned to the protons of the methylene unit of the vinylidene bridge.

The ^{13}C NMR of **4a** includes a low-field singlet at 346.8 ppm, which is assigned to the carbide unit of the vinylidene fragment. An upfield resonance at 7.5 ppm appears as a triplet in the gated-decoupled ^{13}C NMR spectrum ($^1J_{C\text{-}H}$ = 162 Hz) and is assigned to the bridging methylene carbon atom. All other signals appear in relative intensity ratios of 2:1 and are assignable to the pyrazole ring carbons of the Tp' ligand.

While both the ^{1}H and ^{13}C NMR spectra of **4a** indicate a structure possessing mirror symmetry at room temperature, tungsten analogue **4b** displays room temperature spectra which suggests fluxional behavior. The proton spectrum of **4b** displays two downfield signals: a broad singlet at 5.90 ppm integrating for 4H and a sharp singlet integrating for 2H at 5.77 ppm. The upfield region contains a series of broad resonances ascribed to the methyl substituents of the Tp' ligands in the 2.7-1.9 ppm region. Additionally, a broad singlet at 2.20 ppm can be assigned to the two protons of the bridging methylene unit.

Broadened resonances in the ^{13}C NMR of **4b** also indicate molecular motion in solution. A sharp downfield singlet due to the carbide unit appears at 304.4 ppm ($^1J_{W\text{-}C}$ = 45Hz) and two broad resonances from two distinct pairs of terminal carbonyl ligands occur at 232.8 and 225.2 ppm. (^{13}C NMR at -45 °C displays sharp singlets due to these carbonyl ligands at 232.8 ppm ($^1J_{W\text{-}C}$ = 166 Hz) and 225.0 ppm ($^1J_{W\text{-}C}$ = 163 Hz)). The bridging methylene carbon gives rise to an upfield triplet at -3.3 ppm in the gated decoupled ^{13}C NMR ($^1J_{C\text{-}H}$ = 162 Hz).

Upon cooling to -75 °C, the broad ^{1}H NMR signals observed in the room temperature spectrum of **4b** separate, so that the downfield resonances occur as three sharp singlets at 5.95, 5.87 and 5.78 ppm (2H each) and six separate Tp' methyl singlets (6H each) are observed in addition to the bridging methylene singlet. As the sample is warmed, the two lower field aromatic resonances coalesce (T_c = 2.5 °C, Figure 3). The methyl resonances undergo corresponding changes. From the variable temperature data, a $\Delta G^{\ddagger}$ of 13.9 kcal/mol for the fluxional process can be determined. The molybdenum bridging vinylidene behaves similarly, with a T_c of -56 °C, and a $\Delta G^{\ddagger}$ of 10.7 kcal/mol.

At low temperature pairwise equivalence characterizes the six pyrazole rings and four terminal carbonyls in vinylidenes **4a** and **4b**. Citing the structural features of the tungsten dimer in the solid state, and invoking an idealized octahedral geometry at each metal center, a plausible ground state at low temperature would possess a C_2 symmetry axis coincident with the C-C bond of the bridging vinylidene ligand. In this manner, pyrazoles C and C' are equivalent, as are pyrazoles A and A', and B and B'. There will be two pairs of equivalent carbonyls also, each comprised of one carbonyl from each metal.

One process which can explain the observed variable temperature NMR behavior involves a net rotation of the bridging methylene unit about the M-C-M axis by 180° to regenerate an equivalent conformer wherein pyrazoles A and A', and B and B', have exchanged; pyrazoles C and C' are unchanged. As the methylene rotates freely about the M-C-M axis, the four off-axis pyrazoles all become equivalent, as do all four terminal carbonyls; the pyrazoles along the axis remain distinct.

^{13}C NMR spin saturation transfer experiments[21] do not indicate site exchange of the μ-carbide and μ-methylene nuclei, obviating formation of a high-energy μ-η^2-η^2-alkyne complex via reversible 1,2-hydride migration.

Figure 3. ^{1}H NMR Spectra of **4b**.

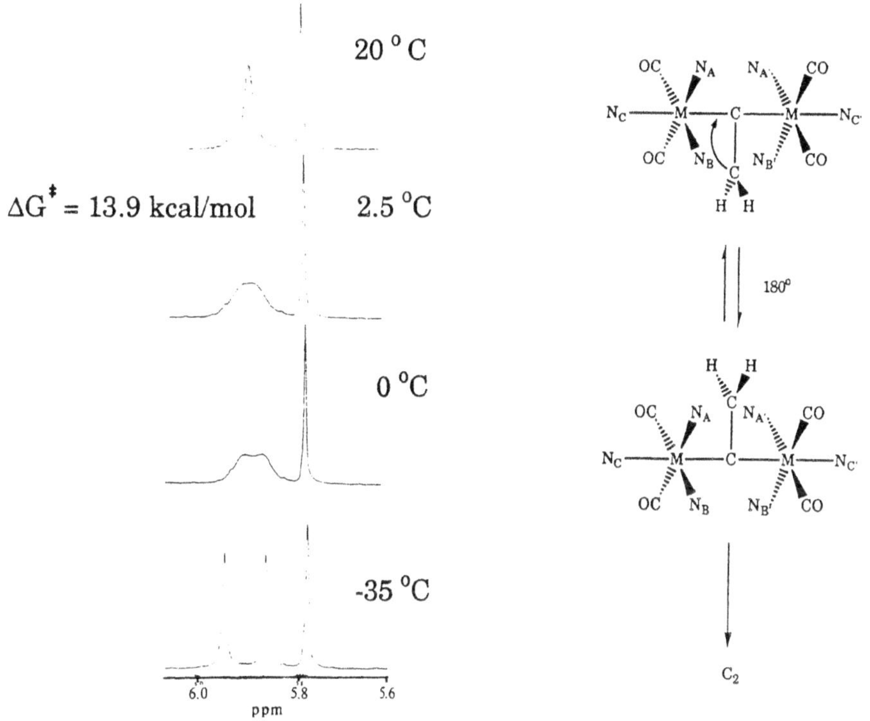

Determination of Homonuclear Carbon-Carbon Coupling(s) in Bridging Vinylidene 4b*.

Isotopically enriched samples of the tungsten derivative, Tp'(*CO)$_2$W(μ-η^2:η^2-*C*CH$_2$)W(*CO)$_2$Tp' (**4b***), reveals several features which were not apparent from NMR studies of unlabeled bridging vinylidenes **4a** and **4b**. A CD_2Cl_2 sample of **4b*** reveals homonuclear couplings between the two distinct types of terminal carbonyls with $^2J_{\underline{C}O\text{-}\underline{C}O}$ = 4 Hz.

Additionally, the upfield carbonyl resonance at 225 ppm shows coupling to an additional carbon nucleus, with a coupling of $J_{C\text{-}C}$ = 10 Hz. A homonuclear 2D COSY experiment[22] unambiguously identifies the coupling interaction between the carbide, C_α, and the pair of terminal carbonyls resonating at 225 ppm. The COSY experiment also indicates that there is homonuclear coupling between the carbide and methylene carbon nuclei on the vinylidene.

INADEQUATE spectroscopy[22b, 23] permits an assignment of $J_{C\alpha\text{-}C\beta}$ = 10.2 Hz. The coupling constant $J_{C\alpha\text{-}C\beta}$ is significantly diminished from that of a formal C-C single bond ($^1J_{C\text{-}C}$ ~ 30-40 Hz for aliphatic systems;[24] $J_{C\text{-}C}$ = 10 Hz for **4b***). It is unlikely that this represents coupling of the methylene carbon to one of the pairs of terminal carbonyls, which should be observable in the ^{13}C-^{13}C COSY experiment.

Also visible was the remaining coupling interaction between the bridging methylene carbon and the two tungsten nuclei ($^1J_{W\text{-}\underline{C}H2}$ = 10.5 Hz).

Electronic Structure of Bridging Vinylidene 4b.

In an attempt to probe the nature of the bonding in the W_2C_2 core, an extended Hückel molecular orbital calculation[25] was carried out on the isoelectronic model compound $[H_3(CO)_2W(\mu\text{-}\eta^2\text{:}\eta^2\text{-}CCH_2)W(CO)_2H_3]^{4-}$ in which hydride ligands were substituted for the Tp' ligands. The geometry of the $[(CO)_2W(\mu\text{-}\eta^2\text{:}\eta^2\text{-}CCH_2)W(CO)_2]$ core was taken from the x-ray structural data of **4b**, and the methylene hydrogen locations were established based on the supposition that the methylene carbon is sp^2-hybridized (planar, 120° H-C-H angle, $r_{C\text{-}H}$ = 1.08 Å).[26]

The results of the calculation verify that the bridging vinylidene should be a stable diamagnetic compound; the model system has a large HOMO-LUMO gap of 4.29 eV.

The nonclassical bonding in the W_2C_2 core can be understood in terms of two orthogonal three-center interactions. A set of molecular orbitals can be identified which involves combinations of tungsten d_{xz} atomic orbitals with the unhybridized p_x orbital at the bridging methylene carbon (refer to the coordinate system in Figure 4). A low-lying filled bonding combination of two out-of-phase metal d_{xz} orbitals with the methylene p_x constitutes a three center-two electron σ-bonding scheme; the nonbonding combination (the molecule's LUMO) and antibonding combination remain unfilled.

The carbide nucleus of the vinylidene bridge can use sp^2 hybrid orbitals for bonds to the methylene carbon and both tungsten nuclei, leaving an unhybridized p_y orbital at the carbide atom. The carbide p_y is of the proper symmetry to interact with the metal d_{xy} orbitals, generating a set of three molecular orbitals which define a three center-four electron π-bond. Both low-lying bonding ($d_{xy} + p_y - d_{xy}$) and nonbonding ($d_{xy} + d_{xy}$) molecular orbitals are doubly occupied, leaving an empty antibonding combination well above the frontier orbitals (Figure 4).

Figure 4. Qualitative Molecular Orbital Scheme for $Tp'(CO)_2M(\mu\text{-}\eta^2\text{:}\eta^2\text{-}CCH_2)M(CO)_2Tp'$

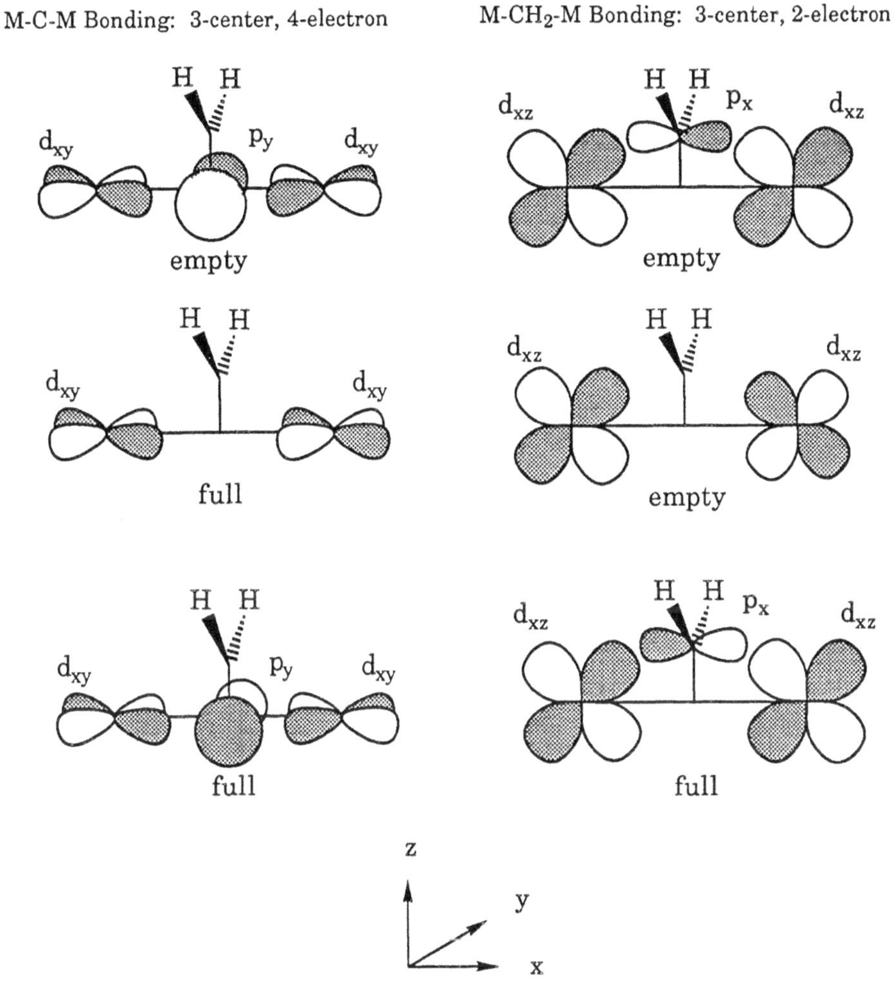

EXPERIMENTAL SECTION

All reactions were run under dry argon with use of standard Schlenk techniques unless otherwise noted. Solvents were dried under nitrogen by standard methods.[27] Dimethylphenylchlorosilane was used as obtained from Hüls-Petrarch. Tetrabutylammonium fluoride was used as obtained from Aldrich Chemicals.

Infrared spectra were obtained with a Mattson Polaris Fourier transform spectrophotometer. ^1H and ^{13}C NMR spectra were recorded on a Varian XL400 instrument. ^{13}C double quantum and saturation transfer NMR experiments were conducted on a Bruker AMX 300 instrument using standard Bruker pulse programs. Elemental and molecular weight analyses were performed by Oneida Research Services, Whitesboro, NY.

Cyclic voltammograms were obtained on a Bioanalytical Systems CV-27 instrument; samples were dissolved in dry THF containing 0.1 M $[Et_4N][PF_6]$ as supporting electrolyte. The voltammograms were obtained at a scan rate of 100 mV/sec, and $E_{1/2}$ values were determined relative to ferrocene/ferrocenium as an internal standard. The electrode array consisted of a saturated calomel reference electrode and platinum disk (working) and wire (auxilliary) electrodes. Potentials were uncorrected for junction effects.

Tp'(CO)$_2$Mo≡C-SiMe$_2$Ph (2a). Identical procedures were used in the preparation of both molybdenum and tungsten silylcarbynes. A $PhMe_2SiLi$ solution in 75 mL of anhydrous THF was generated[28] by stirring 5.20 mL (31.4 mmol) of $PhMe_2SiCl$ vigorously with 1.10 g (160 mmol) of finely cut lithium wire (high sodium content) in THF at 0 °C. This solution was added to 750 mL of anhydrous Et_2O containing $Mo(CO)_6$ (7.75 g, 28.6 mmol) at 0 °C. The purple solution was cooled to -78 °C before slow addition of 4.0 mL (28.3 mmol) of trifluoroacetic anhydride. A degassed methanol solution of KTp' (9.55 g, 28.4 mmol in 100 mL MeOH) was added via cannula to the ethereal molybdenum solution at -78 °C, followed by slow warming to 20 °C overnight. Removal of solvent under vacuum followed by alumina chromatography (4:1 hexanes:CH_2Cl_2) yielded 3.28 g (19% yield) of silylcarbyne **2a** as an analytically pure yellow powder. IR (CH_2Cl_2): 1997, 1911 cm^{-1} (ν_{CO}). ^1H NMR (CD_2Cl_2): δ 7.60, 7.38 (m, 5H, Mo≡C-SiMe$_2$<u>Ph</u>), 5.88, 5.73 (s, 2:1H, Tp' C<u>H</u>), 2.43, 2.34, 2.32, 2.29 (s, 6:6:3:3H, Tp' C<u>H$_3$</u>), 0.47 (s, 6H, Mo≡C-Si<u>Me$_2$</u>Ph). ^{13}C {^1H} NMR(CD_2Cl_2): δ 360.4 (Mo≡<u>C</u>-SiMe$_2$Ph), 227.8 (Mo(<u>C</u>O)), 151.9, 151.3, 146.0, 145.8 (1:2:1:2, Tp' <u>C</u>CH$_3$), 137.1, 134.5, 129.8, 128.3 (Mo≡C-SiMe$_2$<u>Ph</u>), 106.7, 106.6 (1:2, Tp' <u>C</u>H), 16.4, 14.6, 13.0, 12.8 (2:1:2:1, Tp'C<u>C</u>H$_3$), -2.6 (Mo≡C-Si<u>Me$_2$</u>Ph). Anal. Calcd for $C_{26}H_{33}BN_6O_2SiMo$: C, 52.36; H, 5.58; N, 14.09. Found: C, 52.52; H, 5.40; N, 13.67.

Tp'(CO)$_2$W≡C-SiMe$_2$Ph (2b). Tungsten silylcarbyne **2b** was synthesized in a manner similar to that described for **2a**. Product **2b** was isolated in analytically pure form after repeated alumina chromatography in 7% yield. IR (CH_2Cl_2): 1982, 1889 cm^{-1} (ν_{CO}). ^1H NMR (CD_2Cl_2): δ 7.66, 7.40 (m, 5H, W≡CSiMe$_2$<u>Ph</u>), 5.97, 5.81 (s, 2:1H, Tp' C<u>H</u>), 2.50, 2.42, 2.39, 2.34 (s, 6:3:6:3H, Tp' C<u>H$_3$</u>), 0.48 (s, 6H, W≡CSi<u>Me$_2$</u>Ph). ^{13}C {^1H} NMR(CD_2Cl_2): δ 339.0 ($^1J_{W-C}$ = 160 Hz, W≡<u>C</u>SiMe$_2$Ph), 226.1 ($^1J_{W-C}$ = 173 Hz, W(<u>C</u>O)), 153.0, 152.0, 146.2, 145.7 (1:2:1:2, Tp' <u>C</u>CH$_3$), 138.6, 134.4, 129.5, 128.2 (W≡CSiMe$_2$<u>Ph</u>), 107.1, 106.9 (1:2, Tp' <u>C</u>H), 17.1, 15.2, 13.0, 12.8 (2:1:2:1,

Tp'C$\underline{C}H_3$), -2.0 (W≡CSi$\underline{Me}_2$Ph). Anal. Calcd for $C_{26}H_{33}BN_6O_2SiW$: C, 45.63; H, 4.86; N, 12.28. Found: C, 45.79; H, 5.00; N, 11.69.

Tp'(CO)$_2$W≡C-H (3b). An oven-dried Schlenk flask was charged with 1.50 g (2.19 mmol) of **3b** and 75 mL of dry THF. The resulting yellow solution was cooled to -78 °C before 2.40 mL (2.40 mmol) of a 1.0 M solution of Bu_4NF in THF was added via syringe. Monitoring the reaction by IR spectroscopy revealed a slight shift in the carbonyl absorptions from 1981 and 1890 cm^{-1} (Tp'(CO)$_2$W≡C-SiMe$_2$Ph) to 1985 and 1892 cm^{-1}. The mixture was warmed to room temperature and stirred for 30 min., then solvent was removed under vacuum. The resulting residue was washed with dry diethyl ether (5x20 mL), and the combined ether washings were evaporated to dryness under vacuum. Chromatography of the yellow residue on alumina (5:1 hexanes:CH_2Cl_2 eluent) separated a bright yellow band. Evaporation of solvent yielded 0.36 g of Tp'(CO)$_2$W≡C-H as a bright yellow powder in 30% yield. IR (THF): 1986, 1891 cm^{-1} (ν_{CO}). ^{1}H NMR (CD_2Cl_2): δ 8.22 (s, $^2J_{W-H}$ = 83 Hz, W≡C-$\underline{H}$), 5.94, 5.81 (s, 2:1H, Tp' C$\underline{H}$), 2.54, 2.40, 2.37, 2.33 (s, 6:3:6:3H, Tp' C$\underline{H}_3$). ^{13}C {^{1}H} NMR (CD_2Cl_2): δ 280.6 ($^1J_{W-C}$ = 192 Hz, $^1J_{C-H}$ = 142 Hz, W≡$\underline{C}$-H), 224.5 ($^1J_{W-C}$ = 169 Hz, W($\underline{C}$O)), 152.5, 151.6, 145.9, 145.1 (1:2:1:2, Tp' $\underline{C}CH_3$), 106.7, 106.4 (1:2, Tp' $\underline{C}$H), 16.6, 14.9, 12.5 (2:1:3, Tp' C$\underline{C}H_3$). Anal. Calcd for $C_{18}H_{23}BN_6O_2W$: C, 39.3; H, 4.21; N, 15.28. Found: C, 39.39; H, 4.06; N, 14.92. Mol. Wt. Calcd for $C_{18}H_{23}BN_6O_2W$: 550. Found: 552 (3-pt. VPO determination in CH_2Cl_2).

Synthesis of Tp'(CO)$_2$Mo(μ-η^2:η^2-CCH$_2$)Mo(CO)$_2$Tp' (4a). An oven-dried Schlenk flask was charged with **2a** (1.00 g, 1.68 mmol) and 125 mL of dry THF, and cooled to -78 °C. Upon addition of Bu_4NF (1.70 mL, 1.70 mmol) the carbonyl bands of the silylcarbyne (at 1995 and 1910 cm^{-1}) shifted to a dicarbonyl pattern with absorptions at 2001 and 1913 cm^{-1}, ascribed to formation of the molybdenum hydridocarbyne Tp'(CO)$_2$Mo≡CH. As the solution was warmed to 25 °C, the new carbonyl absorptions diminished as bands at 1990, 1934, 1901, and 1855 cm^{-1} increased. The mixture was evaporated under vacuum, and the residue was chromatographed on alumina (hexanes/CH_2Cl_2 eluent). A green-yellow band eluted first. When dissolved in THF this material had an IR spectrum with bands at 1983 and 1894 cm^{-1}. After repeating alumina chromatography this material was identified as Tp'(CO)$_2$Mo≡C-CH$_3$ by comparison of its spectroscopic properties with those of an authentic sample.[29] Upon increasing the solvent polarity an olive-green band eluted. Evaporation and recrystallization of the solid residue from CH_2Cl_2/hexanes gave red Tp'(CO)$_2$Mo(μ-η^2:η^2 CCH$_2$)Mo(CO)$_2$Tp'. IR (KBr): 1993, 1931, 1902, 1849 cm^{-1} (ν_{CO}). ^{1}H NMR (CD_2Cl_2) at -90 °C: δ 5.93, 5.84, 5.76 (s, 2:2:2H, Tp' C$\underline{H}$), 2.62, 2.35, 2.31, 2.19, 1.91 (s, 6:12:6:6:6H, Tp' C$\underline{H}_3$), 2.24 (br s, 2H, Mo(μ-C$\underline{H}_2$)Mo); at -56 °C (T_c of Tp' C$\underline{H}$ signals): δ 5.89 (br, 4H, Tp' C$\underline{H}$), 5.76

(s, 2H, Tp' C$\underline{H}$), 2.64, 2.38, 2.36, 2.22, 1.94 (Tp' CC$\underline{H}_3$), 2.31 (br s, 2H, Mo(μ-C$\underline{H}_2$)Mo); at 20 °C: δ 5.88, 5.75 (s,4:2H Tp' C$\underline{H}$), 2.42, 2.40, 2.25 (s, Tp' CC$\underline{H}_3$), 2.34 (br, 2H, Mo(μ-C$\underline{H}_2$)Mo). ^{13}C {^1H} NMR (CD_2Cl_2) at 20 °C: δ 346.8 (Mo(μ-$\underline{C}$)Mo), 232.0 (br, Mo($\underline{C}$O), 153.3, 152.6, 145.8 (1:2:3, Tp' $\underline{C}CH_3$), 107.4, 107.1 (2:1, Tp' $\underline{C}$H), 15.5, 15.3, 13.2, 13.1 (2:1:1:2, Tp' C$\underline{C}H_3$), 7.5 (Mo(μ-$\underline{C}H_2$)Mo). Anal. Calcd for $C_{36}H_{46}B_2N_{12}O_4Mo_2$: C, 46.77; H, 5.02; N, 18.18. Found: C, 46.88; H, 5.03; N, 17.96.

Isolation and Characterization of Tp'(CO)$_2$W(μ-η^2:η^2-CCH$_2$)W(CO)$_2$Tp' (4b). During purification of hydridocarbyne **3b** by alumina chromatography, the hydridocarbyne separated from an olive-green band, which was subsequently eluted by increasing the eluent polarity. Evaporation of the green fraction yielded complex **4b** which could be recrystallized from CH_2Cl_2/hexanes to give dark red crystals. Complex **4b** was also observed to form from analytically pure solutions of hydridocarbyne **3b** in a number of solvents upon standing. Separation of **3b** and **4b** could be accomplished by alumina chromatography. IR (KBr): 1977, 1916, 1876, 1825 cm^{-1} (ν_{CO}). ^1H NMR (CD_2Cl_2) at -45 °C: δ = 5.95, 5.87, 5.78 (s, 2:2:2H, Tp' C$\underline{H}$), 2.62, 2.42, 2.38, 2.34, 2.22, 1.90 (6:6:6:6:6:6H, Tp' CC$\underline{H}_3$), 2.08 (br s, 2H, W(μ-C$\underline{H}_2$)W); at 3 °C (T_c of Tp' C$\underline{H}$ signals): δ = 5.91 (br, 4H, Tp' C$\underline{H}$), 5.77 (s, 2H, Tp' C$\underline{H}$), 2.66, 2.44, 2.39, 2.24, 1.93 (Tp' CC$\underline{H}_3$), 2.16 (s, 2H, W(μ-C$\underline{H}_2$)W); at 20 °C: δ 5.90 (br, 4H, Tp' C$\underline{H}$), 5.77 (s, 2H, Tp' C$\underline{H}$), 2.64, 2.46, 2.40, 2.25, 2.19 (Tp' CC$\underline{H}_3$), 2.20 (br, 2H, W(μ-C$\underline{H}_2$)W). ^{13}C {^1H} NMR (CD_2Cl_2) at 20 °C: δ 304.4 ($^1J_{W\text{-}C}$ = 45 Hz, W(μ-$\underline{C}$)W), 232.8, 225.2 (br, W($\underline{C}$O)), 154.2, 153.2, 145 5 (Tp' $\underline{C}CH_3$), 107.4, 107.1 (Tp' $\underline{C}$H), 15.7, 12.9, 12.8 (Tp' C$\underline{C}H_3$), -3.3 (W(μ-$\underline{C}H_2$)W). Anal. Calcd for $C_{36}H_{46}B_2N_{12}O_4W_2$: C, 39.30; H, 4.21; N, 15.28. Found: C, 39.26; H, 4.16; N, 15.04.

Synthesis of ^{13}C-Labeled Tp'(*CO)$_2$W(μ-η^2:η^2-*C*CH$_2$)W(*CO)$_2$Tp' (4b*). A procedure identical to that described for **4b** was employed, utilizing partially labeled W(*CO)$_6$ as the tungsten source.[30] Following conversion to, and isolation of, the pure silylcarbyne complex Tp'(*CO)$_2$W≡*C-SiMe$_2$Ph (**2b***), ^{13}C NMR spectroscopy allowed for the determination of 20% enrichment at the terminal carbonyl and methylidyne sites. The labeled silylcarbyne was carried on to give a pure sample of ^{13}C-enriched Tp'(*CO)$_2$W(μ-η^2:η^2-*C*CH$_2$)W(*CO)$_2$Tp' (**4b***).

Collection of Diffraction Data for 4b. A red crystal of dimensions 0.40 x 0.30 x 0.25 mm was selected, mounted on a glass fiber and coated with epoxy. Diffraction data were collected on a Rigaku AFC6S automated diffractometer. Cell parameters were refined by full-matrix least squares from the positions of 25 well-centered reflections found in the region 25.0° < 2θ < 30.0° and indicated a monoclinic cell.

Intensity data were collected in the quadrant ± *h*, *k*, *l* under the conditions specified in Table II. Only data with $I > 2.5\sigma(I)$ were used in the structure and refinement.[31]

Table II. Crystallographic Data Collection Parameters for 4b.

molecular formula	$C_{36.5}H_{47}B_2ClN_{12}O_4W_2$
formula weight, g/mol	1142.62
crystal dimensions, mm	0.40 x 0.30 x 0.25
space group	$P2_1/n$
cell parameters	
a, Å	12.567(3)
b, Å	17.207(3)
c, Å	20.673(4)
β, deg	100.40(2)
vol., $Å^3$	4397(2)
Z	4
d (calc'd), g/cm^3	1.723
Collection and Refinement Parameters	
radiation (wavelength, Å)	MoKα (0.71073)
monochromator	graphite
linear abs. coeff., cm^{-1}	55.1
scan mode	ω/2θ
background	25% of full scan width on both sides
2θ limits, deg	$2 < 2\theta < 45$
quadrant collected	±h, k, l
total no. reflections	5738
data with $I \geq 2.5\sigma(I)$	3909
R	0.059
R_w	0.080
GoF	2.28
no. of parameters	528
largest parameter shift	0.098

Solution and Refinement of the Structure of 4b. The space group $P2_1/n$ was confirmed and the position of the tungsten atom was deduced from the three-dimensional Patterson function. The postitions of the remaining non-hydrogen atoms were determined through subsequent Fourier and difference Fourier calculations.

The non-hydrogen atoms were refined with anisotropic thermal parameters.[32] Hydrogen atom positions were calculated with the use of a C-H distance of 0.96 Å and an isotropic thermal parameter calculated from the anisotropic values for the atoms to which they were connected. Final least squares refinement resulted in residuals of R = 5.9% and R_w = 8.0%.[33] The final difference Fourier map had no peak greater than 1.42 e/$Å^3$.[34]

ACKNOWLEDGEMENT

We thank the United States Department of Energy, Division of Chemical Sciences (Grant No. 85ER13430), for financial support.

REFERENCES

1. Fischer, E. O.; Kreis, G.; Kreiter, C. G.; Müller, J.; Huttner, G.; Lorenz, H. *Angew. Chem., Int. Ed. Engl.* **1973**, *12*, 564-565.
2. For recent reviews see: (a) Kim, H. P.; Angelici, R. J. *Adv. Organomet. Chem.* **1987**, *27*, 51-111. (b) Buhro, W. E.; Chisholm, M. H. *Adv. Organomet. Chem.* **1987**, *27*, 311-369. (c) Nugent, W. A.; Mayer, J. M. *Metal-Ligand Multiple Bonds*; Wiley: New York, 1988. (d) Fischer, H.; Hofmann, P.; Kreissl, F. R.; Schrock, R. R.; Schubert, U.; Weiss, K. *Carbyne Complexes*; Verlag Chemie: Weinheim, Germany, 1988. (e) Mayr, A.; Hoffmeister, H. *Adv. Organomet. Chem.* **1991**, *32*, 227-324.
3. (a) Jeffery, J. C.; Laurie, J. C. V.; Moore, I.; Stone, F. G. A. *J. Organomet. Chem.* **1983**, *258*, C37-C40. (b) Howard, J. A. K.; Jeffery, J. C.; Laurie, J. C. V.; Moore, I.; Stone, F. G. A.; Stringer, A. *Inorg. Chim. Acta* **1985**, *100*, 23-32.
4. Green, M.; Porter, S. J.; Stone, F. G. A. *J. Chem. Soc., Dalton Trans.* **1983**, 513-517.
5. (a) Chisholm, M. H.; Huffman, J. C.; Rothwell, I. P. *J. Am. Chem. Soc.* **1981**, *103*, 4245-4246. (b) Chisholm, M. H.; Folting, K.; Hoffman, D. M.; Huffman, J. C. *J. Am. Chem. Soc.* **1984**, *106*, 6794-6805. (c) Chisholm, M. H.; Conroy, B. K.; Huffman, J. C.; Marchant, N. S. *Angew. Chem., Int. Ed. Engl.* **1986**, *25*, 446-447.
6. Hoffman, D. M.; Hoffmann, R.; Fisel, C. R. *J. Am. Chem. Soc.* **1982**, *104*, 3858-3875, and references cited therein.
7. (a) Bruce, M. I.; Swincer, A. G. *Adv. Organomet. Chem.* **1983**, *22*, 59-128. (b) Bruce, M. I. *Chem. Rev.* **1991**, *91*, 197-257.
8. (a) Doherty, N. M.; Eischenbroich, C.; Kneuper, H.-J.; Knox, S. A. R. *J. Chem. Soc., Chem. Commun.* **1985**, 170-171. (b) Mercer, R. J.; Green, M.; Orpen, A. G. *J. Chem. Soc., Chem. Commun.* **1986**, 567-569. (c) Froom, S. F. T.; Green, M.; Mercer, R. J.; Nagle, K. R.; Orpen, A. G.; Schwiegk, S. *J. Chem. Soc., Chem. Commun.* **1986**, 1666-1668. (d) Boese, R.; Huffman, M. A.; Vollhardt, K. P. C. *Angew. Chem., Int. Ed. Engl.* **1991**, *30*, 1463-1465.
9. (a) Werner, H.; Wolf, J.; Müller, G.; Krüger, C. *Angew. Chem., Int. Ed. Engl.* **1984**, *23*, 431-432. (b) Werner, H.; Wolf, J.; Müller, G.; Krüger, C. *J. Organomet. Chem.* **1988**, *342*, 381-398.
10. Jamison, G. M.; Bruce, A. E.; White, P. S.; Templeton, J. L. *J. Am. Chem. Soc.* **1991**, *113*, 5057-5059.
11. (a) Frauendorfer, E.; Brunner, H. *J. Organomet. Chem.* **1982**, *240*, 371-379. (b) Trofimenko, S. *Inorg. Chem.* **1971**, *10*, 504-507. (c) Curtis, M. D.; Shiu, K. B.; Butler, W. M. *Organometallics* **1983**, *2*, 1475-1477.

(d) Desmond, T.; Lalor, F. J.; Ferguson, G.; Ruhl, B.; Parvez, M. *J. Chem. Soc., Chem Commun.* **1983**, 55-56. (e) Rusik, C. A.; Collins, M. A.; Gamble, A. S.; Tonker, T. L.; Templeton, J. L. *J. Am. Chem. Soc.* **1989**, *111*, 2550-2560. (f) Bruce, A. E.; Gamble, A. S.; Tonker, T. L.; Templeton, J. L. *Organometallics* **1987**, *6*, 1350-1352.

12. Desmond, T.; Lalor, F. J.; Ferguson, G.; Parvez, M. *J. Chem. Soc., Chem. Commun.* **1983**, 457-459.
13. Fischer, E. O.; Hollfelder, H.; Kreissl, F. R. *Chem. Ber.* **1979**, *112*, 2177-2189.
14. (a) Mayr, A.; McDermott, G. A.; Dorries, A. M. *Organometallics* **1985**, *4*, 608-610. (b) McDermott, G. A.; Dorries, A. M.; Mayr, A. *Organometallics* **1987**, *6*, 925-931.
15. (a) Sharp, P. R.; Holmes, S. J.; Schrock, R. R.; Churchill, M. R.; Wasserman, H. J. *J. Am. Chem. Soc.* **1981**, *103*, 965-966. (b) Churchill, M. R.; Rheingold, A. L.; Wasserman, H. J. *Inorg. Chem.* **1981**, *20*, 3392-3399. (c) Holmes, S. J.; Clark, D. N.; Turner, H. W.; Schrock, R. R. *J. Am. Chem. Soc.* **1982**, *104*, 6322-6329. (d) Holmes, S. J.; Schrock, R. R.; Churchill, M. R.; Wasserman, H. J. *Organometallics* **1984**, *3*, 476-484.
16. (a) Davison, A.; McCleverty, J. A.; Wilkinson, G. *J. Chem. Soc.* **1963**, 1133-1138. (b) Faller, J. W.; Anderson, A. S.; Chen, C. C. *J. Chem. Soc., Chem. Commun.* **1969**, 719-720. (c) Keppie, S. A.; Lappert, M. F. *J. Chem. Soc. (A)* **1971**, 3216-3220.
17. Caffyn, A. J. M.; Feng, S. G.; Dierdorf, A.; Gamble, A. S.; Eldredge, P. A.; Vossen, M. R.; White, P. S.; Templeton, J. L. *Organometallics* **1991**, *10*, 2842-2848.
18. Silverstein, R. M.; Bassler, G. C.; Morrill, T. C. *Spectrometric Identification of Organic Compounds*; Wiley: New York, 1981, 4th Ed.
19. Evans, W. J.; Ulibarri, T. A.; Ziller, J. W. *J. Am. Chem. Soc.* **1990**, *112*, 219-223, and references cited therein.
20. (a) Dötz, K. H.; Fischer, H.; Hofmann, P.; Kreissl, F. R.; Schubert, U.; Weiss, K. *Transition Metal Carbene Complexes*; Verlag Chemie: Deerfield Beach, FL, 1983. (b) $r_{W\text{-}C}$ = 2.15 (2) Å in $(CO)_5W{=}C(Ph)_2$: Casey, C. P.; Burkhardt, T. J.; Bunnell, C. A.; Calabrese, J. C. *J. Am. Chem. Soc.* **1977**, *99*, 2127-2134.
21. (a) Dahlquist, F. W.; Longmuir, K. J.; DuVernet, R. B. *J. Magn. Reson.* **1975**, *17*, 406-410. (b) Morris, G. A.; Freeman, R. *J. Magn. Reson.* **1978**, *29*, 433-462.
22. (a) Bax, A.; Freeman, R. *J. Magn. Reson.* **1981**, *44*, 542-561. (b) Sanders, J. K. M.; Hunter, B. K. *Modern NMR Spectroscopy*; Oxford: New York, 1987.
23. Bax, A.; Freeman, R.; Kempsell, S. P. *J. Am. Chem. Soc.* **1980**, *102*, 4849-4851.
24. (a) Bartuska, V. J.; Maciel, G. E. *J. Magn. Reson.* **1971**, *5*, 211-219. (b) Carhart, R. E.; Roberts, J. D. *Org. Magn. Reson.* **1971**, *3*, 139-141.
25. (a) Hoffmann, R. *J. Chem. Phys.* **1963**, *39*, 1397-1412. (b) Parameters were taken from: Kubácek, P.; Hoffmann, R. *J. Am. Chem. Soc.* **1981**, *103*, 4320-4332.

26. *CRC Handbook of Chemistry and Physics*; Weast, R. C.; Astle, M. J., Eds.; CRC: Boca Raton, FL, 1982, 63rd Ed., p. F-180.
27. Gordon, A. J.; Ford, R. A. *The Chemist's Companion*; Wiley: New York, 1972.
28. (a) Gilman, H.; Peterson, D. J.; Whittenberg, D. *Chem. and Ind.* **1958**, 1479-1480. (b) Davis, D. D.; Gray, C. E. *Organomet. Chem. Rev. A* **1970**, *6*, 283-318. (c) Armitage, D. A. in *Comp. Organomet. Chem.*; Wilkinson, G.; Stone, F. G. A.; Abel, E. W., Eds; Pergamon: New York, 1982; Vol. 2, p. 1-397.
29. Brower, D. C.; Templeton, J. L., unpublished results.
30. Darensbourg, D. J.; Darensbourg, M. Y.; Gray, R. L.; Simmons, D.; Arndt, L. W. *Inorg. Chem.* **1986**, *25*, 880-882.
31. Programs used during solution and refinement were from the NRCVAX structure determination package: Gabe, E. J.; Le Page, Y.; Charland, J. P.; Lee, F. L.; White, P. S. *J. Appl. Cryst.*, **1989**, *22*, 384-387.
32. The function minimized was $\Sigma\omega(|F_o| - |F_c|)^2$, where ω is based on counter statistics.
33. $R_{unweighted} = \Sigma(|F_o| - |F_c|)/\Sigma|F_o|$ and $R_{weighted} = [\Sigma\omega(|F_o| - |F_c|)^2/\Sigma\omega F_o^2]^{1/2}$.
34. Scattering factors were taken from the following: Cromer, D. T.; Waber, J. T. in *International Tables for X-Ray Crystallography*; Ibers, J. A. and Hamilton, J. C., Eds., Kynoch: Birmingham, England, 1974, Vol. IV, Table 2.2.

THE ROLE OF NUCLEOPHILES AND ELECTROPHILES IN COUPLING REACTIONS OF ALKYLIDYNE LIGANDS

A. MAYR
Department of Chemistry
State University of New York at Stony Brook
Stony Brook, New York 11794-3400
U. S. A.

ABSTRACT. Alkylidyne ligands undergo coupling with several types of terminal π bonded ligands, such as carbonyl, isocyanide, and alkylidyne ligands, under a variety of conditions. The different coupling mechanisms may be rationalized by a simple qualitative molecular orbital model. In our own work we investigated examples of nucleophile-induced alkylidyne-carbonyl coupling, light-induced alkylidyne-carbonyl coupling, assisted by nucleophiles as well as by electrophiles, proton-induced alkylidyne-isocyanide coupling, and the formal coupling of two alkylidyne ligands.

1. Introduction

In the nearly two decades since the discovery of transition metal carbyne, or alkylidyne, complexes by Fischer and Kreis, the chemistry of this class of compound has developed substantially.[1] In 1976 Kreissl reported the first alkylidyne-carbonyl coupling reaction.[2] Subsequent to this report many additional examples of this and related types of coupling reactions of alkylidyne ligands were described. The full range of this chemistry was treated in a recent review article.[3] In this paper we discuss coupling reactions which were investigated in our laboratory, including previously unpublished results. The reactions are classified by the type of activation that is initiating the coupling process, and mechanistic details are interpreted on the basis of a simple qualitative molecular orbital model for coupling reactions.

2. A Qualitative Molecular Orbital Model for Coupling Reactions of Alkylidyne Ligands

The bonding of alkylidyne metal complexes has been the subject of several theoretical studies.[4] In alkylidyne carbonyl metal complexes the carbonyl ligand forms a significant π interaction only with the d orbital orthogonal to the metal-carbon triple bond (Figure 1a). The carbonyl π* orbital in the plane containing the metal, the alkylidyne ligand, and the carbonyl ligand (coupling plane) is not able to interact significantly with the metal-alkylidyne π bond (Figure 1a and 1b). The different coupling mechanisms (nucleophile-induced, photo-induced, and electrophile-induced) provide ways to increase the direct interaction between the alkylidyne ligand and the carbonyl ligand. In the coupling model only the metal-ligand π orbitals in the coupling plane are considered (Figure 1b).[3] Attack of a nucleophile at the metal center (in the coupling plane) gives rise to the formation of a higher-lying occupied orbital which is localized to a large extent on the alkylidyne carbon

F. R. Kreißl (ed.), Transition Metal Carbyne Complexes, 219–230.

atom and has little metal character. This situation is analogous to the attack of nucleophiles at other multiple bond systems. The "lone pair" on the alkylidyne carbon atom can then directly interact with the carbonyl π* orbital. In photo-induced coupling reactions a similar effect could be achieved by promotion of an electron into the MC π* orbital. In electrophile-induced alkylidyne-carbonyl coupling reactions, attack of the electrophile at the oxygen atom of the carbonyl ligand may be assumed as the first step. As a consequence, the energy of the carbonyl π* orbital is lowered in energy and the system begins

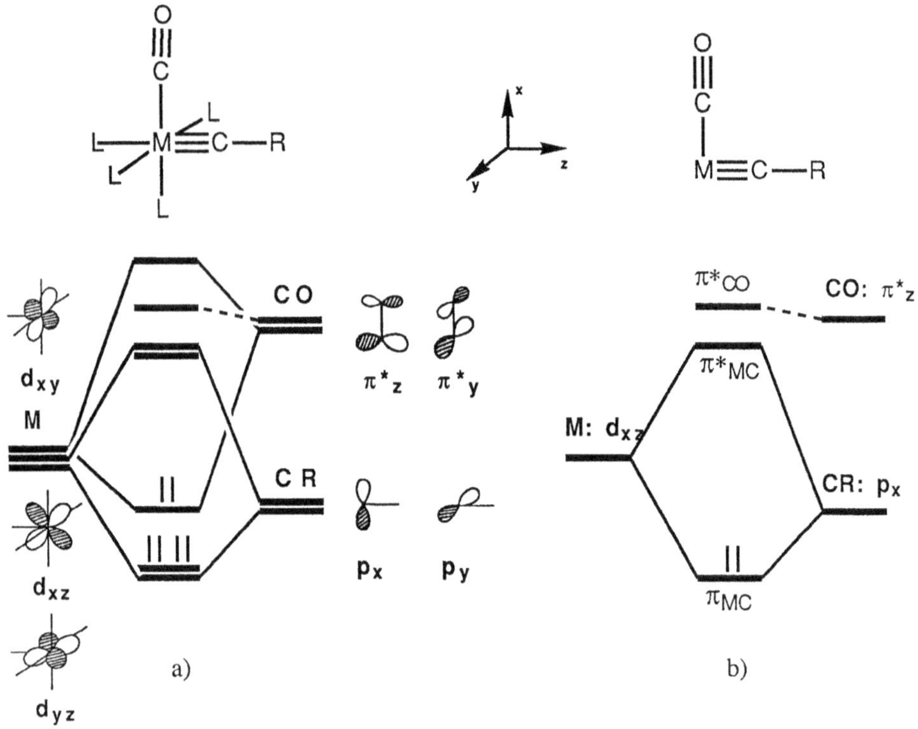

Figure 1. a) Metal-ligand π orbitals of a Fischer-type alkylidyne carbonyl metal complex. b) Metal-ligand π orbitals of the M(CR)(CO) fragment in the coupling plane.

to resemble a metal bis-alkylidyne complex. The coupling of two alkylidyne ligands was predicted by Hoffmann to be a favorable process.[5] Thus in the various types of coupling reactions, different electronic mechanisms are at work to "reduce" the energy gap between the MC π and the CO π* orbitals.

3. Coupling Reactions

3.1. NUCLEOPHILE-INDUCED ALKYLIDYNE-CARBONYL COUPLING

In attempts to synthesize tungsten alkylidyne complexes containing chelating anionic donor ligands we investigated the reactions of the bis-pyridine-substituted complex **1** with dithiocarbamate ligands[6] and with pyrrole-2-carboxaldehyde methylimine in the presence of base.[7] These reactions do not stop after the introduction of one chelate ligand. Instead, a second chelate ligand is taken up and the anionic ketenyl complexes **2** form. Complexes **2** are isolated in the form of tetraethylammonium salts. The presence of an η^2-ketenyl ligand in complexes **2** is recognized by the presence of a characteristic infrared absorption of weak intensity near ν = 1680 cm^{-1} and two characteristic resonances in the ^{13}C NMR spectrum near δ 200 and 180 ppm. The alkylidyne-carbonyl coupling step appears to occur very easily upon coordination of the second anionic chelate ligand. It seems reasonable to conclude from this and other results that alkylidyne-carbonyl coupling becomes more facile with increasing electron density of the metal center. In terms of the bonding model, the higher energy of the d orbitals of an electron-rich metal center would cause an increase of

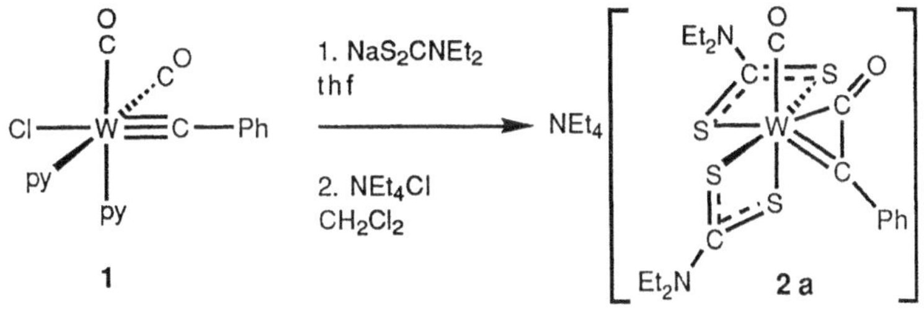

(1)

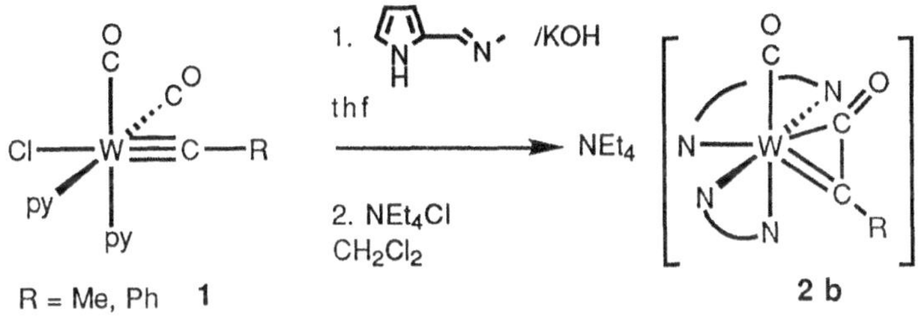

(2)

the energy of the metal-carbon π bonds, thereby improving the preconditions for coupling.[3]

The η^2-ketenyl ligands in **2** react very easily with electrophiles, such as methylating

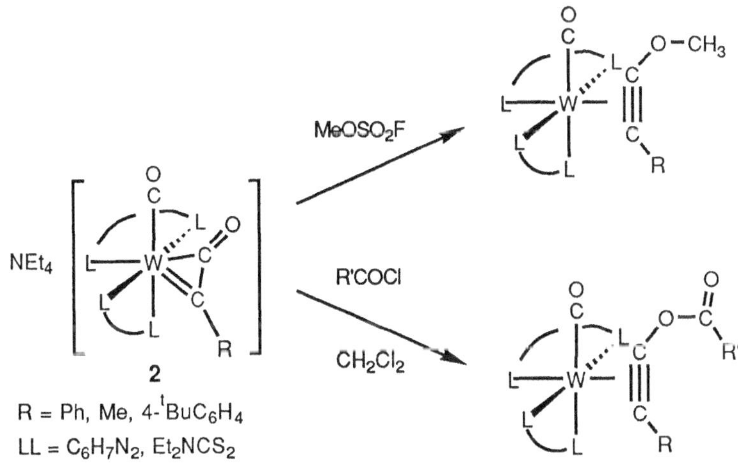

(3a)

(3b)

agents or acyl halides (Eq. 2).[7,8] Several ynolester metal complexes were prepared this way.[8] The high selectivity of η^2-ketenyl ligands towards addition of electrophiles at the oxygen atom is well established.[9]

The cis isomer of the *bis*-trimethylphosphine-substituted tungsten alkylidyne complex **3** reacts with diethyldithiocarbamate to give the ketenyl complex **4**.[10] The same product is obtained when the trans isomer of **3** is used as the starting material. It was not expected that the reactions of the cis and trans isomers of **3** would give the same result. In other investigations it was found that the cis and trans isomers of **3** react differently. In particular, facile substitution of carbon monoxide by donor ligands was observed for the trans isomer, but not for the cis isomer. Therefore, it would have been reasonable to expect the formation of complex **6** in the reaction of *trans*-**3** with diethyldithiocarbamate. The fact that this did not happen indicates that the dithiocarbamate ligand is very effective in inducing the coupling step.

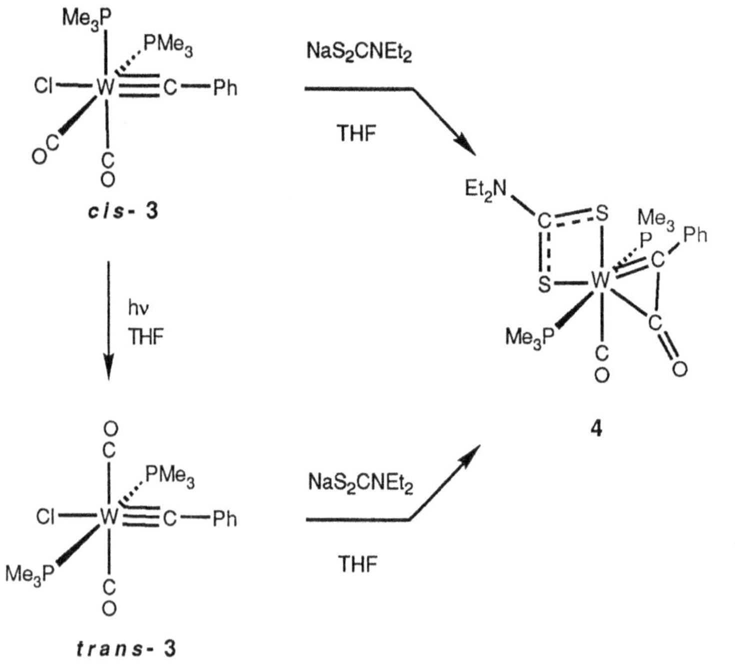

(4a)

(4b)

Complex **6** was obtained by the reaction of the *tris*-trimethylphosphine-substituted complex **5** with diethyldithiocarbamate.[10] Complex **6** exhibits an IR absorption for the

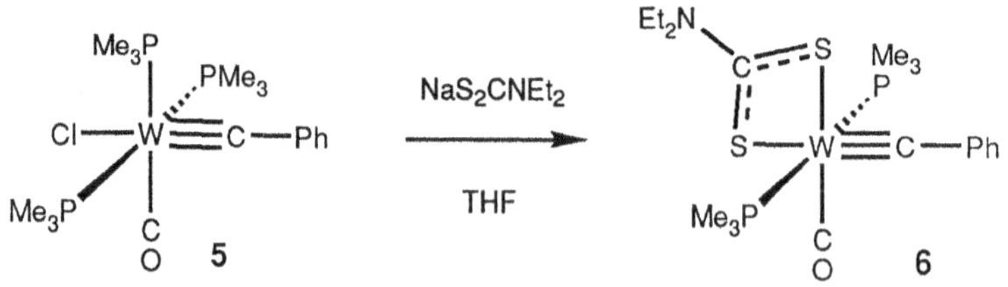

(5)

carbonyl ligand at 1848 cm^{-1} for the carbonyl ligand and a resonance at δ 271 ppm for the alkylidyne carbon in the ^{13}C NMR spectrum. The absence of alkylidyne-carbonyl coupling in this reaction may be due to two reasons. First, the meridional arrangement of the three phosphine ligands in complex **5** is likely to cause repulsive interactions between the phosphine ligands, thereby facilitating substitution of one phosphine ligand. Second, alkylidyne-carbonyl coupling in the monocarbonyl complex **5** would lead to a carbonyl-free ketenyl metal complex as the product. Ketenyl complexes, such as **4**, are d^4 systems, counting the ketenyl ligands as $RCCO^-$. Consequently, the remaining carbonyl ligand plays an important role in stabilizing two occupied metal d orbitals.

The effectiveness of dithiocarbamate ligands in inducing coupling reactions suggested the possibility that the π donor ability of this type of ligand may be, in part, responsible. We were therefore interested in exploring if non-chelating π donor ligands would also have a promotive influence on ligand coupling processes. Several novel tungsten alkylidyne complexes, **7** (**a**: XR = OPh, **b**: XR = $SCMe_3$, **c**: XR = SC_6H_{11}), containing phenoxide and alkylsulfide ligands were synthesized (Eq. 6).[11] All three complexes exhibit very

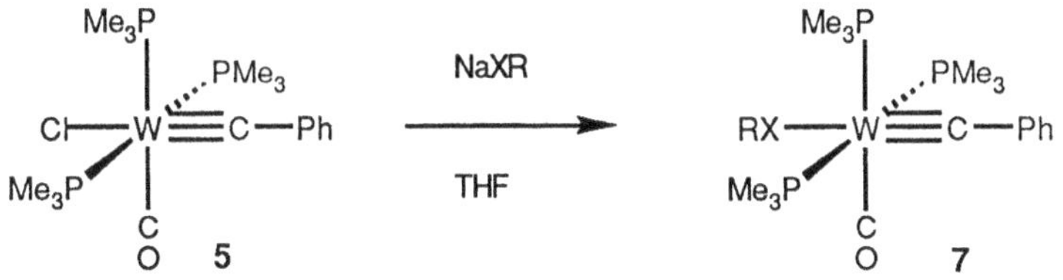

(6)

XR = $SCMe_3$, SC_6H_{11}, OPh

similar spectroscopic data. For example, ν(CO) = 1884 cm^{-1} and δ(CPh) = 259 ppm for the t-butylsulfide complex **7b**. However, preliminary experiments did not reveal a high propensity of complexes **7** to undergo ligand coupling, *e. g.* in reactions with carbon monoxide or trimethylphosphine.

The carbonyl absorptions of complexes **7** are only slightly lower than that of the chloro complex **7** (ν(CO) = 1896 cm^{-1}). Apparently, the introduction of significantly better σ/π donor ligands in the position trans to the alkylidyne ligand does not significantly influence the interaction between the metal and the carbonyl ligand, as measured by the CO stretching frequency. On the other hand, the dithiocarbamate ligand in **6** causes the CO stretching frequency to be as low as 1848 cm^{-1}. It seems reasonable to conclude from these observations that the stretching frequency of the carbonyl ligand does not reflect the overall "electron-richness" of the metal center, but is specifically sensitive to the energy of that metal d orbital with which it is able to engage in π bonding.

3.2. PHOTO-INDUCED ALKYLIDYNE-CARBONYL COUPLING

The ability of light to induce the coupling of alkylidyne and carbonyl ligands was recognized by Geoffroy.[12] We observed the cis-trans isomerization of the *bis*-phosphine-substituted complexes **3** (Eq. 7).[13] To account for this isomerization process we postulated electronically unsaturated, pentacoordinated ketenyl metal complexes **8** as photo-intermediates. Support for the formulation of intermediates **8** comes from the successful isolation of ketenyl metal complexes **9** when irradiation of complexes **3** is conducted in the presence of suitable ligands L (**a**: L = $CNCMe_3$, **b**: L = $CNCH_2Ph$, **c**: L = PMe_3).[14] Complexes **9a** and **9b** are stable and were fully characterized. Characteristic spectroscopic data for **9a** are: ν(CN) = 2165 cm^{-1}, ν(CO) = 1905 cm^{-1}, ν(CCO) = 1684 cm^{-1}, δ(Ph*C*CO) = 223 ppm, δ(PhC*C*O) = 221 ppm.

The nature of the postulated photo-generated intermediates **8**, and the known reactivity of η^2-ketenyl ligands towards electrophiles[9] suggested the possibility of trapping these intermediates with electrophilic reagents. Hence, the oxyacetylene derivatives **10** were obtained by irradiation of complexes **3** in the presence of HCl, acyl chlorides, and silyl chlorides (Eqs. 8 and 9).[15]

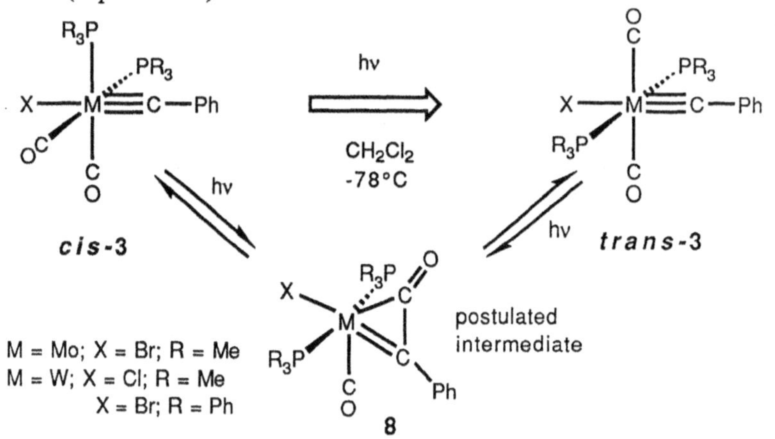

(7)

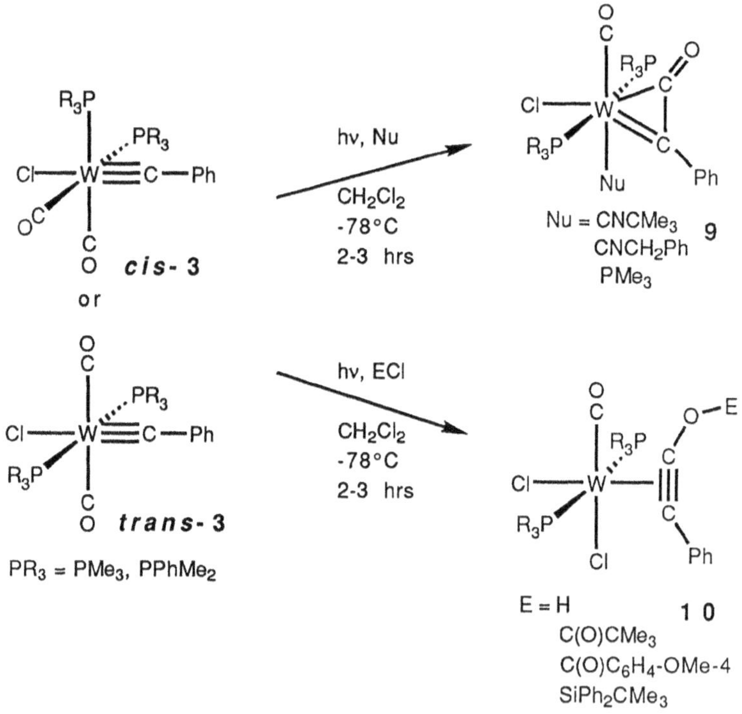

(8)

(9)

All expamples of photo-induced alkylidyne-carbonyl coupling reported in the literature to this date involve systems with aryl substituents on the alkylidyne carbon atom.[3] Some results even suggested that this type of coupling reaction may not be viable with alkyl-substituted systems.[3] For example, Templeton studied the photochemical reactivity of a pair of molybdenum alkylidyne complexes, Mo(CR)(*tris*-dimethylpyrazolylborate)$(CO)_2$, with R = Ph and R = Me.[16] Irradiation of the phenyl-substituted system in acetonitrile gives a ketenyl complex. Irradiation of the methyl-substituted system results in substitution of a carbonyl ligand. In contrast, irradiation of the alkyl derivatives **11** of complexes **3** in the presence of nucleophiles as well as electrophiles was found to result in the formation of the coupling products **12** and **13** in high yield (Eqs. 10 and 11).[17] The efficiency of coupling of the alkyl-substituted systems is, however, lower as indicated by the longer reaction times needed for the reactions shown in Eqs. 10 and 11 to go to completion, compared to the reactions shown in Eqs. 8 and 9.

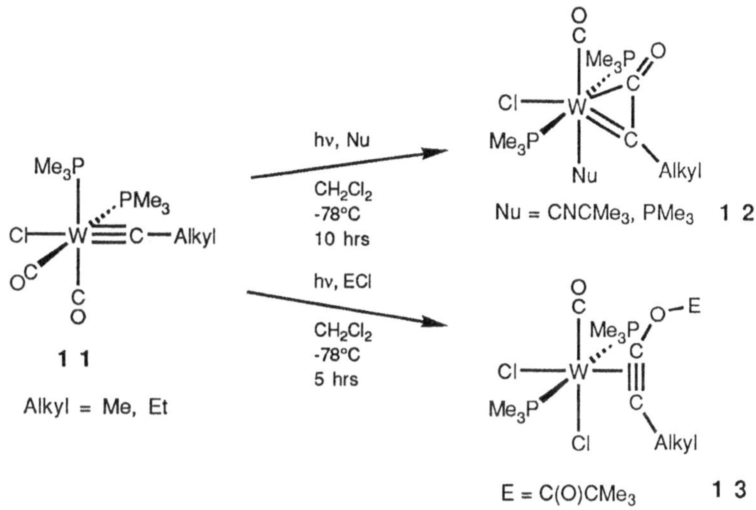

(10)

(11)

3.3. ELECTROPHILE-INDUCED ALKYLIDYNE-CARBONYL COUPLING

Currently, there exists no firmly established example of electrophile-induced alkylidyne-carbonyl coupling, *i. e.* a coupling reaction in which the interaction of an electrophile with the oxygen atom of the carbonyl ligand is essential in the coupling process. There are, however, potential examples in the literature. Schrock reported the formation of $[WCl(R_3AlOCCH)(CO)(PMe_3)_3]$ in the reaction of the methylidyne complex $[W(CH)Cl(PMe_3)_4]$ with CO in the presence of aluminum-based Lewis-acids.[18] In the absence of the Lewis-acids, no coupling products were observed. The "reductive coupling of two carbonyl ligands", developed by Lippard and coworkers, involves the reaction of siloxymethylidyne carbonylmetal complexes of the type $M(COSiR_3)(CO)(dmpe)_2$ (M = V, Nb, Ta) with silyl halides.[19] It seems reasonable to assume that in these electron-rich systems silylation of the carbonyl ligand initiates the coupling process. Attempts in our laboratory to observe electrophile-induced alkylidyne-carbonyl coupling reactions have not yet been successful. However, we did make the observation that electrophiles are capable of increasing the reactivity of alkylidyne complexes. For example, at -78 °C complex **6** does not react with carbon monoxide (1 atm), but in the presence of BPh_3 a green product forms, even though there is no apparent reaction between complex **6** and BPh_3 in the absence of CO.[20] Although the nature of the green product, which is only stable at low temperatures, is not yet known, preliminary data indicate that ligand coupling is not involved in its formation.

The basic types of alkylidyne-carbonyl coupling are represented in abstract form in Scheme 1.

Scheme 1.

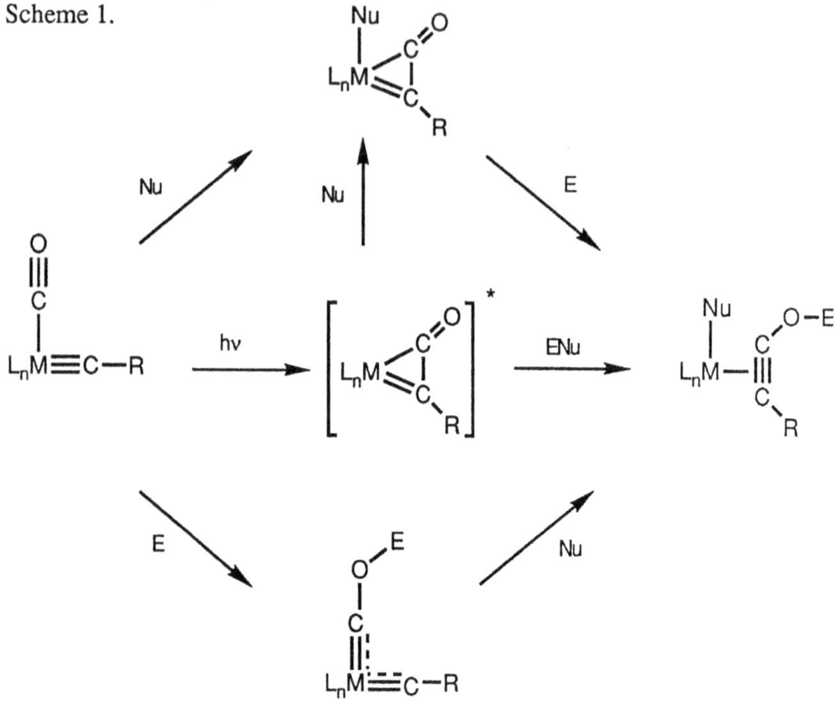

3.4. ALKYLIDYNE-ISOCYANIDE COUPLING REACTIONS

The coupling model for Fischer-type alkylidyne complexes suggests that nucleophile-induced alkylidyne-isocyanide coupling should be more difficult than nucleophile-induced alkylidyne-carbonyl coupling, due to a larger gap between the MC π orbital and the CNR π* orbital.[3] To test this situation we investigated the reaction of complex **14** with pyrrole-2-carboxaldehyde methylimine in the presence of base (Eq. 12) and with diethyldithiocarbamate.[21] The reaction affords the ketenyl complexes **2**. The overall result is substitution of the isocyanide ligand and alkylidyne-carbonyl coupling. The reaction of complex **15** with diethyldithiocarbamate (Eq. 13) does not proceed cleanly, but substitution of the isocyanide ligands and alkylidyne-carbonyl coupling is also observed

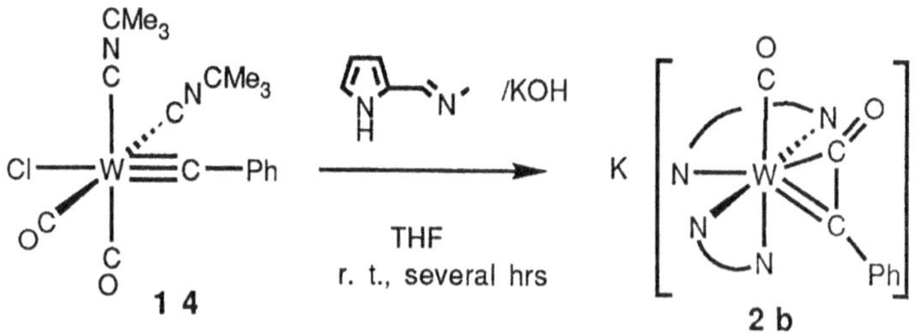

(12)

as the major reaction pathway. These results support the conclusion drawn from the coupling model that isocyanide ligands are less easily induced than carbonyl ligands to undergo coupling with alkylidyne ligands.

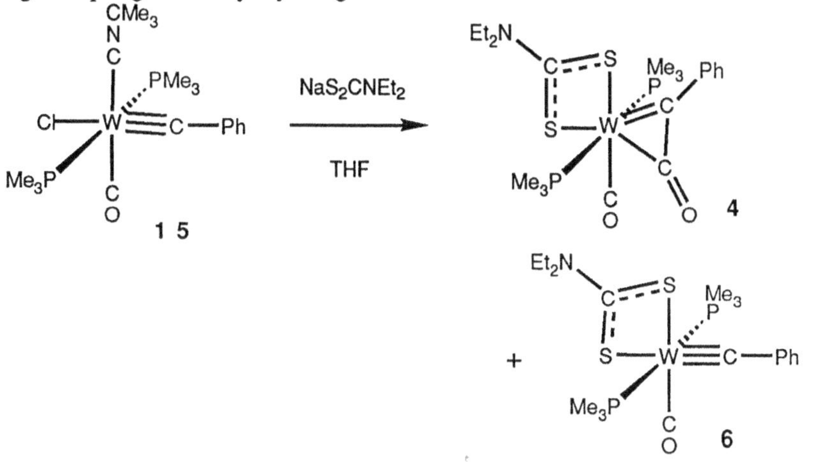

(13)

In contrast, the coupling of alkylidyne and isocyanide ligands in the presence of proton acids is well established.[3,22] Complex **15** reacts with HCl in CH_2Cl_2 to give a mixture of the nitrilium complex **16** and the aminoalkyne complex **17**.[23] The nitrilium ligand may form via double protonation of the alkylidyne ligand and migratory insertion between the formed benzyl ligand and the isocyanide ligand. Formation of the aminoalkyne complex

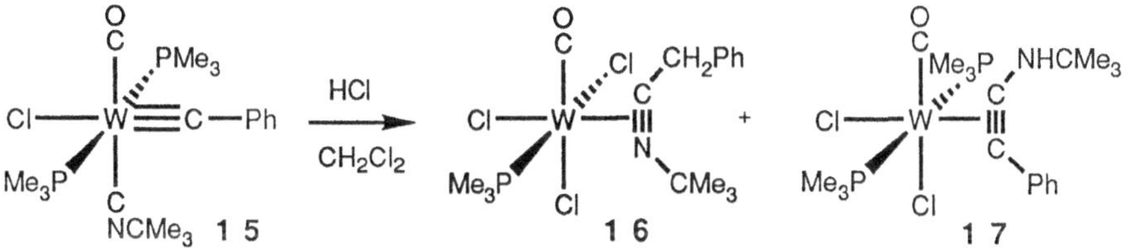

(14)

17 is of interest in the current context. Protonation of complex **15** with CF_3SO_3H was found to occur at the alkylidyne carbon atom (MC π bond) to give **18** (Scheme 2). In the presence of chloride and/or small amounts of methanol or traces of water, the alkylidene complex **18** transforms into the aminoacetylene complexes **17** (Y = Cl or CF_3SO_3). The role of methanol is postulated to help establish the equilibrium between the C-protonated form of the alkylidyne complex, **18**, and the N-protonated form, **19**. The N-protonated form **19** is an alkylidyne aminomethylidyne tungsten complex. In the protonated form the isocyanide ligand is highly activated towards coupling with the alkylidyne ligand (similar to the *bis*-alkylidyne case). The actual ligand coupling step is then induced by addition of the counterion to the metal center. The postulation of intermediate **19** is supported by the recent isolation by Filippou and coworkers of stable bis-(aminocarbyne) tungsten complexes.[24] These compounds were found to undergo ligand coupling upon the addition of nucleophilic ligands.

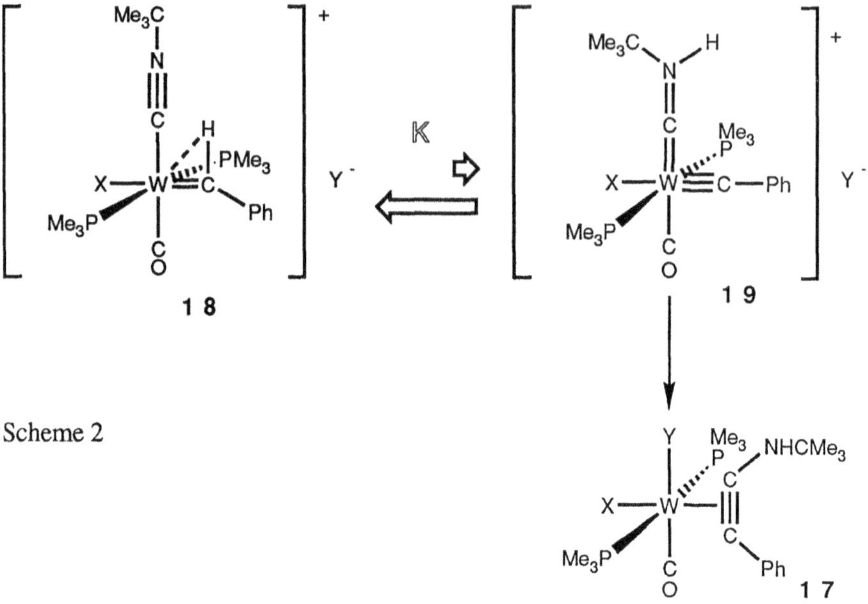

Scheme 2

18 19 17

3.5. FORMAL COUPLING OF TWO ALKYLIDYNE LIGANDS

The formal coupling of two alkylidyne ligands was achieved by the reaction sequence shown in Eqs. 15 and 16.[25,26] In this reaction sequence, the steps involved in the transformation of a carbonyl ligand into an alkylidyne ligand, *i. e.* addition of a nucleophile to a carbonyl ligand and oxide abstraction from the generated acyl ligand by oxalyl bromide, was applied to an alkylidyne carbonylmetal complex. The alkylidyne carbonyl

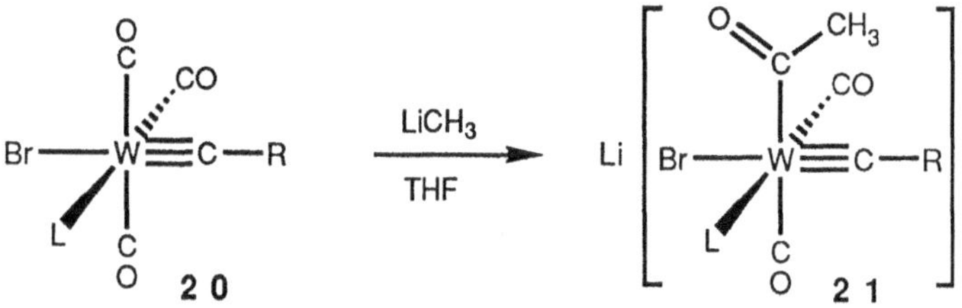

20 21 (15)

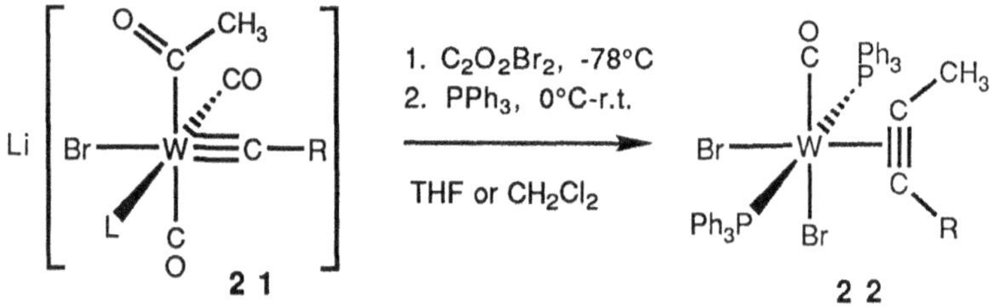

21 22 (16)

L = CO: R = C_6H_4-OMe-4, C_6H_4-NMe_2-4

L = PPh_3: R = Ph, C_6H_4-NMe_2-4

tungsten complexes **20** react with methyl lithium to give the alkylidyne acyl tungsten complexes **21** (Eq. 15). Treatment of complexes **21** with oxalyl bromide and addition of PPh_3 results in the formation of the alkyne complexes **22** (Eq. 16). The alkyne ligand in complexes **22** is formally derived from the coupling of two alkylidyne units. The available experimental evidence does not provide any information regarding the actual coupling step. The immediate coupling precursor is believed to be an alkylidyne alkylidene metal complex of type **23**. Dissociation of the bromooxalato group from the alkylidene ligand probably initiates the coupling step. Whether a bis-alkylidyne species forms as a short-lived intermediate, **24**, or whether carbon-carbon bond formation occurs simultaneously with the dissociation of the bromooxalate entity, **25**, is not known. We favor the latter

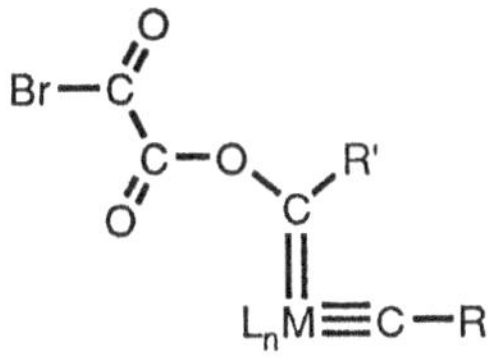

23

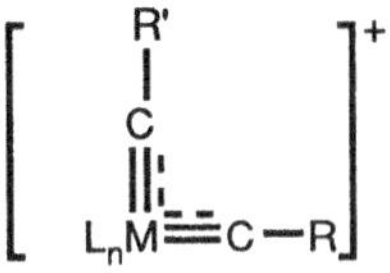

24

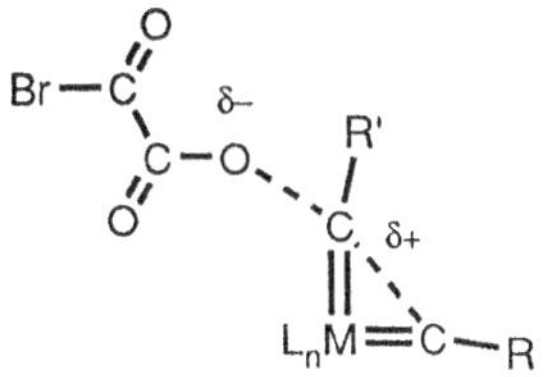

25

mechanistic possibility, since a metal center in octahedral coordination geometry cannot form two independent metal-ligand triple bonds.

Acknowledgement

I wish to thank my coworkers for their enthusiasm, skillful work, and for many stimulating discussions. Support for this work by the National Science Foundation and by the donors of the Petroleum Research Fund, administered by the American Chemical Society, is gratefully acknowledged.

References

(1) E. O. Fischer, G. Kreis, C. G. Kreiter, J. Müller, G. Huttner, and H. Lorenz, *Angew. Chem.* **85**, 618 (1973); *Angew. Chem. Int. Ed. Engl.* **12**, 564 (1973).

(2) F. R.Kreissl, A. Frank, U. Schubert, T. L. Lindner, and G. Huttner, *Angew. Chem.*, **88**, 649-650 (1976); *Angew. Chem. Int. Ed. Engl.*, **15**, 632 (1976).

(3) A. Mayr and C. M. Bastos, Prog. Inorg. Chem. **40**, 1 (1992).

(4) P. Hofmann in: H. Fischer, P. Hofmann, F. R. Kreissl, R. R. Schrock, U. Schubert, and K. Weiss, *Carbyne Complexes*, VCH Publishers, Weinheim, Germany, 1988.

(5) C. N. Wilker, R. Hoffmann, and O. Eisenstein, *Nouveau Journal de Chimie*, **7**, 535 (1983).

(6) A. Mayr, G. A. McDermott, A. M. Dorries, A. K. Holder, W. C. Fultz, and A. L. Rheingold, *J. Am. Chem. Soc.*, **108**, 310 (1986).

(7) A. Mayr, G. A. McDermott, A. M. Dorries, and D. Van Engen, *Organometallics*, **6**, 1503 (1987).

(8) K. A. Belsky, M. F. Asaro, S. Y. Chen., and A. Mayr, *Organometallics*, **11**, 1926 (1992).

(9) F. R. Kreissl, in *Organometallics in Organic Synthesis* (A. de Meijere and H. tom Dieck, Eds.), Springer-Verlag Berlin Heidelberg, pp. 105-109, 1987.
(10) A. Mayr, M. A. Kjelsberg, G. A. McDermott, D. Van Engen, R. T. Chang, and T.-Y. Lee, in preparation.
(11) A. Mayr, T.-Y. Lee, and M. A. Kjelsberg, in preparation.
(12) J. B. Sheridan, D. B. Pourreau, G. L. Geoffroy, and A. L.Rheingold, *Organometallics*, **7**, 289 (1988).
(13) A. Mayr, M. A. Kjelsberg, K. S. Lee, M. F. Asaro, and T. Hsieh, *Organometallics*, **6**, 2610 (1987).
(14) C. M. Bastos, "Coupling Reactions of Alkylidyne Ligands with Carbonyl, Isocyanide, and Alkylidyne Ligands on Tungsten", Ph. D. Thesis. State University of New York at Stony Brook, 1991.
(15) A. Mayr, C. M. Bastos, R. T. Chang, J. X. Haberman, and K. S. Robinson, D. A. Belle-Oudry, *Angew. Chem.* **104**, 802 (1992); *Angew. Chem. Int. Ed. Engl.* **31**, 747 (1992).
(16) D. C. Brower, M. Stoll, and J. L. Templeton, *Organometallics*, **8**, 2786 (1989).
(17) A. Mayr, C. M. Bastos, and R. T. Chang, in preparation.
(18) S. J. Holmes, R. R. Schrock, M. R. Churchill, and H. J. Wasserman, *Organometallics*, **3**, 476 (1984).
(19) R. N. Vrtis, S. Liu, Ch. P. Rao, S. G. Bott, and S. J. Lippard, *Organometallics*, **10**, 275 (1991).
(20) O. Cheung and T.-Y. Lee, unpublished results.
(21) A. Mayr, S. M. Holmes, and C. M. Bastos, *Organometallics*, in press.
(22) A. C. Filippou and W. Grünleitner, *Z. Naturforsch.*, **44b** , 1023 (1989).
(23) A. Mayr and C. M. Bastos, *J. Am. Chem. Soc.*, **112**, 7797 (1990).
(24) A. C. Filippou, W. Grünleitner, C. Völkl, and P. Kiprof, *Angew. Chem.* **103**, 1188 (1991); *Angew. Chem. Int. Ed. Engl.* **19**, 1167 (1991).
(25) G. A. McDermott and A. Mayr, *J. Am. Chem. Soc.*, **109**, 580 (1987).
(26) A. Mayr, C. M. Bastos, N. Daubenspeck, and G. A. McDermott, *Chem. Ber.*, **125**, 1583 (1992).

Dicarbonyl(η^5-cyclopentadienyl)carbyne Complexes of Molybdenum and Tungsten as Building Blocks

F. R. Kreißl, J. Ostermeier, W. Schütt, C. M. Stegmair, N. Ullrich, W. Ullrich
Anorganisch-chemisches Institut der Technischen Universität München
Lichtenbergstraße 4, Garching,, Germany

ABSTRACT. Dicarbonyl(η^5-cyclopentadienyl)carbyne complexes of molybdenum and tungsten prove to be a valuable synthetic tool: Reaction with phosphines provides substituted carbyne complexes and leads via an intramolecular CC-coupling to η^1- or η^2-ketenyl complexes respectively. Electrophiles attack the metal carbyne triple bond forming hetero- and acyclic carbene complexes, η^2-acyl compounds, η^3-ketene complexes and metalla-dithia-bicyclobutane cations. Dithiocarboxylates are formed in reaction of these dicarbonyl(η^5-cyclo-pentadienyl)carbyne complexes with sulfur or cyclohexene sulfide.

Introduction

In 1973, E. O. Fischer reported the first transition metal complex containing a metal-carbon triple bond [1]. Since this discovery many other carbyne complexes have been synthesized by the classical route of E. O. Fischer, by Schrock's method or via new preparative approaches. The exploration of the metal carbon triple bond has fascinated many research groups and carbyne complexes have become more and more important in theory, synthesis and catalytic research. Carbyne complexes can be conveniently classified into Fischer-type complexes with the metal not in its highest possible oxidation state and into Schrock-type complexes with the metal in a considerably high oxidation state. On that basis dicarbonyl(η^5-cyclopentadienyl)carbyne molybdenum or tungsten $Cp(CO)_2M{\equiv}C\text{-}R$ complexes may be regarded as lying in between these two classes.

Results

When trans-halo-tetracarbonylcarbyne complexes of molybdenum and tungsten, $X(CO)_4M{\equiv}CR$, are treated with sodium cyclopentadienyl, not only the displacement of the halide, but an additional elimination of two carbonyl ligands is observed, affording dicarbonyl(η^5-cyclopentadienyl)carbyne complexes [2,3]. A different synthetic approach converts vinylidene ligands into carbyne ligands to yield comparable bisdimethylphosphite substituted complexes of molybdenum [4].

F. R. Kreißl (ed.), Transition Metal Carbyne Complexes, 231–238.

Carbonyl-Phosphine-Substitution.

Under specific reaction conditions dicarbonyl-(η^5-cyclopentadienyl)carbyne complexes of molybdenum and tungsten (R = CH_3, C_6H_5, $C_6H_4CH_3$) react with trimethylphosphine via substitution

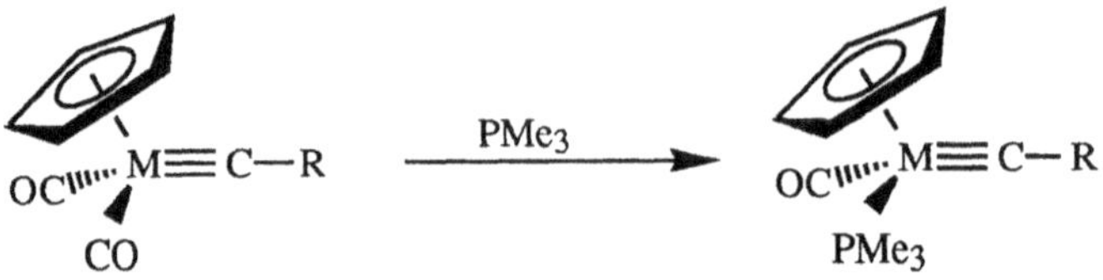

of one carbonyl ligand [3,5-7]. With triphenylphosphine substitution can be achieved in boiling thf [8].

Carbonyl Carbyne Coupling - Ketene Reaction.

If, however, carbonyl(cyclopentadienyl)(trimethylphosphine)carbyne complexes of molybdenum and tungsten are allowed to react with carbon monoxide, carbonylation at the carbyne carbon takes place forming η^1-ketenyl complexes in nearly quantitative yields [3,9].

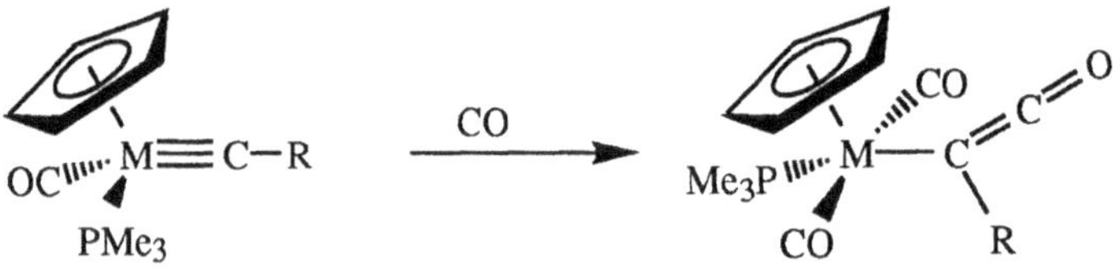

The first well established and characterized example of such a carbonyl-carbyne coupling reaction

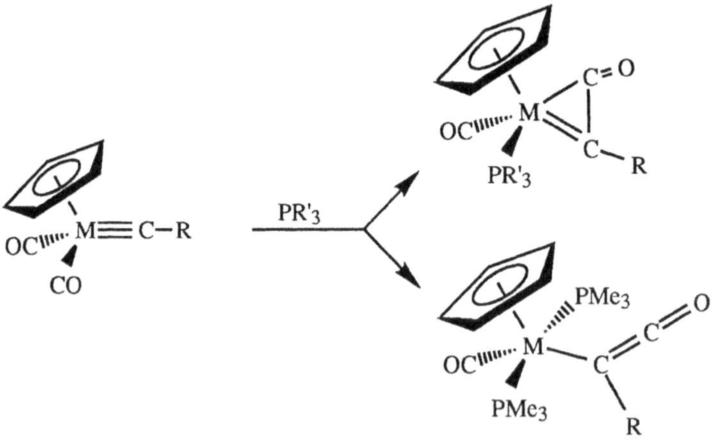

occurs on treating dicarbonyl(cyclopentadienyl)carbyne complexes with trialkylphosphines or triphenylphosphine. Depending on the applied reaction conditions $Cp(CO)L_2M[\eta^1\text{-}C(R)CO]$ and/or η^2-ketenyl $Cp(CO)LM[\eta^2\text{-}C(R)CO]$ complexes (M = Mo, W; R = CH_3, C_3H_5, C_5H_7, $C_5H_4FeC_5H_5$, C_6H_5, $C_6H_4CH_3$, $C_6H_2(CH_3)_3$, $Si(C_6H_5)_3$; R´= Me, Ph) are formed in high yields [3,10-12].

The application of the nucleophilic or electrophilic induced carbonyl carbyne coupling reaction has been perfectly demonstrated by E. O. Fischer [13,14], R. R. Schrock [15], J. L. Templeton [16,17], F. G. A. Stone [18], Angelici [19,20], A. Mayr [21,22], R. J. G. L. Geoffroy [23] and S. J. Lippard [24], preparing a variety of interesting ketenyl, alkyne and binuclear complexes involving η^2-ketenyl species as possible intermediates.

Carbyne Carbene Rearrangement - Protonation of the Carbyne Carbon.

Fenske´s theoretical studies on cyclopentadienyl-substituted carbyne complexes indicate a remarkable charge on the carbyne carbon atom [25]. This existence of a nucleophilic carbyne carbon becomes evident in the protonation of different carbyne complexes of molybdenum [26], tungsten [27-29]) and osmium [30].

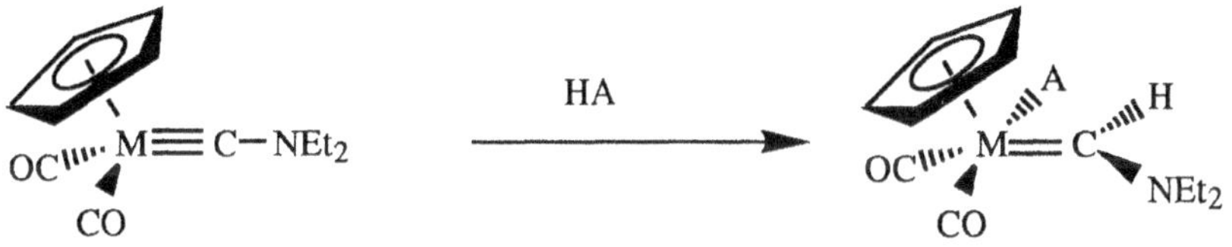

For the title compounds, $Cp(CO)_2M{\equiv}C\text{-}R$, protonation (HCl, CF_3COOH) of the carbyne carbon was easily accomplished. In the case of the dicarbonyl(η^5-cyclopentadienyl)diethylaminocarbyne complexes the corresponding diethylaminocarbene moieties were formed for the first time [31]. Treatment of $Cp(CO)_2M{\equiv}C\text{-}R$ (R = Me, Tol) with aqueous hydrogen iodide affords the corresponding iodo carbene complex [32]

Carbyne Acyl Rearrangement.

Contrary to Stone´s synthesis of $Cp(CO)_2IW{=}CHR$ [32], reaction of dicarbonyl(η^5-cyclopentadienyl)organylcarbyne complexes of molybdenum and tungsten with hydrogen chloride, trifluoro

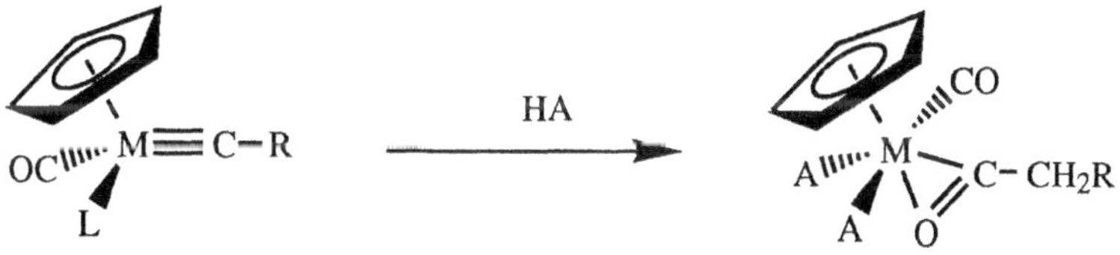

acetic acid or trichloroacetic acid provides, via a combined protonation and carbon carbon

coupling reaction, η^2-acyl complexes [33,34].
It has been proposed that these carbyne-acyl rearrangements start with the protonation of the carbyne carbon atom to initially give a carbene complex. For the next step two different pathways have been proposed: **i**) An intramolecular carbene carbonyl coupling to yield a π-ketene complex,

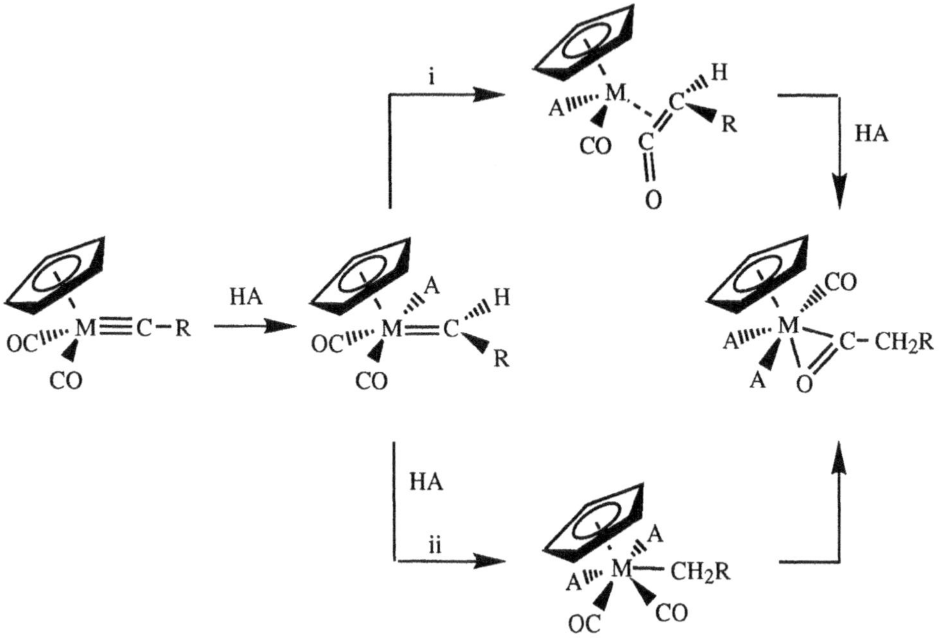

which undergoes further protonation to give the final acyl compound. **ii**) Subsequent protonation of the intermediate carbene carbon to form an alkyl ligand, which yields the η^2-acyl compound in a pseudo carbonyl insertion reaction [35].

Carbyne η^3-Ketene Conversion.

Dicarbonyl(η^5-cyclopentadienyl)carbyne complexes of tungsten (R = CH_3, C_6H_5, $C_6H_4CH_3$) react with halophosphines ($ClPMe_2$, $ClPPh_2$), dimethyliodoarsine ($IAsMe_2$) or 2-nitrobenzenesulphenyl chloride ($ClSC_6H_4NO_2$), via an electrophilic induced carbon carbon coupling reaction,to

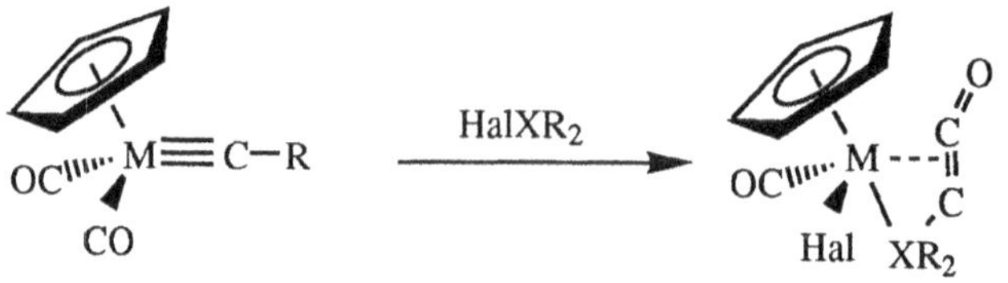

give the corresponding η^3-coordinated phosphino-, arsino- and thioketene complexes. Upon treatment of these products with trimethylphosphine, triphenylphosphine or trimethylphosphite, phosphinoketene and 1-tungsta-2-phospha-5-oxa-cyclopentene-3 complexes are formed [36].

Carbyne Chalcogenocarboxylate Conversion.

Reaction of certain cyclopentadienyl substituted carbyne complexes with sulfur or selenium affords dithio- or diselenocarboxylates [37]. Efforts to add only one sulfur atom to the metal carbon

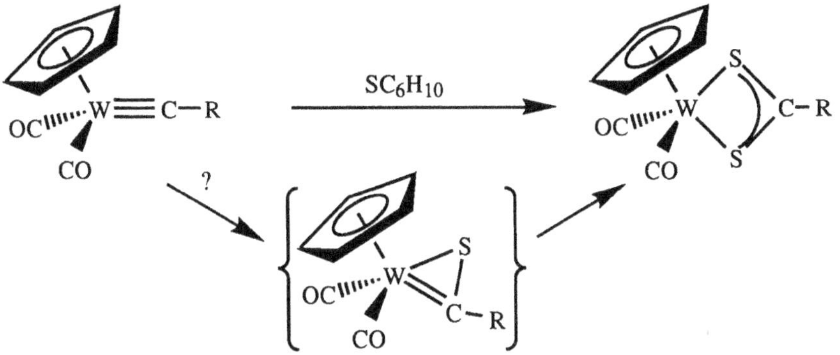

triple bond of dicarbonyl(η^5-cyclopentadienyl)carbyne complexes of tungsten (R = CH_3, C_6H_5, $C_6H_4CH_3$), using cyclohexene sulfide again result in the formation of dithiocarboxylate complexes [38].

Stepwise Addition of SMe^+ to the Metal Carbon Multiple Bond.

In a reaction sequence, characterized by a twofold electrophilic attack on the carbon atom of a metal carbon multiple bond, η^5-cyclopentadienylcarbyne complexes of molybdenum and tungsten Cp(CO)LM≡C-R (R = CH_3, C_6H_5, $C_6H_4CH_3$; L = CO, PMe_3; n = 0, 1) react in a stepwise, easily observed manner with dimethyl(methylthio)sulfonium tetrafluoroborate [Me_2S-SMe][BF_4], to give first the cationic η^2-thiocarbene and then a dicationic metalla-dithiabicyclo[1.1.0]butane moiety [39-41]. In contrast, the diethylamino-substituted carbyne complex $Cp(CO)_2M$≡C-NEt_2

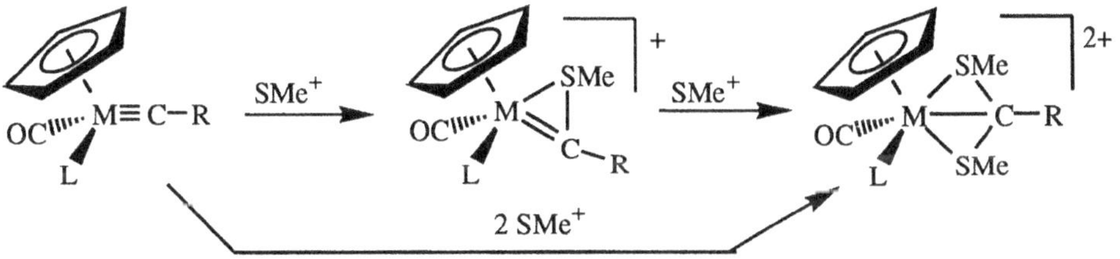

adds only one SMe^+ cation - presumably due to electron delocalization. In a previous experiment

Angelici has demonstrated that protonation of the methylthiocarbyne complex $HBpz_3(CO)_2W \equiv CSMe$ affords the cationic Fischer-type η^2-thiocarbene complex $[HBpz_3(CO)_2W\text{-}\eta^2\text{-}(=C(H)SMe)][CF_3SO_3]$ [42]. However, the comparable cyclopentadienyl substituted η^2-thiocarbene complexes $[Cp(CO)_2M\text{-}\eta^2\text{-}(=C(R)SMe)][BF_4]$ still contain a nucleophilic carbene carbon whereas an amphiphilic behavior is found for the trimethylphosphine-substituted derivatives Thus. nucleophiles as well as electrophiles add to the carbene carbon of $[Cp(CO)(PMe_3)M\text{-}\eta^2\text{-}(=C(R)SMe)][BF_4]$ [41].

[3+2]-Cycloaddition of Azides to the Metal-Carbon Triple Bond.

The reaction of dicarbonyl(η^5-cyclopentadienyl)-carbyne complexes $Cp(CO)_{2-n}(PMe_3)_nM{\equiv}C\text{-}R$ (M = Mo, W; n = 0, 1; R = Me, Ph, Tol) with azides, N_3R' (R´ = CO_2CH_3, $CH_2CO_2CH_3$), results in a [3+2]-cycloaddition to the electron rich metal carbon triple bond, leading to neutral 1-metalla-2,3,4-triazole complexes in high yields [43,44].

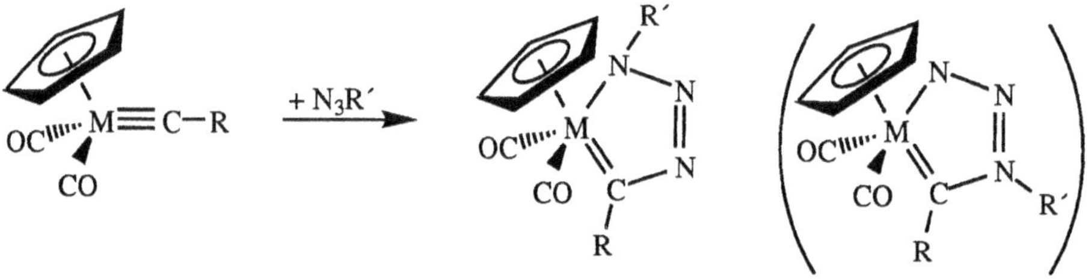

According to X-ray investigations, the metalla-triaza-cyclopentadiene ring system is nearly planar; the ester substituent is found to be at the nitrogen closest to the metal. The other possible isomer, with the ester function at the nitrogen atom in the neighborhood of the carbene carbon, is not formed. This result is in accordance to Regitz´s addition of an organic azide to a phosphinoalkyne [45] or to the postulated intermediate in the reaction of aromatic azides to a rhodium-carbonyl-bond [46]; in both cases only one isomer is observed.

Synthesis of Di- and Trinuclear Metal Complexes

The synthetic potential of transition metal carbyne complexes for preparing di-, tri- and polynuclear complexes has been impressively demonstrated by F. G. A. Stone [47-50]. In the

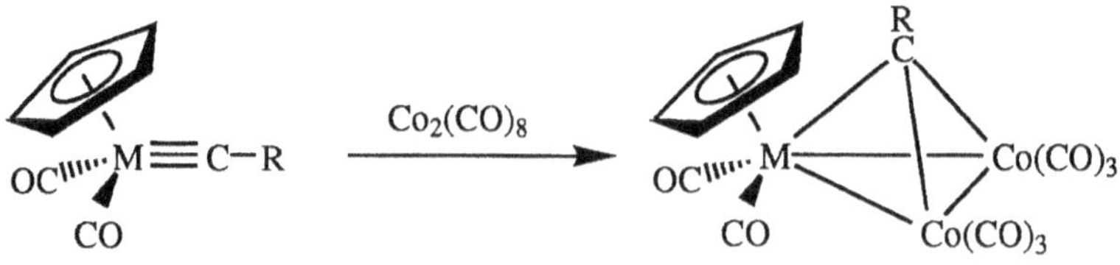

"Stone Reaction" mainly dicarbonyl(η^5-cyclopentadienyl)carbyne complexes are reacted with

other metal complexes of titanium, zirconium, vanadium, chromium, molybdenum, tungsten, manganese, rhenium, iron, ruthenium, osmium, cobalt, rhodium, iridium, nickel, palladium, platinum, copper, silver and gold to synthesize a wide variety of interesting new compounds. It utilizes Hoffmann's [51] isolobal relationship between the C≡C bond and the metal-carbon triple bond.

Acknowledgements

Support from the Deutsche Forschungsgemeinschaft and the Fonds der Chemischen Industrie is gratefully acknowledged. We thank Dr. E. Herdtweck, Prof. G. Huttner, Dr. P. Kiprof, Prof. G. Müller and Prof. U. Thewalt for X-ray structural studies.

References

[1] Fischer, E. O., Maasböl, A. (1964), Angew. Chem. 76, 645; (1964), Angew. Chem., Int. Ed. Engl. 3, 580.
[2] Fischer, E. O., Lindner, T. L., Kreißl, F. R. (1976), J. Organomet. Chem. 112, C27.
[3] Uedelhoven, W., Eberl, K., Kreißl, F. R. (1979), Chem. Ber. 112, 3376.
[4] Beevor, R. G., Green, M., Orpen, A. G., Williams, I. D. (1983), J. Chem. Soc., Chem. Commun. 673.
[5] Byrne, P. G., Garcia, M. E., Hoa Tran-Huy, N., Jeffery, J. C., Stone, F. G. A., J. Chem. Soc. (1987), Dalton Trans. 1243.
[6] Kreißl, F. R., Uedelhoven, W., Neugebauer, D. (1988), J. Organomet. Chem. 344, C27.
[7] Stegmair, C. M. (1991), Dissertation TU München.
[8] Dossett, S. J., Hill, A. F., Jeffery, J. C., Marken, F., Sherwood, P., Stone, F. G. A. (1988), J. Chem. Soc., Dalton. Trans. 2453.
[9] Kreißl, F. R., Uedelhoven, W., Eberl, K. (1978), Angew. Chem. 90, 908; (1978), Angew. Chem., Int. Ed. Engl. 17, 859.
[10] Kreißl, F. R., Frank, A., Schubert, U., Lindner, T. L., Huttner, G. (1976), Angew. Chem. 88, 649; (1976), Angew. Chem., Int. Ed. Engl. 15, 632.
[11] Kreißl, F. R., Friedrich, P., Huttner, G. (1977), Angew. Chem. 89, 110; (1977), Angew. Chem., Int. Ed. Engl. 16, 102.
[12] Sieber, W. J., Wolfgruber, M., Tran-Huy, N. H., Schmidt, H. R., Heiss, H., Hofmann, P., Kreißl, F. R. (1988), Chem. Ber. 340, 341.
[13] Fischer, E. O., Friedrich, P. (1979), Angew. Chem. 91, 345; (1979), Angew. Chem., Int. Ed. Engl. 18, 327.
[14] Fischer, E. O., Filippou, A. C., Alt, H. G., Ackermann, K. (1983), J. Organomet. Chem. 254, C21.
[15] Holmes, S. J., Schrock, R. R., Churchill, M. R., Wasserman H. J. (1984), Organometallics 3, 476.
[16] Birdwhistell, K. R., Tonker, T. L., Templeton J. L. (1985), J. Am. Chem. Soc. 107, 4474.
[17] Gambale, A. S., Birdwhistell, K. R., Templeton J. L. (1990), J. Am. Chem. Soc. 112, 1818.
[18] James, A. P., Stone, F. G. A. (1986), J. Organomet. Chem. 310, 47.
[19] Kim, H. P., Kim, S., Jacobsen, R. A., Angelici, R. J. (1986), Organometallics 5, 2481.

[20] Doyle, R. A., Angelici, R. J. (1989), Organometallics 8, 2207.
[21] Mayr, A., McDermott, G. A., Dorries, A. M., Holder, A. K., Fultz, W. C., Rheingold, A. L (1986), J. Am. Chem. Soc. 108, 310.
[22] Mayr, A., McDermott, G. A., Dorries, A. M., Van Engen D. (1987), Organometallics 6, 1503.
[23] Sheridan, J. B., Geoffroy, G. L., Rheingold, A. L. (1986), Organometallics 5, 1514.
[24] Vrtis, R. N., Liu, S., Rao, Ch. P., Bott, S. G., Lippard, S. J. (1991), Organometallics 10, 275.
[25] Kostic, N. M., Fenske, R. F. (1981), J. Am. Chem. Soc. 103, 4677.
[26] Green, M., Orpen, A. G., Williams, I. D. (1982), J. Chem. Soc., Chem. Commun. 493.
[27] Holmes, S. J., Schrock, R. R. (1981), J. Am. Chem. Soc. 103, 4599.
[28] Freudenberger, J. H., Schrock, R. R. (1985), Organometallics 4, 1937.
[29] Kim, H. P., Kim, S., Jacobson, R. A., Angelici, R. J. (1984), Organometallics 3, 1124
[30] Clark, G. R., Marsden, K., Roper, W. R., Wright, L. J. (1980), J. Am. Chem. Soc. 102, 6570.
[31] Kreißl, F. R., Sieber, W. J., Wolfgruber, M. (1984), J. Organomet. Chem. 270, C45.
[32] Howard, J. A. K., Jeffery, J. C., Laurie, J. C. V., Moore, I., Stone, F. G. A., Stringer, A. (1985), Inorg. Chim. Acta 100, 23.
[33] Kreißl, F. R., Sieber, W. J., Wolfgruber, M., Riede, J. (1984), Angew. Chem. 96, 618; (1984), Angew. Chem., Int. Ed. Engl. 23, 640.
[34] Kreißl, F. R., Sieber, W. J., Keller, H., Riede, J., Wolfgruber, M. J. (1987), Organomet. Chem. 320, 83.
[35] Kreißl, F. R., Keller, H., Schütt, W. J. Organomet. Chem. in press.
[36] Wolfgruber, M., Stegmair, C. M., Kreißl, F. R. (1989), J. Organomet. Chem. 376, 45.
[37] Gill, D. S., Green, M., Marsden, K., Moore, I., Orpen, A. G., Stone, F. G. A., Williams, I. D., Woodward, P. (1984), J. Chem. Soc., Dalton. Trans. 1343.
[38] Kreißl, F. R., Ullrich, N. (1989), J. Organomet. Chem. 361, C30.
[39] Kreißl, F. R., Keller, H. (1986), Angew. Chem. 89, 924; (1986), Angew. Chem., Int. Ed. Engl. 25, 904.
[40] Ullrich, N., Stegmair, C., Keller, H., Herdtweck, E., Kreißl, F. R. (1990), Z. Naturforsch. B45, 921.
[41] Kreißl, F. R., Stegmair, C. M. (1991), Chem. Ber. 124, 2747.
[42] Kim, H. P., Kim, S., Jacobsen, R. A., Angelici, R. J. (1984), Organometallics 3, 1124.
[43] Stegmair, C. M., Ullrich, W., Schütt, W., Kiprof, P., Kreißl, F. R. (1992), Chem. Ber. 125, 1571.
[44] Stegmair, C. M., Schütt, W., Ullrich, W., Kiprof, P., Ostermeier, J., Kreißl, F. R. J. Organomet. Chem., in press.
[45] Rösch, W., Regitz, M. (1984), Angew. Chem., 96, 898; (1984), Angew. Chem., Int. Ed. Engl. 23, 900.
[46] Iqbal, A. F. M. (1972), Helv. Chim. Acta 55, 2637.
[47] Stone, F. G. A. (1983), 'Inorganic Chemistry: towards the 21st century' (Chisholm, M. H., ed.) Am. Chem. Soc., Symp. Ser. 211, 383.
[48] Kim, H. P., Angelici, R. J., (1987), 'Transition Metal Complexes with Terminal Carbyne Ligands', Adv. Organomet. Chem., 27, 51.
[49] Kreißl, F. R., (1988), 'Selected Reactions of Carbyne Complexes', (Fischer, H., Hofmann, P., Kreißl, F. R., Schrock, R. R., Schubert, U., Weiss, K. ed.) Carbyne Complexes, VCH Verlag.
[50] Mayr, A., H. Hoffmeister (1991), Adv. Organomet. Chem, 32, 227.
[51] Hoffmann, R. (1982), Angew. Chem. 10, 725; Angew. Chem., Int. Ed. Eng. 21, 711.

DIVERSIONS *EN ROUTE* TO ALKYLIDYNE COMPLEXES OF IRON

A.F. HILL
Department of Chemistry
Imperial College of Science, Technology and Medicine
South Kensington
London SW7 2AY
United Kingdom

ABSTRACT. The Mayr carbyne synthesis is readily extendable to aminomethylidyne complexes of group 6 metals, e.g., $[M(\equiv CN^iPr_2)(O_2CCF_3)(CO)_3(PPh_3)]$ (M = Cr, Mo, W) result from successive treatment of $[M(CO)_6]$ with LiN^iPr_2, $(CF_3CO)_2O$ and PPh_3. The same approach with $[Fe(CO)_5]$ however, provides the trifluoromethyl-carbamoyl complex $[Fe(\eta\text{-}OCN^iPr_2)(CF_3)(CO)_2(PPh_3)]$, whilst replacing LiN^iPr_2 with $LiC_6H_3Me_2$-2,6 provides the dinuclear complex $[Fe_2(\mu\text{-}OCC_6H_3Me_2\text{-}2,6)_2(CO)_5(PPh_3)]$. Reaction of $[M(\equiv CN^iPr_2)(O_2CCF_3)(CO)_3(PPh_3)]$ with $K[HB(pz)_3]$ (pz = pyrazol-1-yl) provides $[M(\equiv CN^iPr_2)(CO)_2\{HB(pz)_3\}]$ whilst similar treatment of $[Fe(\eta\text{-}OCN^iPr_2)(CF_3)(CO)_2(PPh_3)]$ provides the unusual metallaoxetene complex $[Fe\{C(N^iPr_2)OCF_2\}(CO)\{HB(pz)_3\}]$ *via* fluoride abstraction.

1. Introduction

1.1 LATE TRANSITION METAL ALKYLIDYNE COMPLEXES

Given the isoelectronic relationship between $[CR]^+$ and $[NO]^+$ and the ubiquity of this latter ligand in the coordination chemistry of later transition metals, the scarcity of mononuclear alkylidyne complexes of metals from groups 8-10 is surprising [1-4]. Isolated examples have been reported for iron [5], cobalt [6], ruthenium [4,7], osmium [4,8-9] and iridium [10]. Most of the examples known employ routes with extensive precedent in early transition metal systems, i.e., either electrophilic attack at the β-atom of a hetero carbonyl (CS [5], CTe [4], or $C{=}CH_2$ [10]) or the Lewis-acid assisted abstraction of an alkoxide group from a carbene precursor [5] (Scheme 1). The one approach which is, too date, peculiar to group 8 metals involves reduction of a divalent dichlorocarbene complex by lithium aryls [4]. The limitation of this procedure to ruthenium and osmium is presumably not a feature of these metals but rather a result of the present lack of synthetic routes to suitable dihalocarbene precursor complexes of earlier metals.

F. R. Kreißl (ed.), Transition Metal Carbyne Complexes, 239–248.

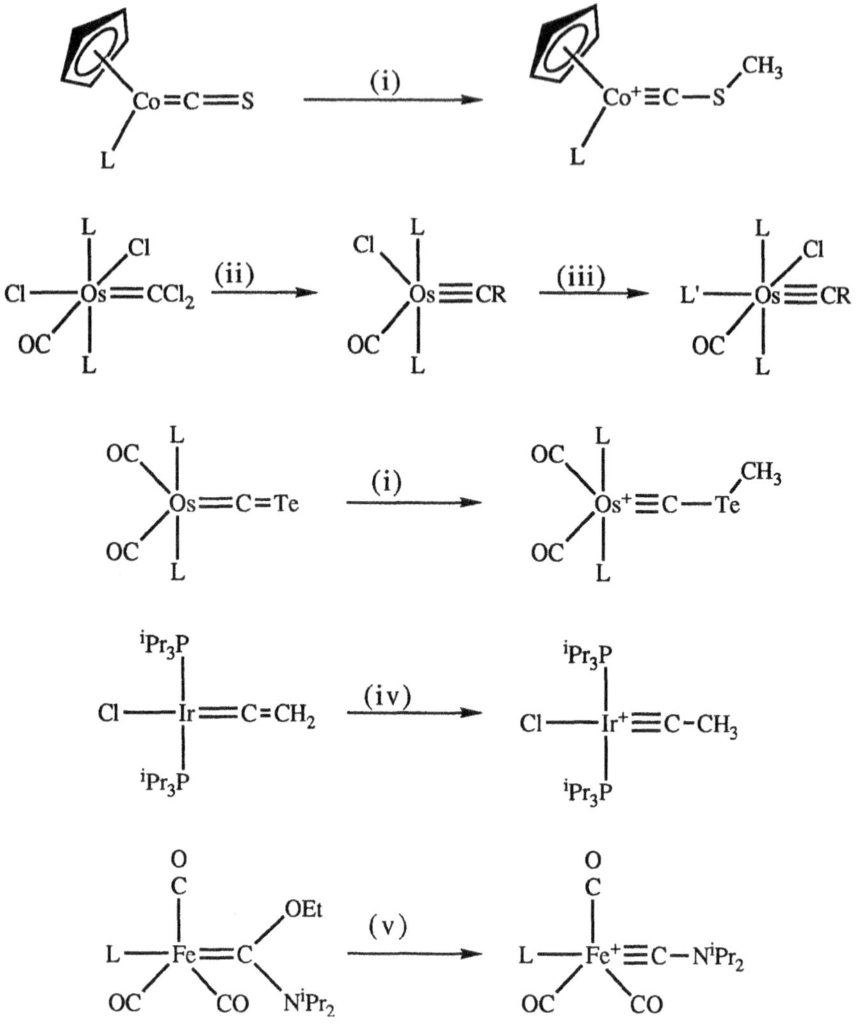

Scheme 1. L = PPh_3; (i) CH_3^+; (ii) 2 LiC_6H_4R-4, R = H, OMe, Me, NMe_2; (iii) O_2, H^+, L', L' = CNC_6H_4Me-4$^{(+)}$, SCN; (iv) H^+; (v) BCl_3.

1.2 STABILISATION OF ALKYLIDYNE LIGANDS

The work described herein is directed towards the synthesis of alkylidyne complexes of the later transition metals, specifically iron. Two approaches present themselves for the synthesis of alkylidyne complexes which might otherwise be unstable, *viz* steric or electronic stabilisation. The first approach involves the accumulation of steric bulk in the vicinity of the metal-carbon multiple bond, an effect easily acheived for benzylidyne complexes by 2,6-disubstitution of the aryl ring [11]. Alternatively, electronic stabilisation is afforded by alkylidyne substituents capable of hyperconjugative donation into the alkylidyne acceptor orbital. This approach is best illustrated by aminomethylidynes which may be better described as 2-aza-vinylidenes (Scheme 2).

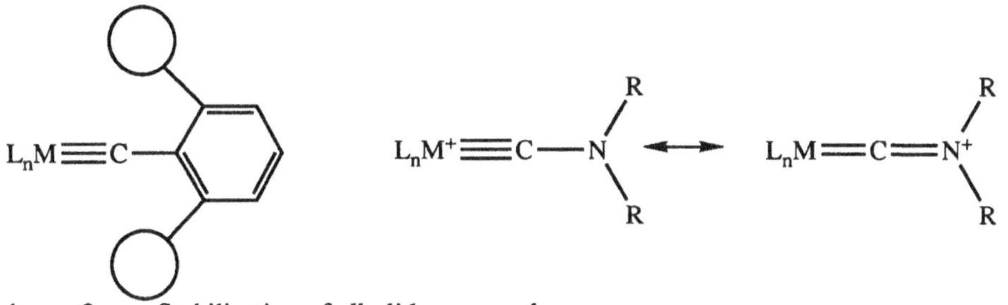

Scheme 2. Stabilisation of alkylidyne complexes.

2. Results and Discussion

2.1 GROUP 6 AMINOMETHYLIDYNE COMPLEXES

Mayr has reported a convenient synthetic approach to the large-scale preparation of thermally stable aryl and alkyl methylidyne complexes of group 6 metals based on the abstraction of oxide from suitable metal carbonyl acylates [12]. This approach has been extended to provide alkylidyne complexes bearing alkynyl [13], ferrocene-diyl [14], cymantrenyl [15] and silyl [16] substituents (Scheme 3).

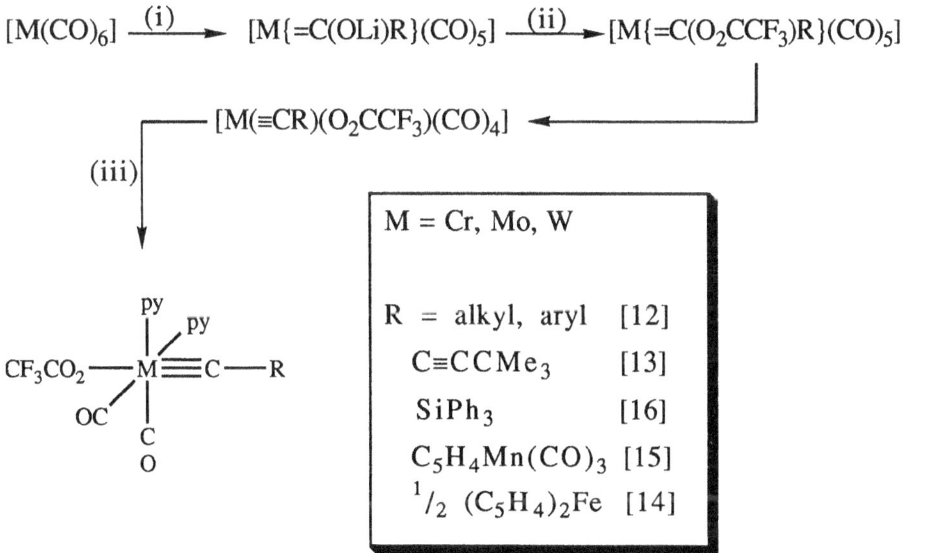

Scheme 3. Synthesis of alkylidyne complexes by oxide abstraction. (i) LiR; (ii) $(CF_3CO)_2O$; (iii) py = pyridine.

We find that the oxide abstraction approach also works well for aminomethylidyne complexes of chromium, molybdenum and tungsten providing the complexes $[M(\equiv CN^iPr_2)(O_2CCF_3)(CO)_3(PPh_3)]$ (M = Cr, Mo, W) in high yield (Scheme 4).

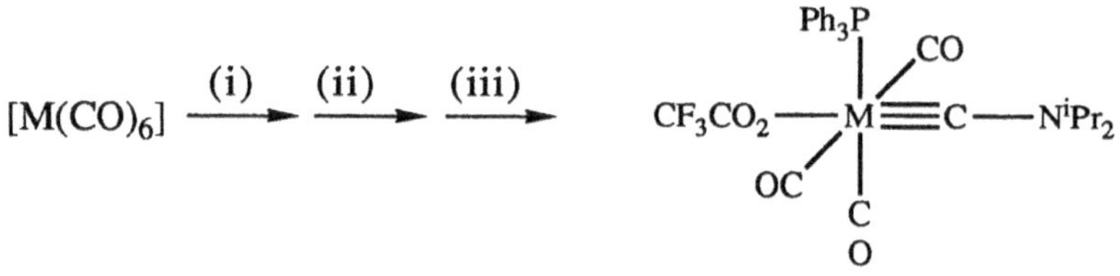

Scheme 4. Synthesis of group 6 aminomethylidyne complexes. M = Cr, Mo, W; (i) LiN^iPr_2; (ii) $(CF_3CO)_2O$; (iii) PPh_3.

These complexes are thermally and comparatively air stable and as such provide convenient precursors for the synthesis of further aminomethylidyne derivatives (Scheme 5).

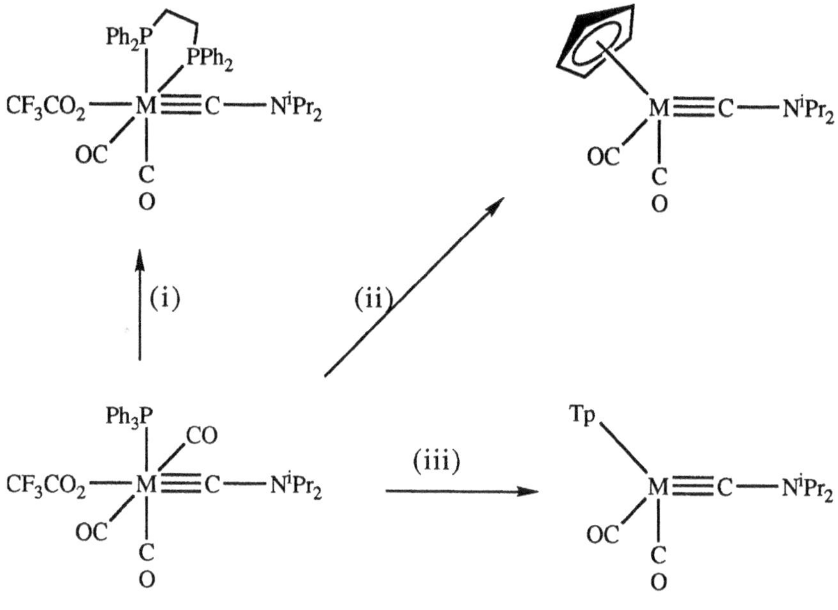

Scheme 5. Ligand exchange reactions for the complexes $[M(\equiv CN^iPr_2)(O_2CCF_3)(CO)_3(PPh_3)]$ (M = Cr, Mo, W); (i) 1,2-bis(diphenylphosphino)ethane; (ii) NaC_5H_5; (iii) $K[HB(pz)_3]$; Tp = κ^3-$HB(pz)_3$.

Thus, for example, reaction of the chromium derivative with $K[HB(pz)_3]$ (pz = pyrazol-1-yl) provides the crystallographically characterised hydrotris(pyrazolyl)borato derivative $[Cr(\equiv CN^iPr_2)(CO)_2\{HB(pz)_3\}]$ in high yield (Figure 1). Similar reaction with sodium cyclpentadienide with the molybdenum derivative provides the 'half-sandwich' complex $[Mo(\equiv CN^iPr_2)(CO)_2(\eta\text{-}C_5H_5)]$ and 1,2-bis(diphenylphosphino)ethane (dppe) provides the complex $[Mo(\equiv CN^iPr_2)(O_2CCF_3)(CO)_2(dppe)]$. In all these cases the triphenylphosphine is lost in preference to a carbonyl ligand. It is noteworthy that the molybdenum complex $[Mo(\equiv CN^iPr_2)(CO)_2\{HB(pz)_3\}]$ also results, in very low yield, from the reaction of the carbamoyl complex $[Mo(\eta^2\text{-}OCCN^iPr_2)(CO)_2\{HB(pz)_3\}]$ with sodium ethoxide in a manner analogous to the alkoxide mediated conversion of $[Mo(\eta^2\text{-}OCMe)(CO)_2\{HB(pz)_3\}]$ to $[Mo(\equiv CMe)(CO)_2\{HB(pz)_3\}]$ described by Templeton and and co-workers [17]. The reduced yield presumably results from the reduced suceptibility of the carbyne carbon in the aminomethylidyne complex towards nucleophilic attack by alkoxide.

Figure 1. Molecular structure of $[Cr(\equiv CN^iPr_2)(CO)_2\{HB(pz)_3\}]$

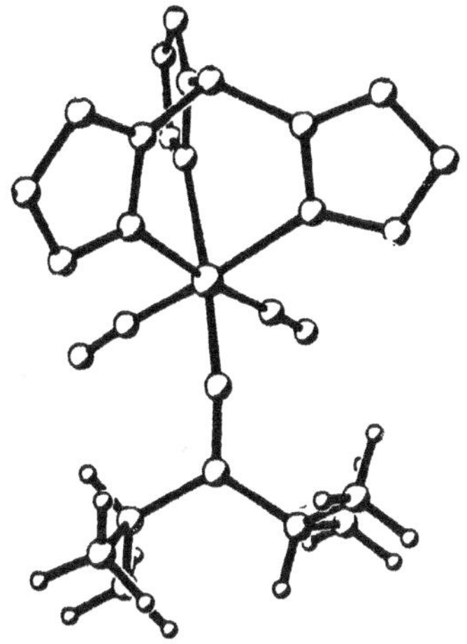

2.2 CARBAMOYL COMPLEXES OF IRON

The convenience of this route into the chemistry of aminomethylidyne complexes suggested that it might also provide access to such complexes of the later transition metals. Accordingly iron pentacarbonyl was treated under similar conditions. The product of the reaction was however found to not be an aminomethylidyne complex but rather a trifluoromethyl-carbamoyl derivative of iron(II) characterised on the basis of spectroscopic data and X-ray diffraction (Figure 2). A proposed mechanism for the formation of the carbamoyl complex (Scheme 6) involves the initial formation of a carbamoylate complex which, by way of contrast with the octahedral group 6 analogues, is five coordinate at iron and therefore suceptibile to electrophilic attack at either the carbamoyl oxygen or the electron rich iron centre. Electrophilic attack by the hard electrophile $[Et_3O]BF_4$ has been shown by Fischer [18] to occur at the carbamoyl oxygen to provide an alkoxy - carbene complex. In the

Figure 2. Molecular structure of $[Fe(OCN^iPr_2)(CF_3)(CO)_2(PPh_3)]$

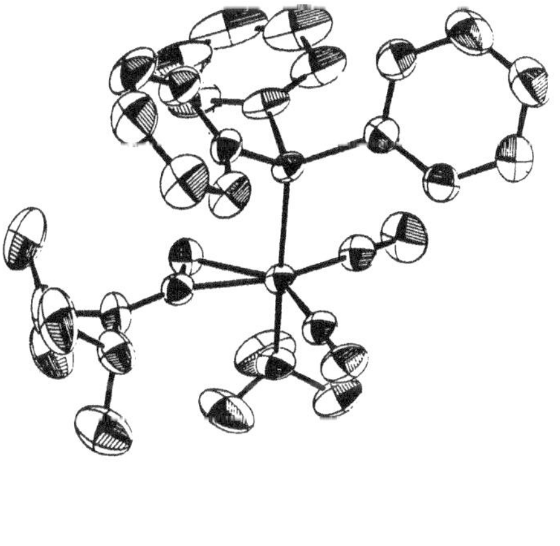

present case, however, trifluoroacetic anhydride would appear to attack at the iron centre to provide an unstable trifluoroacyl complex which may decarbonylate under the reaction conditions to provide a trifluoromethyl ligand, although it is surprising that this occurs under such mild condition (-10 °C). Further decarbonylation at the iron centre allows coordination of the carbamoyl oxygen and one triphenylphosphine. Further phosphine coordination is possible if the complex is treated with dppe whereby the complex [Fe(η^2-OCN^{i}Pr$_2$)(CF$_3$)(CO)(dppe)] is obtained.

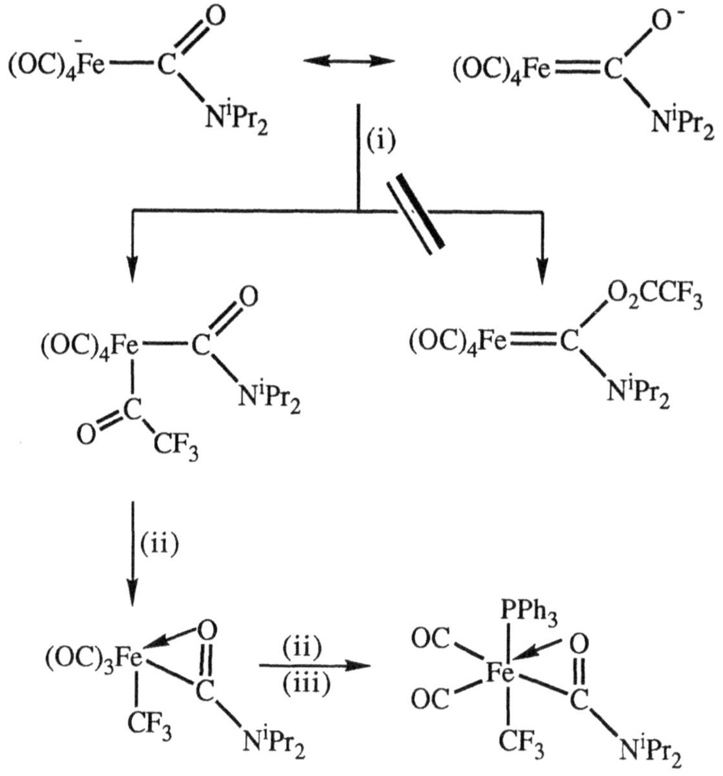

Scheme 6. Synthesis of [Fe(η^2-OCN^{i}Pr$_2$)(CF$_3$)(CO)$_2$(PPh$_3$)]. (i) (CF$_3$CO)$_2$O; (ii) -CO; (iii) PPh$_3$.

The extensive synthetic utility of trifluoromethyl ligands has only recently been realised [19] and derives from the nucleofugicity of the fluoride substituents. Thus, with moist Lewis acids the CF$_3$ group of [Fe(η^2-OCN^{i}Pr$_2$)(CF$_3$)(CO)$_2$(PPh$_3$)] is hydrolytically converted into a carbonyl ligand (Scheme 7).

The intermediacy of a cationic difluorocarbene complex in this reaction has considerable precedent, however given the typical electrophilicity of this ligand and the established nucleophilicity of carbamoyl ligands, we were somewhat surprised not to observe any form of interaction between these two ligands. Such an interaction was however observed in the unexpected reaction of the trifluoromethyl complex with K[HB(pz)$_3$]. Rather than forming an anionic carbamoyl complex, [Fe{η^1-C(O)N^{i}Pr$_2$} (CF_3)(CO){HB(pz)$_3$}]K, the final product isolated is a neutral ferraoxetene derived from the coupling of the carbamoyl ligand with a putative difluorocarbene ligand, *viz* [Fe{C(N^{i}Pr$_2$)OCF$_2$)(CO){HB(pz)$_3$}] (Scheme 8).

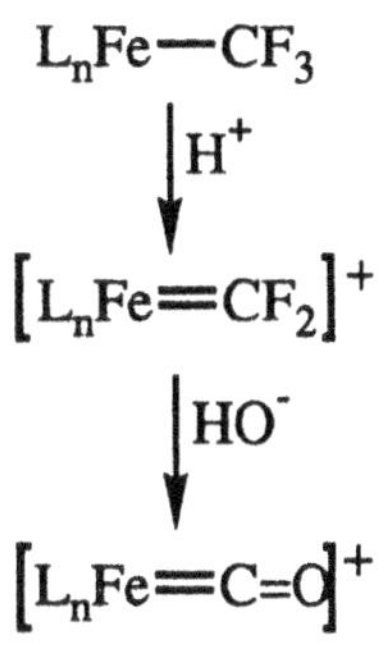

Scheme 7.

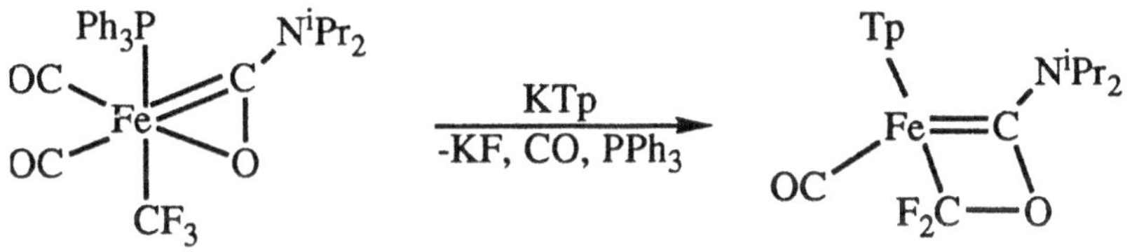

Scheme 8. Synthesis of a ferraoxetene. Tp = HB(pz)$_3$

The characterisation of the ferraoxetene included a single crystal structure determination which whilst confirming the metallacycle formulation revealed considerable departure from the structural parameters of the only other example of such a metallacycle. The ruthenaoxetene [Ru(CH_2OCPh)(CO)$_2$(PPh$_3$)$_2$]$^+$ [20] described by Roper *et al* shows significant difference between the two ruthenium carbon bond lengths which approach values typical of single and double Ru-C bonds. In contrast the metallacycle of the ferraoxetene shows essentially equal Fe-C bond lengths consistent with a reduction in Fe-C bond order for the amino substituted carbon. This is clearly due to considerable pπ donation from the diisopropylamino group, the nitrogen atom of which is trigonally coordinated. The mechanism for the oxetene formation has yet to be resolved, however it is somewhat surprising that the ferraoxetene does not result from the reaction of K[HB(pz)$_3$] with [Fe(η^2-OCN^{i}Pr$_2$)(CF_3)(CO)$_3$] (generated *in situ*).

Figure 3.
Molecular structure of [Fe(CF_2OCN^{i}Pr$_2$)(CO){HB(pz)$_3$}]

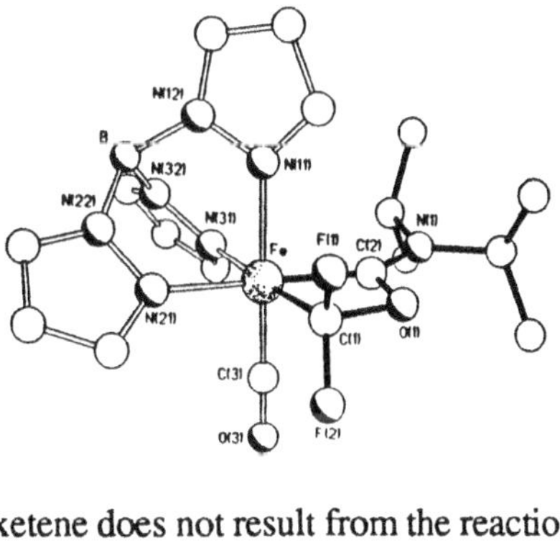

2.3 μ-BENZOYL COMPLEXES OF IRON

With the failure of the diisopropylamino group to favour formation of alkylidyne iron complexes we next turned our attention to the possibility of using steric factors to favour alkylidyne formation. The reaction of [Fe(CO)$_5$] with simple aryl or alkyl lithium reagents (LiR; R = Me, tBu, C_6H_4Me-4, C_6H_4OMe-4) followed by trifluoroacetic anhydride and triphenylphosphine lead in all cases to the exclusive formation of [Fe(CO)$_3$(PPh$_3$)$_2$] in high yield (Scheme 9).

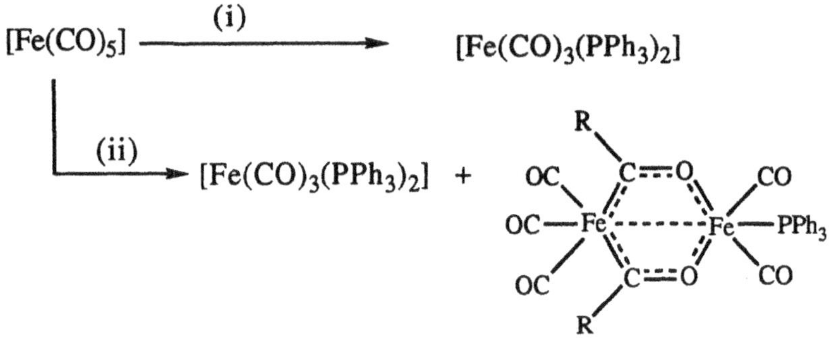

Scheme 9. Synthesis of [Fe$_2$(μ-OCC$_6$H$_3$Me$_2$-2,6)$_2$(CO)$_5$(PPh$_3$)] *(Fe-Fe)*. (i) LiR, [R = Me, ^{t}Bu, C_6H_4R'-4, R' = H, OMe, Me], (CF$_3$CO)$_2$O, PPh$_3$; (ii) LiC$_6$H$_3$Me$_2$-2,6, (CF$_3$CO)$_2$O, PPh$_3$.

If however more sterically congested 2,6-dimethylphenyl lithium is used in this sequence, in addition to [Fe(CO)$_3$)(PPh$_3$)$_2$] a binuclear complex may be isolated in moderate yield. The nature of the product does not follow unambiguously from spectroscopic data, however a single crystal structure determination reveals that the molecule has two chemically distinct iron centres with the two bridging benzoyl ligands being bound through carbon to one iron and oxygen to the other iron centre. The bonding in the Fe$_2$C$_2$O core may be described by three valence bond contributors (Scheme 10), two of which require an Fe->Fe dative bond to satisfy the EAN requirements of both metal

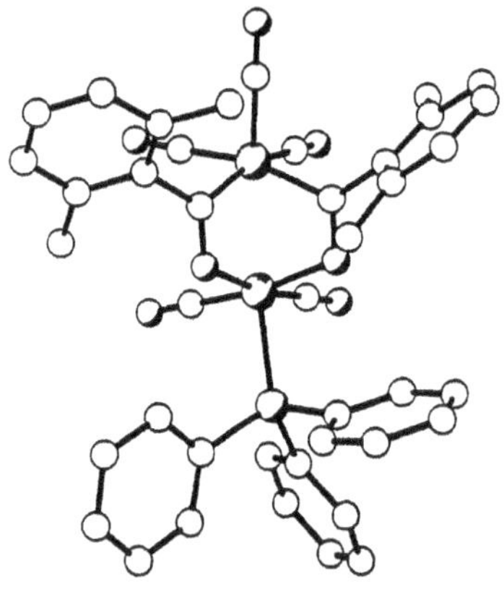

Figure 4. Molecular structure of [Fe$_2$(μ-σ,σ'-OCC$_6$H$_3$Me$_2$)$_2$(CO)$_5$-(PPh$_3$)]

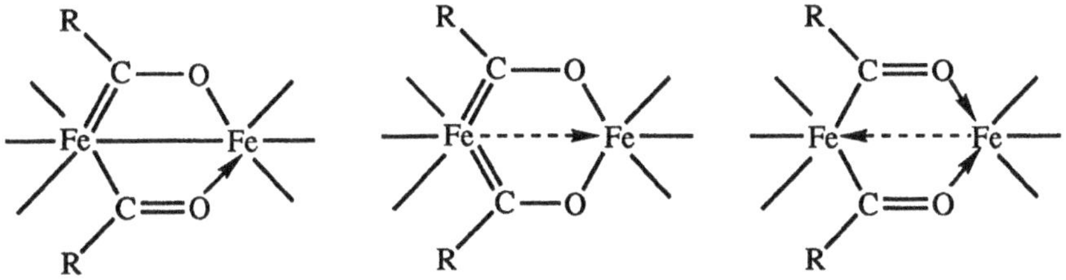

Scheme 10. Canonical forms for the bonding in $[Fe_2(\mu\text{-}OCC_6H_3Me_2)_2(CO)_5(PPh_3)]$

3. Concluding Remarks

The present study has not led to stable alkylidyne complexes of iron, however the problems encountered here represent diversions encountered in applying synthetic strategies developed for group 6 metals. The complex $[Fe(\equiv CN^iPr_2)(CO)_3(PPh_3)]^+$ has been reported by Fischer et al [5] and indicates that given suitable synthetic methodology the chemistry of this class of compounds may yet be developed.

4. Acknowledgements

We gratefully acknowledge the award of a studentship by Science and Engineering Research Council (U.K.) to S.A. who carried out the synthetic work described herein. We wish to thank George R. Clark, David J. Williams, Alexandra M.Z. Slawin and Graham Hogarth for the crystallographic studies.

5. References

1 Kim, H.P.; Angelici, R.J.; *Adv. Organomet. Chem.,* 27, 1987, 51.

2 Mayr, A.; Hoffmeister, H., *Adv. Organomet. Chem.*, 32, 1991, 227.

3 Gallop, M.A.; Roper, W.R., *Adv. Organomet. Chem.*, 21, 1986, 121.

4 Roper, W.R.; *J. Organomet. Chem.,* 300, 1986, 167

5 Fischer, E.O.; Schneider, J.; Neugebauer, D., *Angew. Chem., Int. Ed .Engl.,* 23, 1984, 820

6 Fortune, J.; Manning, A.R., *Organometallics,* 2, 1983, 1719. The possibility that the compound described here is actually a cationic thioacyl complex must also be considered in the absence of crystallographic data and the in view of results obtained for rhodium and iridium: Faraone, F.; Tresoldi, G.; Loprette, G.A., *J. Chem. Soc., Dalton Trans.,* 1979, 933; Tresoldi, F.G.; Faraone, F.; Piraino, P., *ibid.,* 1979, 1053.

7 Clark, G.R.; Marsden, K.; Roper, W.R.; Wright, L.J., *J. Am. Chem. Soc.,* 102, 1980, 6570.

8 Vogler, A.; Kisslinger, J.; Roper, W.R., *Z. Naturforsch.*, 32B, 1977, 473.

9 Clark, G.R.; Edmonds, N.R.; Pauptit, R.A.; Roper, W.R.; Waters, J.M.; Wright, A.H., *J. Organomet. Chem.,* 244, 1983, C57.

10 Höhn, A.; Werner, H., *Angew. Chem., Int. Ed. Engl.,* 25, 1986, 737; Höhn, A.; Werner, H., *J. Organomet. Chem.*, 382, 1990, 255.

11 Dossett, S.J.; Hill, A.F.; Jeffery, J.C.; Marken, F.; Sherwood, P.; Stone, F.G.A., *J. Chem Soc. Dalton Trans.,* 1988, 2453; Dossett, S.J.; Hill, A.F.;

Howard, J.A.K.; Nasir, B.A.; Spaniol, T.P.; Sherwood, P.; Stone, F.G.A., *ibid.*, 1989, 1871
12 Mayr, A.; McDermottt, G.A., *Organometallics,* 3, 1985, 608.
13 Hart, I.J.; Hill, A.F.; Stone, F.G.A., *J. Chem. Soc., Dalton Trans.*, 1989, 2261.
14 Davies, S.J.; Hill, A.F.; Pilotti, M.U.; Stone, F.G.A., *Polyhedron,* 8, 1989, 2265.
15 Hill, A.F.; Stone, F.G.A., unpublished results.
16 Jamieson, G.M.; Bruce, A.E.; White, P.S.; Templeton, J.L., *J. Am. Chem. Soc.,* 113, 1991, 5057.
17 Brower, D.C.; Stoll, M.; Templeton, J.L., *Organometallics,* 8, 1989, 2786.
18 Fischer, E.O.; Schneider, J.; Ackermann, K., *Z. Naturforsch.*, 39B, 1984, 468.
19 Hoskins, S.V.; Pauptit, R.A.; Roper, W.R.; Waters, J.M., *J. Organomet. Chem.,* 269, 1984, C55; For a recent review of this aspect of metal perfluoroalkyl chemistry see P.J. Brothers, W.R. Roper, *Chem. Rev.*, 88, 1988, 1293
20 Bohle, D.S.; Clark, G.R.; Rickard, C.E.F.; Roper, W.R.; Shepard, W.E.B.; Wright, L.J., *J. Chem. Soc., Chem. Commun.*, 1987, 563

ELECTRON-RICH TUNGSTEN AMINOCARBYNE COMPLEXES WITH Cp* LIGANDS SYNTHESIS AND PROTONATION REACTIONS

B. LUNGWITZ and A. C. FILIPPOU*

Anorganisch-chemisches Institut der Technischen Universität München, Lichtenbergstr. 4, W-8046 Garching (Germany)

ABSTRACT. Efficient methods for the synthesis of electron-rich aminocarbyne complexes of the type $Cp^*(CO)_n(L)_{2-n}W\equiv CNEt_2$ ($Cp^* = \eta^5\text{-}C_5Me_5$; n = 0, 1; L = PMe_3, RNC) are reported starting from $Cp^*(Br)_2(CO)W\equiv CNEt_2$. Protonation of $Cp^*(CO)_n(L)_{2-n}W\equiv CNEt_2$ occurs either at the carbyne-carbon or the metal center affording aminocarbene or hydrido(aminocarbyne) complexes of the type $Cp^*(X)(CO)_n(L)_{2-n}W=C(H)NEt_2$ and $[Cp^*(H)(CO)_n(L)_{2-n}W\equiv CNEt_2]X$ (X = Br, I, CN). Both types of products are susceptible to further protonation yielding hydrido(aminocarbene) complexes of the type $[Cp^*(H)(X)(CO)_n(L)_{2-n}W=C(H)NEt_2]X$.

INTRODUCTION

Electron-rich carbyne complexes have been recently shown to play a central role in several coupling reactions of terminal two-faced π-acceptor ligands such as carbon monoxide [1] and isocyanides [2] undergoing a variety of electrophile-induced CC-bond forming reactions [3]. The course of these reactions is strongly dependent on the site of the electrophilic attack. This has initiated the following studies on the protonation of electron-rich tungsten aminocarbyne complexes containing Cp* ligands.

RESULTS AND DISCUSSION

Two routes have been developed to electron-rich aminocarbyne complexes of the type $Cp^*(CO)_n(L)_{2-n}W\equiv CNEt_2$ (n = 0, 1; L = PMe_3, RNC) (**2a-c**, **4a-c**). Synthesis of the monocarbonyl derivatives $Cp^*(CO)(L)W\equiv CNEt_2$ (**2a-c**) is achieved by reductive dehalogenation of $Cp^*(Br)_2(CO)W\equiv CNEt_2$ (**1**) with Na/Hg in the presence of the ligand L (scheme 1). In comparison, the carbonyl-free derivatives $Cp^*(L)_2W\equiv CNEt_2$ (**4a-c**) are obtained from **1** in two steps.

F. R. Kreißl (ed.), Transition Metal Carbyne Complexes, 249–254.

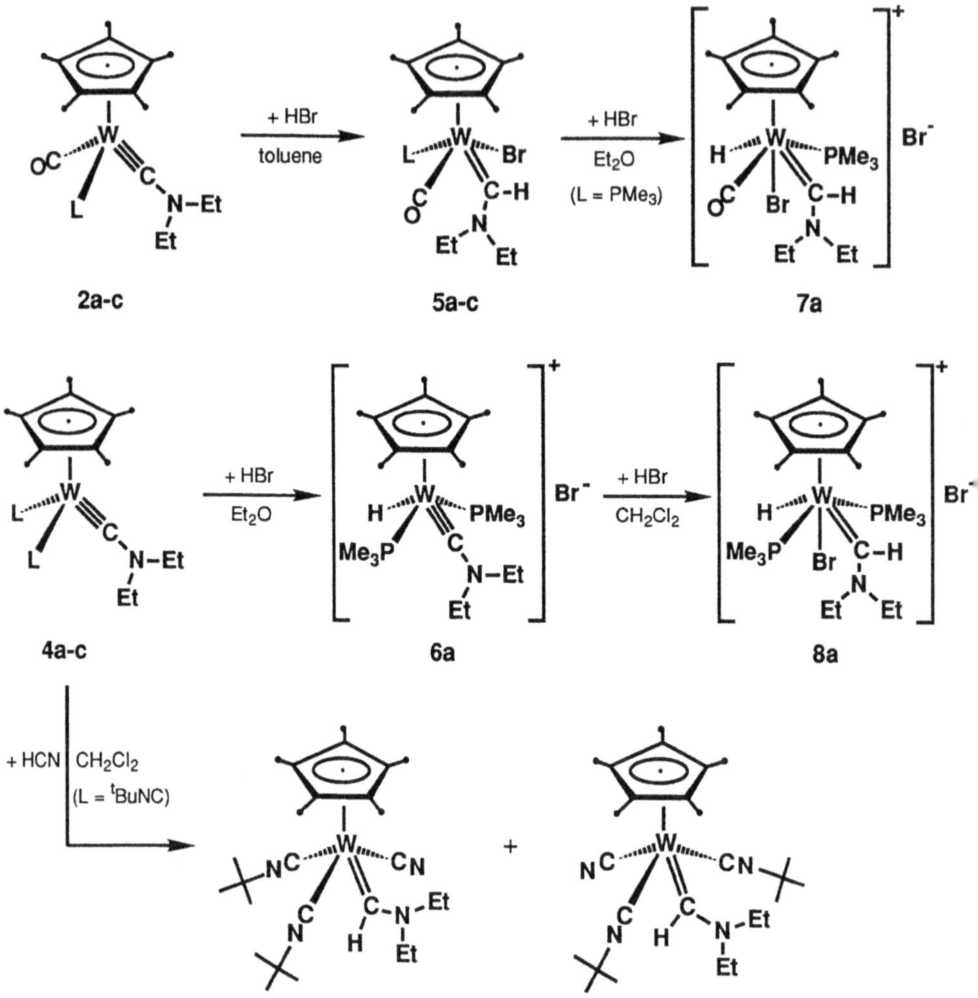

scheme 2

Both types of products are susceptible to further protonation, allowing the selective synthesis of hydrido(aminocarbene) complexes (scheme 2). Thus, treatment of **5a** with HBr gives, after protonation at the metal center, [Cp*(H)(Br)(CO)(PMe_3)W=C(H)NEt_2]Br (**7a**). In comparison, protonation of **6a** occurs selectively at the carbyne-carbon to give [Cp*(H)(Br)(PMe_3)$_2$W=C(H)NEt_2]Br (**8a**) [4d].

The first step involves oxidative decarbonylation of **1** with Br_2 to give the 2-azavinylidene complex $Cp^*(Br)_4WCNEt_2$ (**3**). This is followed by the reductive dehalogenation of **3** with Na/Hg, which affords in the presence of L the desired aminocarbyne complexes **4a-c** (scheme 1). Reduction of **3** to give $Cp^*(^tBuNC)_2W{\equiv}CNEt_2$ (**4c**) is accompanied by the formation of two CC-coupling products, the diaminoacetylene complexes $Cp^*(CN)(^tBuNC)W[(^tBu)(H)NC{\equiv}CNEt_2]$ and $[Cp^*(^tBuNC)_2W[(^tBu)(H)NC{\equiv}CNEt_2]]Br$ [4a].

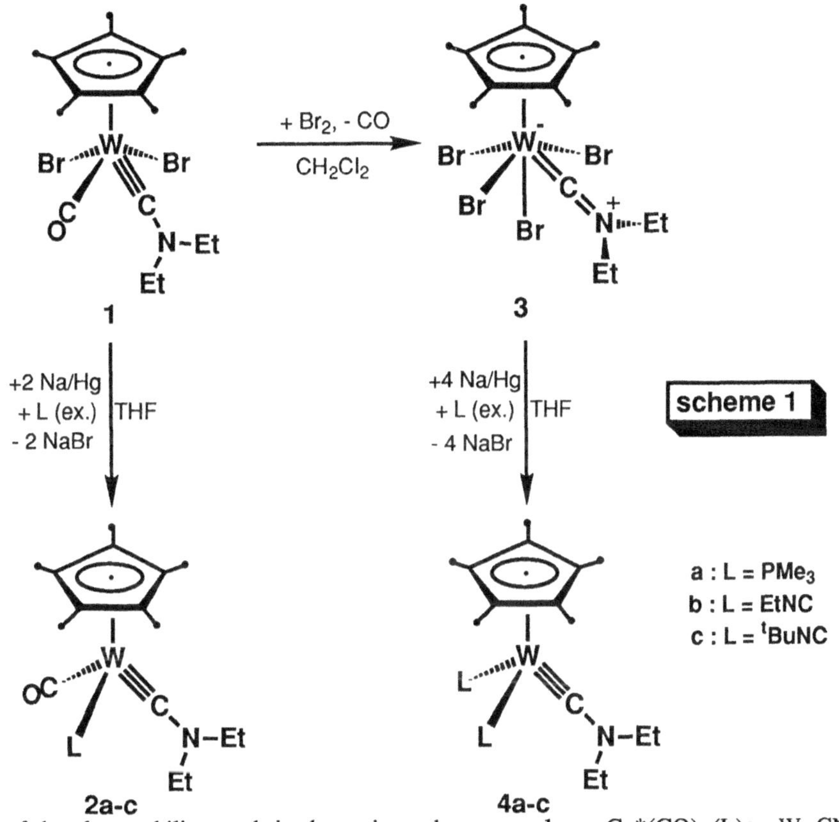

The site of the electrophilic attack in the aminocarbyne complexes $Cp^*(CO)_n(L)_{2-n}W{\equiv}CNEt_2$ (**2a-c**, **4a-c**) is strongly dependent on the ligand sphere. Thus, protonation of **2a-c** with HBr occurs exclusively at the carbyne-carbon to give the electron-rich aminocarbene complexes $Cp^*(Br)(CO)(L)W{=}C(H)NEt_2$ (**5a-5c**) (scheme 2). Analogous reactions of **2b** and **2c** with HI afford the corresponding iodo derivatives $Cp^*(I)(CO)(L)W{=}C(H)NEt_2$ (**5b´**, **5c´**). A similar reactivity pattern is found also for the dicarbonyl derivatives $(\eta^5\text{-}C_5R_5)(CO)_2W{\equiv}CNEt_2$ (R = H, Me) [4a-c]. In contrast, complex **4a** is protonated by HBr at the metal center to yield the hydrido(aminocarbyne) complex $[Cp^*(H)(PMe_3)_2W{\equiv}CNEt_2]Br$ (**6a**) (scheme 2) [4d].

Reaction of $Cp^*(^tBuNC)_2W{\equiv}CNEt_2$ (**4c**) with HCN affords a mixture of the aminocarbene isomers *cis*- and *trans*-$Cp^*(CN)(^tBuNC)_2W{=}C(H)NEt_2$ (**9c**, **9c´**) (scheme 2) [4a, c]. This reaction again indicates the strong effect of the ligand sphere on the chemoselectivity of the protonation reactions of $Cp^*(CO)_n(L)_{2-n}W{\equiv}CNEt_2$.

Complexes **5a**-**5c´** are convenient starting materials for the synthesis of cationic aminocarbene complexes as demonstrated by the reactions of $Cp^*(I)(CO)(EtNC)W{=}C(H)NEt_2$ (**5b´**) with EtNC and ^tBuNC to give $[Cp^*(CO)(EtNC)_2W{=}C(H)NEt_2]I$ (**10**) and $[Cp^*(CO)(EtNC)(^tBuNC)W{=}C(H)NEt_2]I$ (**11**, **11´**) respectively (scheme 3).

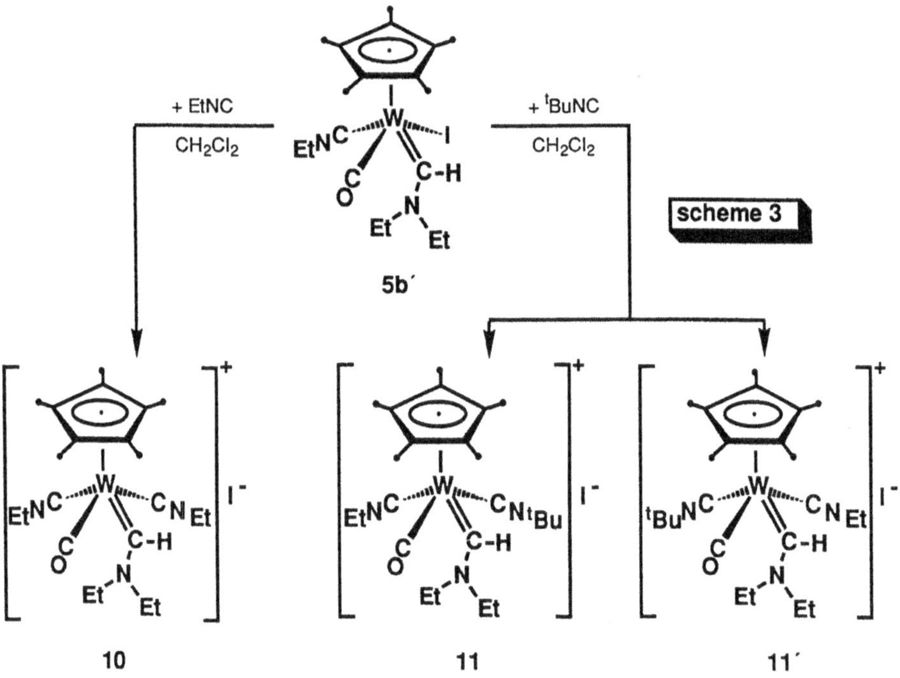

Similarly, bromide abstraction from $Cp^*(Br)(CO)_2W{=}C(H)NEt_2$ (**12**) by $TlPF_6$ affords in the presence of RNC the cationic aminocarbene complexes $[Cp^*(CO)_2(RNC)W{=}C(H)NEt_2]PF_6$ (**13**: R = Et; **14**: R = ^tBu) (eq. 1). Studies are currently carried out to elucidate, whether coupling of the carbene with the isocyanide ligand can be induced in complexes **9c**-**14**.

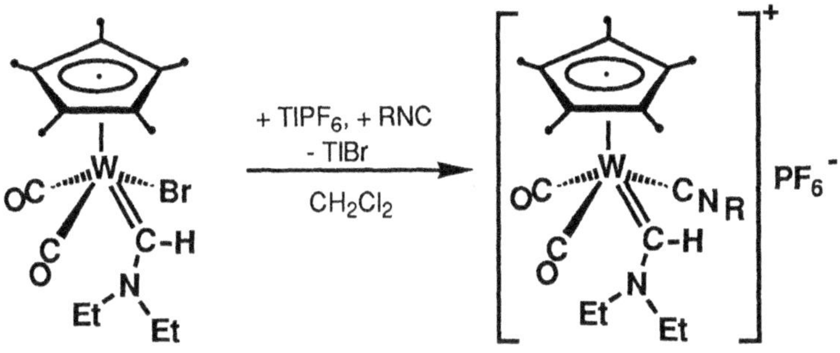

(eq. 1)

12 13, 14

CONCLUSIONS

Electron-rich aminocarbyne complexes of the type $Cp^*(CO)_n(L)_{2-n}W{\equiv}CNEt_2$ (n = 0, 1; L = PMe_3, EtNC, ^tBuNC) are susceptible to successive protonation by HX (X = Br, I, CN). The site of the electrophilic attack depends strongly on the ligand sphere allowing the selective synthesis of a variety of new compounds with high synthetic potential.

REFERENCES

1. (a) R. N. Vrtis and S. J. Lippard, Isr. J. Chem., **30** (1990) 331; (b) R. N. Vrtis, S. Liu, Ch. P. Rao, S. G. Bott and S. J. Lippard, Organometallics, **10** (1991) 275; (c) J. D. Protasiewicz and S. J. Lippard, J. Am. Chem. Soc., **113** (1991) 6564.
2. (a) A. C. Filippou and W. Grünleitner, J. Organomet. Chem., **393** (1990) C10; (b) A. C. Filippou and W. Grünleitner, Z. Naturforsch, **46b** (1991) 216; (c) E. M. Carnahan and S. J. Lippard, J. Chem. Soc., Dalton Trans., (1991) 699; (d) A. C. Filippou, W. Grünleitner, C. Völkl and P. Kiprof, Angew. Chem. Int. Ed. Engl., **30** (1991) 1167.
3. (a) F. R. Kreißl, W. J. Sieber, M. Wolfgruber and J. Riede, Angew. Chem. Int. Ed. Engl., **23** (1984) 640; (b) S. J. Holmes, R. R. Schrock, M. R. Churchill and H. J. Wasserman, Organometallics, **3** (1984) 476; (c) A. C. Filippou and W. Grünleitner, Z. Naturforsch., **44b** (1989) 1023; (d) A. C. Filippou, Polyhedron, **9** (1990) 727; (e) A. Mayr and C. M. Bastos, J. Am. Chem. Soc., **112** (1990) 7797; (f) A. C. Filippou, C. Völkl, W. Grünleitner and P. Kiprof, J. Organomet. Chem., **434** (1992) 201; (g) A. Mayr and C. M. Bastos, Prog. Inorg. Chem., **40** (1992) 1.

4. (a) W. Grünleitner, Dissertation, Technische Universität München, 1991; (b) F. R. Kreißl, W. Sieber and M. Wolfgruber, J. Organomet. Chem., **270** (1984) C45; (c) A. C. Filippou, Habilitationsschrift, Technische Universität München, 1992; (d) B. Lungwitz, Diplomarbeit, Technische Universität München, 1992.

Index

GPSR Compliance

The European Union's (EU) General Product Safety Regulation (GPSR) is a set of rules that requires consumer products to be safe and our obligations to ensure this.

If you have any concerns about our products, you can contact us on ProductSafety@springernature.com

In case Publisher is established outside the EU, the EU authorized representative is:

Springer Nature Customer Service Center GmbH
Europaplatz 3
69115 Heidelberg, Germany

Batch number: 10761697

Printed by Printforce, the Netherlands